30.00
C38

Cognitive Neurochemistry

ALSO PUBLISHED BY OXFORD UNIVERSITY PRESS

Fast and slow chemical signalling
in the nervous system

Edited by

L.L. Iversen and E.C. Goodman
Neuroscience Research Centre
Merck Sharp & Dohme Research Laboratories
Harlow, Essex

Cognitive Neurochemistry

EDITED BY

S.M. STAHL, S.D. IVERSEN
AND E.C. GOODMAN

Neuroscience Research Centre
Merck Sharp & Dohme Research Laboratories
Harlow, Essex

Oxford   New York   Tokyo
OXFORD UNIVERSITY PRESS
1987

Oxford University Press, Walton Street, Oxford OX2 6DP
Oxford   New York   Toronto
Delhi   Bombay   Calcutta   Madras   Karachi
Petaling Jaya   Singapore   Hong Kong   Tokyo
Nairobi   Dar es Salaam   Cape Town
Melbourne   Auckland
and associated companies in
Beirut   Berlin   Ibadan   Nicosia

Oxford is a trade mark of Oxford University Press

© The contributors listed on pp. vii–ix, 1987

All rights reserved. No part of this publication may be reproduced,
stored in a retrieval system, or transmitted, in any form or by any means,
electronic, mechanical, photocopying, recording, or otherwise, without
the prior permission of Oxford University Press

British Library Cataloguing in Publication Data
Cognitive neurochemistry.
1. Dementia   2. Mental illness —— Physiological aspects
I. Stahl, S.M.   II. Iversen, Susan D. III. Goodman, E.C.   IV. Series
616.89   RC524
ISBN 0-19-854225-9

Library of Congress Cataloging in Publication Data
Cognitive neurochemistry.
Based on the proceedings of a symposium held in Oct. 1986 to celebrate the opening of the
Clinical Neuroscience Research Centre established by Merck Sharp & Dohme Research
Laboratories at Harlow, UK.
1. Cognition disorders——Congresses.   2. Neurochemistry
——Congresses.   2. Neuropsychology——Congresses.
3. Neurochemistry——Congresses.
I. Stahl, S.M.   II. Iversen, Susan D., 1940-
III. Goodman, E.C. (Elisabeth C.)   IV. Neuroscience
Research Centre (Merck Sharp & Dohme).   Clinical
Neuroscience Research Unit.
[DNLM: 1. Brain Chemistry——congresses.
2. Cognition——congresses. WL 300 C676 1986]
RC394.C64C6   1987   616.8   87–11300
ISBN 0-19-854225-9

Set by Colset Private Limited, Singapore
Printed in Great Britain by
St Edmundsbury Press Ltd
Bury St Edmunds, Suffolk

Contents

Contributors

YVES AGID, *Clinique de Neurologie et de Neuropsychologie, and INSERM U. 289, CHU Pitié-Salpêtrière, 91 Boulevard de l'Hôpital, 75651 Paris Cedex 13, France.*

HARVEY J. ALTMAN, *Department of Psychiatry, Wayne State University School of Medicine, Detroit, MI 48207, USA.*

FLOYD E. BLOOM, *Scripps Clinic and Research Foundation, 10666 Torrey Pines Road, La Jolla, CA 92037, USA.*

ROSE-MARIE BLUTHE, *Laboratoire de Psychobiologie des Comportements Adaptifs, INSERM U. 259, Université Bordeaux II, Domaine de Carreire, Rue Camille Saint-Saëns, 33077 Bordeaux Cedex, France.*

PAUL BROKS, *Neuroscience Research Centre, Merck Sharp & Dohme Research Laboratories, Terlings Park, Eastwick Road, Harlow, Essex CM20 2QR.*

NELSON BUTTERS, *Psychology Service (116B), San Diego Veterans Administration Medical Center, 3350 La Jolla Village Drive, La Jolla, CA 92161, USA.*

DANIEL COLLERTON, *Department of Psychiatry, University of Newcastle upon Tyne, The Royal Victoria Infirmary, Queen Victoria Road, Newcastle upon Tyne NE1 4LP.*

TIMOTHY J. CROW, *Division of Psychiatry, Clinical Research Centre, Northwick Park Hospital, Watford Road, Harrow, Middlesex HA1 3UJ.*

ROBERT DANTZER, *Laboratoire de Psychobiologie des Comportements Adaptifs, INSERM U. 259, Universite Bordeaux II, Domaine de Carreire, Rue Camille Saint-Saëns, 33077 Bordeaux Cedex, France.*

BRUNO DUBOIS, *Clinique de Neurologie et de Neuropsychologie, and INSERM U. 289, CHU Pitié-Salpetrière, 91 Boulevard de l'Hopital, 75651 Paris Cedex 13, France.*

JOHN L. EVENDEN, *Department of Experimental Psychology, University of Cambridge, Downing Street, Cambridge CB2 3EB.*

BARRY J. EVERITT, *Department of Anatomy, University of Cambridge, Downing Street, Cambridge CB2 3EB.*

S. FREEDMAN, *Merck Sharp & Dohme Research Laboratories, Neuro-*

science Research Centre, Terlings Park, Eastwick Road, Harlow, Essex CM20 2QR.

ELAINE FUNNELL, *Psychology Department, Birkbeck College, University of London, Malet Street, London WC1E 7HX.*

DAVID GAFFAN, *Department of Experimental Psychology, University of Oxford, South Parks Road, Oxford OX1 3UD.*

SAMUEL GERSHON, *Department of Psychiatry, Wayne State University School of Medicine, Detroit, MI 48207, USA.*

PATRICIA S. GOLDMAN-RAKIC, *Department of Neuroscience, Section of Neuroanatomy, Yale University School of Medicine, 333 Cedar Street, New Haven, CT 06510, USA.*

ERIC GRANHOLM, *Psychology Service (116B), San Diego Veterans Administration Medical Center, 3350 La Jolla Village Drive, La Jolla, CA 92161, USA.*

JEFFREY A. GRAY, *Department of Psychology, Institute of Psychiatry, De Crespigny Park, Denmark Hill, London SE5 8AF.*

WILLIAM HEINDEL, *Psychology Service (116B), San Diego Veterans Administration Medical Center, 3350 La Jolla Village Drive, La Jolla, CA 92161, USA.*

S.D. IVERSEN, *Neuroscience Research Centre, Merck Sharp & Dohme Research Laboratories, Terlings Park, Harlow, Essex CM20 2QR.*

EILEEN M. JOYCE, *Department of Psychiatry, Institute of Psychiatry, De Crespigny Park, Denmark Hill, London SE5 8AF.*

GEORGE F. KOOB, *Scripps Clinic and Research Foundation, 10666 North Torrey Pines Road, La Jolla, CA 92037, USA.*

MICHAEL D. KOPELMAN, *Section of Neuropsychiatry, Department of Psychiatry, Institute of Psychiatry, De Crespigny Park, London SE5 8AF.*

MICHEL LE MOAL, *Laboratoire de Psychobiologie des Comportements Adaptifs, INSERM U. 259, Universite Bordeaux II, Domaine de Carreire, Rue Camille Saint-Saëns, 33077 Bordeaux Cedex, France.*

LAUREN LYON, *Psychology Service (116B), San Diego Veterans Administration Medical Center, 3350 La Jolla Village Drive, La Jolla, CA 92161, USA.*

ROBIN G. MORRIS, *Department of Experimental Psychology, University of Cambridge, Downing Street, Cambridge CB2 3EB.*

HOWARD J. NORMILE, *Department of Psychiatry, Wayne State University School of Medicine, Detroit, MI 48207, USA.*

BERNARD PILLON, *Clinique de Neurologie et de Neuropsychologie, and INSERM U. 289, CHU Pitié-Salpetrière, 91 Boulevard de l'Hopital,*

75651 Paris Cedex 13, France.

CLIFF PRESTON, *Neuroscience Research Centre, Merck Sharp & Dohme Research Laboratories, Terlings Park, Eastwick Road, Harlow, Essex CM20 2QR.*

J.N.P. RAWLINS, *Department of Experimental Psychology, South Parks Road, Oxford OX1 3UD.*

TREVOR W. ROBBINS, *Department of Experimental Psychology, University of Cambridge, Downing Street, Cambridge CB2 3EB.*

M. ROSSOR, *Department of Neurology, St Mary's Hospital, Praed Street, London W2 1NY.*

MERLE RUBERG, *Clinique de Neurologie et de Neuropsychologie, and INSERM U. 289, CHU Pitié-Salpetrière, 91 Boulevard de l'Hopital, 75651 Paris Cedex 13, France.*

N.M.J. RUPNIAK, *Neuroscience Research Centre, Merck Sharp & Dohme Research Laboratories, Terlings Park, Harlow, Essex CM20 2QR.*

BARBARA J. SAHAKIAN, *Section of Old Age Psychiatry, Department of Psychiatry, Institute of Psychiatry, De Crespigny Park, Denmark Hill, London SE5 8AF.*

DAVID P. SALMON, *Psychology Service (116B), San Diego Veterans Administration Medical Center, 3350 La Jolla Village Drive, La Jolla, CA 92161, USA.*

M. TRAUB, *Merck Sharp & Dohme Research Laboratories, Neuroscience Research Centre, Terlings Park, Eastwick Road, Harlow, Essex CM20 2QR.*

DAVID M. WARBURTON, *Department of Psychology, University of Reading, Whiteknights, Reading RG6 2AL.*

Abbreviations

AII	angiotensin II
ACh	acetylcholine
ACTH	adrenocorticotrophic hormone
AD	Alzheimer-type dementia
AK	alcoholic Korsakoff's syndrome
AVP	arginine vasopressin
CAT	computerized axial tomography
CCK	cholecystokinin
ChAT	choline acetyltransferase
CNS	central nervous system
CRF	corticotrophin releasing factor
CBF	cerebrospinal fluid
CT	computerized tomography
CTA	conditional taste aversion
CVA	cerebral vascular accident
EEG	electroencephalogram
DA	dopamine
DAT	dementia of the Alzheimer type
dDAVP	d-des-arginnie vasopressin
DDC	diethyldithiocarbamate
DA-AVP	desglycinamide-lysine vasopressin
DNAB	dorsal noradrenergic ascending bundle
DOPAC	dihydroxyphenylacetic acid
DRS	Dementia Rating Scale
FDA	Food and Drug Administration (USA)
GABA	gamma-amino-butyric acid; gamma-aminobutyrate
GAD	glutamic acid decarboxylase
HD	Huntington's disease
5-HIAA	5-hydroxy-indole acetic acid
HPLC	high performance liquid chromatography
HVA	homovanillic acid
ITI	intertrial interval
LVP	lysine vasopressin
MHPG	3-methyl-4-hydroxy-phenethylene glycol
MPTP	3-methyl-4-phenyl-1,2,3,4-tetrahydropyridine

MQ	memory quotient
NA	noradrenaline
nbM	nucleus basalis of Meynert
NPY	neuropeptide Y
6-OHDA	6-hydroxydopamine
PI	proactive interference
PD	Parkinson's disease
PET	positron emission tomography
QNB	quinuclidinyl benzilate
SDAT	senile dementia of the Alzheimer type
SMA	spontaneous motor activity
S/N	signal-to-noise ratio
SRIF	somatostatin
THIP	4,5,6,7-tetrahydroisoxazola(5,4-c)pyridin-3-ol
TTP	thyamine pyrophosphate
THC	9-tetrahydrocannabinol
VDU	visual display unit
VIP	vasoactive intestinal peptide
WAIS	Wechsler Adult Intelligence Scale
WAIS-R	Wechsler Adult Intelligence Scale—Revised
WMS	Weschler Memory Scale

Introduction

S.M. STAHL AND S.D. IVERSEN

The chapters in this volume arise from the proceedings of a symposium held in October 1986 to celebrate the opening of the Clinical Neuroscience Research Unit of the Neuroscience Research Centre established by Merck Sharp & Dohme Research Laboratories at Harlow in the UK.

The new Clinical Neuroscience Research Unit will be entirely devoted to conceptual and development studies in man, applying new knowledge of brain biochemistry to the discovery of novel treatments for CNS disorders.

The symposium grew out of a wish to bring two areas of scientific enquiry together: cognitive psychology and neurochemistry. An area of priority for clinical neuroscience programmes dedicated to the discovery of new drugs will inevitably focus on the cognitive problems of ageing, such as senile dementia of the Alzheimer type. We hoped that by bringing together the fields of cognitive psychology and neurochemical pharmacology, the perspectives and disciplines offered by each could be applied in a more powerful way to understand and begin to solve the problem of cognitive decline.

Thus, this symposium embodied a number of strategies which our new research group will hope to apply: for example, multidisciplinary teams, preclinical studies closely interfaced with clinical studies, use of drugs as tools for posing conceptual questions, as well as use of drugs for development of treatment strategies. This specific topic arose in large part after discussion with a friend and mentor, Dr Daniel X. Freedman, who suggested the format and the specific topic which we later incorporated into the symposium. We thank him for his input and as usual, his catalytic effects.

Chapters in Part I of this volume focus on the cognitive neuropsychology issues. Hopefully, the language and terminology utilized by each of the authors will be understandable to scientists whose major interests are in the fields of neurochemistry or neuropharmacology.

The chapters in Part II of this volume focus on the neurochemical and neuropharmacological aspects of cognition, and failures of cognition. Hopefully, also, the terminology and language in this section will not create scientific barriers—but help to bridge them. If we can succeed in doing this, the volume will have fulfilled its projected role.

We are very grateful to all who contributed to the book and to the symposium. Thanks especially to Elisabeth Goodman and to Oxford University Press for their help and input in editing the manuscripts for publication.

Part I

NEUROPSYCHOLOGICAL
APPROACHES TO COGNITIVE
DISORDERS

1

Neuropsychological differentiation of amnesic and dementing states

NELSON BUTTERS, DAVID P. SALMON,
ERIC GRANHOLM, WILLIAM HEINDEL, AND
LAUREN LYON

The first author of this paper has previously discussed the potential contributions of experimental studies of amnesia and dementia for our clinical understanding of these disorders (Butters 1984). It was noted that little had been learned about the nature of the memory impairments seen in amnesic and demented patients using standardized tests such as the Wechsler Memory Scale (WMS). Patients with alcoholic Korsakoff's syndrome (AK), Huntington's disease (HD), and dementia of the Alzheimer type (DAT) often achieved similar memory quotients (MQ's) despite obvious differences in their encoding, storage and retrieval deficiencies. Our knowledge as to how these patient populations processed new information had originated primarily from the application of the concepts and experimental procedures of cognitive psychology to the assessment of severely defective memories. Butters (1984) concluded that a symbiotic relationship between clinical neuropsychology and cognitive psychology should be encouraged if the assessment of patients was to advance beyond superficial statements concerning lesion sites and severity of deficiency. Any hope of a successful extension of clinical neuropsychology into the realm of pharmacological therapies and rehabilitation would depend upon a deeper understanding of the processes underlying the observed impairments.

To illustrate the value of cognitive psychology to the study of impaired memory, Butters (1984) reviewed a series of studies comparing the memory disorders of patients with diencephalic (i.e. AK patients) and basal ganglia (i.e. HD patients) damage. It was noted that the anterograde amnesia of AK patients involved a failure in storage due to an increased sensitivity to proactive interference (PI) and limited encoding, whereas the severe deficit of HD patients on recall measures of learning was related to an inability to initiate systematic retrieval processes. The memory failures of AK patients,

3

but not those of HD patients, could be attenuated by the introduction of procedures which reduced PI (e.g. distributed rather than massed learning trials). In contrast, only the HD patients performed at almost normal levels when recognition rather than recall measures of learning were employed.

Besides this distinction between storage and retrieval problems, the AK and HD patients appeared to differ in their ability to acquire a visuomotor skill (Martone *et al*. 1984). Using the reading of mirror-reflected words as a measure of skill learning, Cohen and Squire (1980) had shown that AK patients were capable of normal learning and retention of this skill (as measured by reduction in the temporal durations necessary to read mirror-reflected word triads) despite a total inability to recognize the specific words used to train the skill. When Martone and her colleagues (1984) extended the mirror reading paradigm to HD patients, a double dissociation between recognition memory and skill learning emerged. This group of alcoholic Korsakoff patients performed like those described by Cohen and Squire (1980). However, the HD patients were significantly impaired in the acquisition of the visuomotor skill, although they did demonstrate normal recognition of the words employed on the test. On the basis of these findings, Martone *et al*. (1984) suggested that the learning of motor skills might depend upon the integrity of the basal ganglia (especially the caudate nucleus) whereas the storage of factual (i.e. data-based) materials involved limbic-diencephalic regions.

Since the publication of Butters' (1984) paper several studies from this laboratory have provided additional evidence that the memory disorders of HD patients can be characterized by failures both in initiating systematic retrieval strategies and in the acquisition of motor skills. In addition to comparisons between amnesic and HD patients, the investigations focusing upon the acquisition of general skills have included patients with Alzheimer's disease. The findings of these recent experiments have proven of some relevance for Cummings and Benson's (1984) distinction between 'cortical' and 'subcortical' dementia as well as the neuroanatomical bases of what has been termed 'procedural' memory (Cohen and Squire 1980; Squire 1986).

The patients participating in our recent studies are similar to those described by Butters (1984). The patients with HD have a genetically transmitted disorder that results in a progressive atrophy of the basal ganglia, especially the caudate nucleus. The most common behavioural symptoms include involuntary choreiform movements, a progressive dementia, and in most cases marked personality changes (e.g. depression, increased irritability). These HD patients had a mean age of 46 years and had been diagnosed from 3 months to 19 years prior to testing. The median full-scale IQ (WAIS-R) and MQ (WMS) of these patients were in the middle 80's and high 70's, respectively. Although some of the HD patients had moderate-to-severe choreiform movements, very few were institutionalized and none were considered in the terminal stages of the illness.

Most of the amnesic patients in these studies were alcoholics with Korsakoff's syndrome. They were male military veterans with a mean age of 58 years. They all had 20–30-year histories of alcohol addiction accompanied by malnutrition prior to the onset of their Wernicke–Korsakoff syndrome. At the time of testing, all of the Korsakoff patients were residing in a Veterans Administration facility or nursing home. They had severe antero-grade and retrograde amnesias, as measured by the WMS and on the basis of clinical assessment, but their general intellectual functioning, as measured by the WAIS (or WAIS-R), was within normal limits (mean = 100). Their MQ's (mean = 79) were a minimum of 18 points lower than their full-scale IQ's. Although it is generally assumed that the severe amnesia in these patients is related to haemorrhagic lesions in the medial diencephalon (Victor *et al.* 1971), there is some evidence that alcoholic Korsakoff patients may also have a significant loss of neurons in various structures of the basal forebrain (Arendt *et al.* 1983).

The patients with Alzheimer's disease were diagnosed using the clinical cri-teria developed by the National Institute on Neurological and Communi-cative Disorders and Stroke, and the Alzheimer's Disease and Related Dis-orders Association (McKann *et al.* 1984). All patients scored at or above 104 out of a possible 144 points on the Dementia Rating Scale (DRS) (mean = 118), a mental status examination which assesses a broad spectrum of cogni-tive functions (Mattis 1976). In addition, the patients averaged 10 out of a possible 33 errors on the Blessed mental status scale (Blessed *et al.* 1968), and they earned 21 correct responses out of a possible 30 on the Mini-Mental State examination (Folstein *et al.* 1975).

RETRIEVAL PROCESSES

Evidence consistent with the retrieval hypothesis was found in a study com-paring the performance of HD and amnesic patients on a list learning task (Butters *et al.* 1985). The Rey Auditory Verbal Learning Test (Lezak 1983) was administered with a *recall* and *recognition* procedure to nine amnesic (six AK patients, two post-encephalitics, one patient with a neoplasm located in the medial diencephalic region), 10 HD, and 14 normal control subjects. For the *recall* condition, five presentation-recall trials were followed by a single delayed-recall trial 30 minutes later. On each of the five presentation-recall trials, a list of 15 common words was read to the subject at the rate of one word per second. Immediately after the presentation of the 5th word, the subjects were asked to recall as many of the words as possible.

For the *recognition* procedure, five presentation-recognition learning trials were followed 30 minutes later by a single delayed-recognition test. On

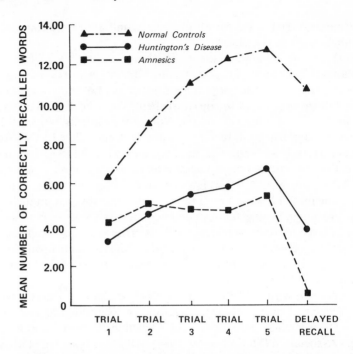

FIG. 1.1. Recall performance of normal control subjects, HD patients, and amnesics on a 15-word list (Rey Auditory Verbal Learning Test). Adapted from Butters *et al.* (1985).

each of the five presentation-recognition learning trials, a 15-word list (different from the one used for recall testing) was read to the subjects at the rate of one word per second. Immediately following the reading of the 15th word, a 30–item yes-no recognition test was administered orally. Thirty words (15 words from the presentation list, 15 distractor words that had not been presented previously) were read to the subject sequentially, and he or she was asked to indicate whether each word *was* or *was not* on the previously presented list.

The results of this study are shown in Figs. 1.1 and 1.2. On the verbal recall test (Fig. 1.1), both the demented HD and amnesic patients were severely impaired in their acquisition and delayed recall of the 15-word lists. Statistical analyses failed to uncover any significant differences between the HD and amnesic patients on the five presentation-recall trials, but on the delayed-recall trials the amnesic patients were significantly worse than were the HD patients. This failure to differentiate the HD and amnesic patients on the acquisition trials of the recall test stands in marked contrast to their performance on the verbal recognition test (Fig. 1.2). Although both the HD and amnesic patients were again impaired in comparison to the normal control

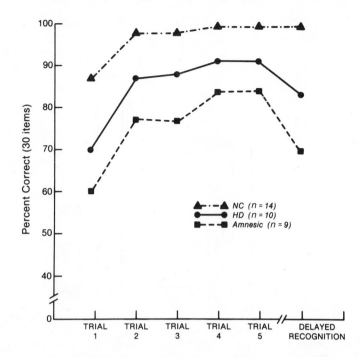

FIG. 1.2.   Recognition performance of normal control subjects, HD patients, and amnesics on a 30-word list (15 target words and 15 distractors). Adapted from Butters *et al.* (1985).

subjects, the recognition scores of the HD patients on the five acquisition trials and the delayed-recognition test were significantly better than those of the amnesic patients. As suggested by the results of Martone *et al.*'s (1984) skill learning study, HD patients demonstrated far better performance when recognition rather than recall measures were employed.

It is obvious from an inspection of Fig. 1.2 that the HD patients' recognition performance, although superior to that of the amnesic patients, was not normal. The major reason for the deficit becomes evident when examining the number of *false positive* and *false negative* errors compiled on the recognition test (Fig. 1.3). Unlike the amnesic patients who made many errors of both types, the HD patients differed from the normal controls primarily in the number of *false positive* errors they compiled during testing. Although HD patients usually detected a word that had been presented on the list, they did tend to say 'yes' when some of the distractor words were presented (i.e. *false positive* errors). Thus, the HD patients recognized words that had been presented, but when they were uncertain about some of the distractor words, they tended to adopt rather liberal judgmental criteria. As one

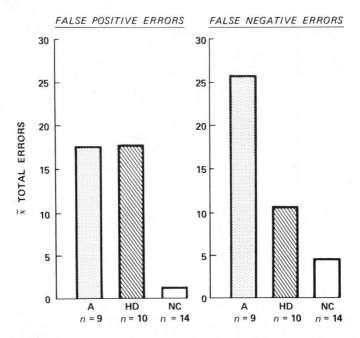

FIG. 1.3.   Mean total false positive and false negative errors made by amnesic patients, HD patients and normal control subjects. Adapted from Butters *et al.* (1985).

might expect, amnesic patients made numerous *false negative* errors (i.e. failed to detect a word that had been presented), a finding which again emphasizes their inability either to recognize or to recall words recently presented to them.

Another experiment concerned with the HD patients' deficiencies in initiating retrieval processes involved the comparison of HD and AK patients on the recall and recognition of prose passages (Butters *et al.* 1986). Based upon the previous findings for word lists, it was anticipated that when the two patient groups were equally impaired in their recall of organized prose passages, the HD patients would perform better than would the AK patients on a recognition test. Since there is abundant evidence that AK patients are more sensitive to PI than are HD patients (for review, see Butters 1984), the AK patients were expected to produce more prior-story intrusion errors (i.e. an element or bit of information from one passage intruding into the recall of a subsequent story) than the patients with HD.

Thirty-one subjects participated in this study: 7 AK patients, 12 patients with HD, and 12 neurologically intact normal control subjects. To ensure that any differences in memory performance were not related to overall degree of dementia the AK and HD patient groups were matched for scores on the Dementia Rating Scale. The materials for this study consisted of two sets of thematically neutral stories (four stories per set). Each story contained 23 phrases or units of information, and was similar in format and length to the Logical Memory Passages of the WMS. One set of stories was used for the recall condition, one for the recognition. To develop a recognition test format, a series of 10 written, multiple-choice questions was constructed for each of the eight stories. For each recognition question there were four alternative answers.

On the recall test, the four different stories of set 1 (or set 2) were read aloud sequentially. The subjects were instructed to listen closely and remember the story they were about to hear. Following the presentation of each story, the subjects were asked to count backwards from 100 by 3's for 30 seconds, and then were asked to recite (i.e. recall) as much of the story as they could remember. When the subjects indicated that they were unable to recall any additional story ideas, they were allowed a 10-second rest interval before the next story presentation. All stories were scored according to a verbatim scale which gave one point credit for each verbatim informational unit recalled by the subject. In addition to prior-story intrusion errors, the number of extra-story intrusion errors (i.e. ideas recalled by the subject which were never presented in any story) were also recorded.

On the recognition test, the four stories of set 2 (or set 1) were presented. After the examiner read aloud each passage, he immediately instructed the subjects to count backwards from 100 by 3's for 30 seconds and then presented them with the 10 written multiple-choice questions about the story. If subjects were uncertain about the answers to some questions, the examiner urged them to guess.

The results for the recall test are shown in Figs. 1.4 and 1.5. Both the HD

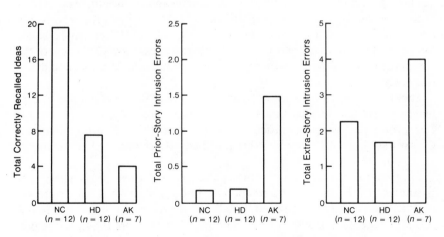

FIG. 1.4. Performance of AK and HD patients, and normal control subjects (NC) on the recall of four prose passages. Adapted from Butters *et al.* (1986).

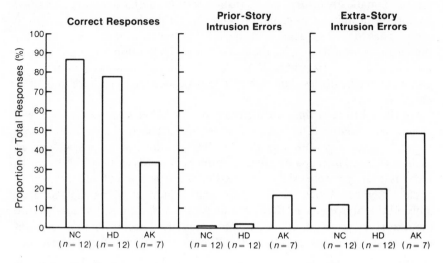

Fig. 1.5.   Recall of prose passages presented as a proportion of the total number of recall responses. Adapted from Butters *et al.* (1986).

and AK patients recalled significantly fewer phrases than did the normal control subjects, and the difference between the two patient groups did not approach statistical significance. Despite this similarity in the number of correct responses produced, the HD and AK groups demonstrated different patterns of errors. The HD patients made significantly fewer prior- and extra-story intrusion errors than did the AK patients.

This significant tendency for the AK patients to produce intrusion errors becomes most evident when performance is calculated as proportions (i.e. percentage) of total responses (Fig. 1.5). Although the HD patients recalled few items, most of them were correct. In contrast, less than 40 per cent of the AK patients' recall was correct due to the large number of prior- and extra-story intrusion errors (Fig. 1.4).

The results for the recognition procedure showed that both patient groups had lower recognition scores than did the normal control group but that the HD group performed significantly better than the AK group (Normal Control subjects, 92 per cent; HD patients, 70 per cent; AK patients, 50 per cent).

The pattern of results noted for the patients' recall of stories is highly consistent with the predictions based upon the findings with word lists (Butters *et al.* 1985) and upon the well-established role of proactive interference in AK and HD patients' anterograde memory disorders (Butters 1984). In the present study, despite recall impairments of equivalent severity, HD patients were again shown to be superior to the AK patients when recognition procedures were employed. Since a recognition test requires less systematic searching for stored information than does recall, this pattern of recall-recognition deficits is consistent with the notion that in patients with HD the verbal memory disorder is primarily attributable to a deficiency in the initiation of adequate retrieval strategies. Again in confirmation of early studies, the frequent intrusion of ideas from one story into the recall of a subsequent passage demonstrates the disruptive influence of PI in anterograde memory processes of Korsakoff patients (e.g. Cermak *et al.* 1974; Warrington and Weiskrantz 1973).

The numerous extra-story intrusion errors made by AK patients on the recall test deserve comment. It is not possible to determine if these extra-story intrusions represent the preservation of ideas learned some time prior to the administraion of the test passages. However, if these errors are not intrusions from prior experiences, they may represent the patients' continuing tendency to confabulate when confronted with lapses in their knowledge. Although confabulation has long been described as a major symptom of patients in the acute stages of Korsakoff's syndrome (Talland 1965; Victor *et al.* 1971), it tends to disappear from the spontaneous everyday behaviour of the typical chronic AK patient after 3–6 months. Thus, the generating of numerous extra-story intrusion errors noted in the present study suggests that years after the onset of the Wernicke–Korsakoff syndrome patients under the stress of a formal test situation may again revert to confabulation as a way of coping with their lack of memory.

Perhaps the most convincing evidence that the memory disorders of AK and HD patients involve different underlying processes can be found in a third experiment which focuses upon the patients' performance on a letter fluency task (Butters *et al.* 1986). Since such verbal tests place great demands upon the subjects' ability to search systematically their semantic (i.e. long-term) memories (Martin and Fedio 1983), HD patients should retrieve fewer words than do AK patients. However, in view of the AK patients' increased sensitivity to PI, they should emit more incorrect perseverative errors (i.e. repetitions of words previously emitted) than should patients with HD. To assess this predicted dissociation, the letter fluency test (FAS) developed by Benton and his colleagues (Benton 1968; Borkowski *et al.* 1967) was administered to 9 HD patients in the early stages of the disease, 11 HD patients in the middle (i.e. advanced) stages, 9 alcoholic Korsakoff patients and 26 normal control subjects. The subjects were read the letters 'F', 'A', and 'S' successively and asked to produce 'as many words as they could think of' that began with the given letter. They were instructed that giving proper nouns or the same word with a different suffix was not permitted. For each of the three letters, the subjects were allowed 60 seconds to generate words. Orally all of the subjects' responses (correct and incorrect) were recorded by the examiner.

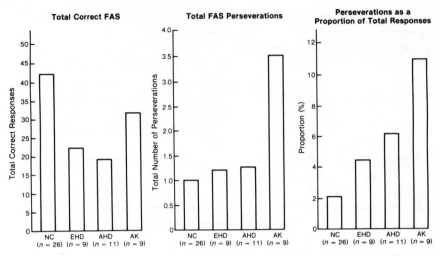

FIG. 1.6.   Performance of AK patients, patients in the early (EHD) and advanced (AHD) stages of Huntington's disease and of normal control (NC) subjects on the letter fluency task. Adapted from Butters *et al.* (1986).

The results are shown in Fig. 1.6. In terms of total correct responses, both the early and advanced HD groups generated fewer words to the three letters than did the AK patients, and normal control subjects. The AK patients' performance, although superior to that of the two HD patient groups, was impaired in comparison to the scores of the normal control subjects. The results for perseverative errors and for perseverations as a proportion (percentage) of total responses indicated that the AK patients were more likely to commit such repetitions than were the other three subject groups. Thus, although the HD patients were able to retrieve fewer correct words than were the AK patients, the latter were the most likely to make perseverative errors.

The results for this letter fluency task have relevance for the interpretation of the performance of HD patients on recognition tests. HD patients have demonstrated relatively preserved verbal recognition memory in a number of experimental paradigms. Their performance could be explained by assuming that recognition tests are simply easier (i.e. less complex) and consequently less sensitive than recall tasks. That is, one could argue persuasively that AK and HD patients share a qualitatively similar memory disorder which differs only in severity. The performance of HD patients on the letter fluency task undermines the appeal to task complexity as an explanation for the pattern of results. The tendency of HD patients to produce fewer correct responses and a lower proportion of perseverative errors than do AK patients suggests that retrieval impairments and increased sensitivity to proactive interference are differentially influencing the verbal memory capacities of these two patient groups. Furthermore, when the results for the various recognition tests and the letter fluency test are considered together, a double dissociation between tasks and patient groups emerges. Although the HD patients demonstrate a greater difference between recognition and recall memory than do the AK patients, the AK patients retrieve significantly more correct words than do the HD patients on the fluency test. This double dissociation indicates that the memory deficiencies of HD patients represent a unique pattern of retrieval and perhaps other processes, rather than a mild form of the amnesic syndrome.

The results for the early and advanced HD patients are consistent with previous reports (Brandt and Butters 1986; Butters *et al.* 1978; Josiassen *et al.* 1983) that the memory disorder associated with HD occurs very early in the disease process, and shows only a moderate progression as the patients' functional disabilities and general dementia worsen. On the letter fluency task, the performances of early and advanced HD patients were virtually identical, with both groups retrieving fewer words than the amnesic AK patients. It appears then that the severe deficits in the initiation of retrieval associated with HD is evident when the functional disability is only mild-to-moderate, and changes relatively little as the abnormal movements and general intellectual deterioration progress.

SKILL LEARNING

It is important to note that the learning of motor skills is a single member of a class of diverse memory tasks that remain preserved in amnesic patients. For example, amnesics can not only learn and retain motor skills such as mirror reading, but also evidence normal classical conditioning, perceptual learning, and verbal priming. Although Squire (1984) has suggested that all of these preserved memory capacities represent the learning of general rules or procedures (i.e. procedural knowledge) rather than specific facts (i.e. declarative memory), the enunciation of a clear operational definition of procedural memory has to-date proven elusive. Also, whether the various types of procedural memory are mediated by a single or by different neuroanatomical systems remains unknown.

Based upon Martone *et al.*'s (1984) findings, Heindel *et al.* (in press) examined the ability of HD patients to acquire the motor skills underlying a pursuit rotor task. This classic test of skill learning has the advantage of avoiding some of the methodological problems encountered on the mirror reading task. For example, the HD patients' impairments in reading mirror-reflected words may have been more attributable to the eye movement problems associated with this disease rather than to a general impairment in skill learning. Also, the interpretation of Martone *et al.*'s (1984) results is further confounded by the failure of the HD patients to begin the mirror reading task at the same initial level of performance as the other subjects (i.e. they were consistently slower than the Korsakoff patients and normal controls). Different levels of initial performance can often result in ceiling or floor effects which reduce the meaningfulness of the results. In Heindel *et al.*'s (in press) study, initial performance levels were equated by manipulating the difficulty (i.e. speed of rotation of the disk) of the pursuit rotor task. The use of this tracking task rather than a mirror reading test also eliminated any effects attributable to the HD patients' abnormal eye movements.

A total of 45 patients participated in this study: 10 HD patients, 10 patients with DAT, 4 amnesic patients of mixed aetiologies, and 20 intact control subjects. The three patient groups were matched for overall degree of dementia with the DRS. The small group of amnesics was included to confirm previous findings of spared motor learning in amnesia (Cermak *et al.* 1973; Corkin 1968). The inclusion of Alzheimer as well as HD patients provided an opportunity to compare the performances of patients with 'cortical' and 'subcortical' dementias. Based upon a recent report by Eslinger and Damasio (1986) concerned with pursuit rotor learning in Alzheimer patients, it was anticipated that patients with DAT would demonstrate normal acquisition of this motor skill. Both the HD and Alzheimer groups were also administered a verbal recognition span test (Moss *et al.* 1986) to establish the severity of their deficits in learning new verbal information.

On the pursuit rotor task subjects were told to try to maintain contact between a stylus held in their preferred hand and a small metallic disc (2 cm in diameter) on a rotating turntable (25 cm in diameter). The turntable could be adjusted to rotate at 15, 30, 45, or 60 rotations per minute (rev/min) for a given 20-second trial. All subjects were tested over three sessions of eight trials each, with each session separated by approximately 30 minutes of other psychometric testing. Within each test session, subjects were also allowed a 1-minute rest interval between the fourth and fifth trials, thereby creating six blocks of four trials each. The total time on target was recorded for each 20-second trial.

The first test session was preceded by a block of practice trials to determine the appropriate speed of rotation of the turntable for each subject. On each successive practice trial the speed of the turntable was increased. The turntable was then set for the remainder of the subject's testing to that speed which was associated with a score (i.e. time on target) closest to 5 seconds. In this manner, the initial level of performance of all subject groups was equated.

The verbal recognition span test has been described in detail elsewhere (Moss *et al.* 1986). It involves a delayed non-matching-to-sample procedure to estimate the longest span of words a subject can recognize before making

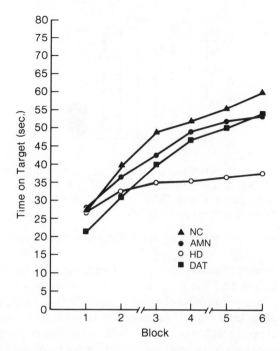

FIG. 1.7. Performance of normal control (NC) subjects, amnesia (AMN) and Huntington's disease (HD) patients, and patients with dementia of the Alzheimer type (DAT) on the pursuit rotor task. Adapted from Heindel *et al.* (in press).

an error and to assess immediate and delayed (2 minutes) recall of the words presented on the recognition test.

The results for the pursuit rotor task are shown in Fig. 1.7. It is evident that all four groups began at the same level of performance (approximately 25 per cent time on target) and that three of the four groups showed systematic skill learning over the six blocks of testing. The Alzheimer and amnesic patients and normal control subjects improved their performance to approximately 52 per cent time on target on block 6, whereas the HD patients maintained contact between the stylus and the disc for only 36 per cent of the time on this last test block. When difference scores (Block 6-Block 1) were calculated to measure the amount of skill acquisition (Fig. 1.8), the HD patients showed significantly less learning than the other three groups. As anticipated, the amnesic and Alzheimer patients did not differ from the normal control subjects on any measure of skill acquisition.

On the recognition span test both the Alzheimer and HD patients were severely impaired in terms of length of verbal span, and of immediate and delayed incidental recall. However, although the HD and Alzheimer patients did not differ on recognition span, the patients with DAT showed less inci-

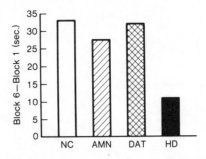

FIG. 1.8.   Difference in performance between the last and first test blocks on the pursuit rotor task. NC = normal control subjects; AMN = amnesic patients; HD = patients with Huntington's disease; DAT = patients with dementia of the Alzheimer type. Adapted from Heindel *et al.* (in press).

dental recall and more rapid forgetting (i.e. immediate-delayed recall) than did the HD patients.

In agreement with Martone *et al.*'s (1984) findings, the results of our study support the notion that the basal ganglia (especially the neostriatum) are involved in the acquisition of motor skills. Since the four subject groups were matched for initial level of performance on the pursuit rotor task, the impairment of the HD patients cannot be attributed to ceiling or floor effects or to the motor limitations associated with the HD patients' choreiform movements. It should also be noted that the matching of the three patient groups in terms of overall level of dementia with the DRS reduces the possibility that the differences in the learning of the motor skill might reflect differences in degree of overall cognitive loss (i.e. dementia).

Despite their profound inability to learn a motor skill, the HD patients were not as impaired on the verbal memory test as the Alzheimer patients were. This pattern of results is similar to the double dissociation noted previously between HD and amnesic patients on a visuomotor skill (i.e. reading of mirror-reflected words) and a verbal recognition test (Martone *et al.* 1984). The ability of the HD patients to retain more information over a 2-minute delay compared to the patients with DAT suggests that their respective memory impairments may be due to different processing deficits. Patients with degenerative diseases affecting medial temporal structures (e.g. DAT) may be unable to store new factual information for more than a brief temporal period while patients with dementias due to dysfunction of the basal ganglia (e.g. HD) may have a memory impairment characterized by deficient retrieval of relatively well-stored information.

Although the linkage between the learning of factual data-based material (i.e. declarative memory) and hippocampal-diencephalic structures appears firmly established (Squire 1986), knowledge of the neuroanatomical substrate mediating the acquisition of motor skills, and the learning of general

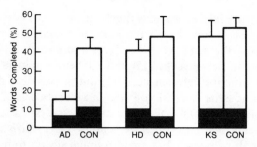

FIG. 1.9.   Lexical priming ability as measured by the word completion test. Patients with Alzheimer's disease (AD), patients with Huntington's disease (HD), alcoholic Korsakoff patients (KS) and normal control (CON) subjects were shown words (e.g., MOTEL) and later asked to complete three-letter word stems (e.g., MOT) with the first words that came to mind. Bars represent percentage of previously-presented words that were used to complete word stems (total number of study words = 20). Dark area represents baseline guessing performance, i.e., the tendency to complete the same words under conditions when they had not been presented. Performance of each patient group is shown next to its age-matched control group. Adapted from Shimamura *et al.* (in press).

procedures and rules is extremely limited. One possibility is that different examplars of procedural memory are mediated by distinct neural systems. The results of our investigation of rotor pursuit learning, in conjunction with other recent findings on verbal priming in HD and Alzheimer patients (Shimamura *et al.* in press) lend credence to this proposed dissociability of various forms of procedural memory (Fig. 1.9). Priming has been defined as the temporary facilitation of performance via prior exposure to stimuli and has been proposed as a form of procedural memory (Squire 1984). For example, it has been shown that both amnesic patients and intact control subjects have a strong tendency (relative to chance) to complete three-letter word stems (e.g. MOT) with previously presented words (e.g. MOTEL) despite the failure of the amnesic patients to recall or recognize these words on standard memory tests (Graf *et al.* 1984). Although patients with DAT in the present study showed intact learning of a motor skill, those in Shimamura *et al.*'s (in press) study were severely impaired when compared to control and amnesic subjects on a stem-completion priming task. The reverse relationship was found for HD patients who were severely impaired on Heindel *et al.*'s in press pursuit rotor task yet exhibited normal levels of stem-completion priming.

This double dissociation between HD and Alzheimer patients on the stem-completion priming and pursuit rotor tasks suggests that various types of procedural memory do depend upon different neuroanatomical substrates. The HD patients' impairments on pursuit rotor and mirror reading tests are consistent with the proposed association between the acquisition of motor skills and the neostriatum, whereas the Alzheimer patients' deficiencies on verbal priming tasks may be attributable to the cortical neuropathology reported in DAT (Blessed *et al*. 1968; Terry and Katzman 1983).

The dissociation between the HD and Alzheimer patients also has relevance for the distinction between 'cortical' and 'subcortical' dementias made by Cummings and Benson's (1984). Patients with subcortical dementias (e.g. HD, Parkinson's disease) usually have much less dysphasia and dyspraxia than do patients with cortical dementias (e.g. Alzheimer's disease, Creuzfeldt–Jakob's disease), but are also much slower to initiate and complete most cognitive as well as motor processes than are the typical patients with cortical degenerative disease. With regard to memory deficits, patients with cortical and subcortical dementias are supposed to have deficiencies in storage and retrieval, respectively. Our data are consistent with this storage-retrieval dichotomy, but they also point to another distinction between these two generic forms of dementia. Patients with cortical dementias have a preserved capacity to acquire and retain motor skills, although other forms of procedural knowledge which depend upon the intactness of the association cortex in the dominant hemisphere may be severely impaired. In contrast, patients with subcortical dementias appear very limited in their ability to learn motor skills despite their normal performance on procedural memory tasks mediated by verbal processes (e.g. stem-completion priming). Further investigations employing various types of cortical and subcortical dementias, and larger batteries of procedural and rule learning tasks will determine the validity of this generalization.

CONCLUSIONS

The studies reviewed in this paper are consistent with the notion that the memory deficits of HD can be differentiated from those of other amnesic and dementing disorders. The recall-recognition, fluency, pursuit rotor, and verbal priming studies indicate that HD patients are impaired primarily in the initiation of systematic retrieval strategies and in the learning of motor skills. However, their deficits in the acquisition of motor skills do not appear to generalize to all forms of 'procedural' or rule-based knowledge. Verbal priming as assessed with a stem-completion paradigm is intact in HD as well as in amnesic patients, but is seriously impaired in patients with Alzheimer's disease. Such findings have considerable import for our understanding of the role of the basal ganglia in memory processes and for the proposed distinction between cortical and subcortical dementias.

From a clinical perspective, it is important to stress again that the present findings emanate from the application of concepts borrowed from cognitive psychology and from careful analyses of the processes underlying the patients' achievements and deficits. These demonstrations of the utility of experimental concepts with pathological populations also provides the constructs themselves with a form of validity and legitimacy unavailable through studies limited to normal subjects. In our view, the mutual benefits that have been described represent an ideal model for the interaction between experimental and clinical approaches to neuropsychology.

ACKNOWLEDGEMENTS

The research reported in this manuscript was supported by funds from the Medical Research Service of the Veterans Administration, by NIAAA grant AA-00187 to Boston University, NIA grant AG-05131 to the University of California at San Diego, and NINCDS grant NS-16367 to Massachusetts General Hospital.

REFERENCES

Arendt, T., Bigl, V., Arendt, A., and Tennstedt, A. (1983). Loss of neurons in the nucleus basalis of Meynert in Alzheimer's Disease. *Acta Neuropathologica* **61**, 101–8.

Benton, A.L. (1968). Differential behavioural effects in frontal lobe disease. *Neuropsychologia* **6**, 53–60.

Blessed, G., Tomlinson, B.E., and Roth, M. (1968). The association between quantitative measures of dementia and of senile change in the cerebral grey matter of elderly subjects. *British Journal of Psychiatry* **114**, 797–811.

Borkowski, J.G., Benton A.L., and Spreen, O. (1967). Word fluency and brain damage. *Neuropsychologia* **5**, 135–40.

Brandt, J. and Butters, N. (1986). The neuropsychology of Huntington's Disease. *Trends in Neuroscience* **9**, 118–20.

Butters, N. (1984). The clinical aspects of memory disorders: Contributions from experimental studies of amnesia and dementia. *Journal of Clinical Neuropsychology* **6**, 17–36.

——, Sax, D.S., Montgomery, K., and Tarlow, S. (1978). Comparison of the neuropsychological deficits associated with early and advanced Huntington's Disease. *Archives of Neurology* **35**, 585–9.

——, Wolfe, J., Martone, M., Granholm, E., and Cermak, L.S. (1985). Memory disorders associated with Huntington's Disease: Verbal recall, verbal recognition and procedural memory. *Neuropsychologia* **6**, 729–44.

——, ——, Granholm, E., and Martone, M. (1986). An assessment of verbal recall, recognition and fluency abilities in patients with Huntington's Disease. *Cortex* **22**, 11–32.

Cermak, L.S., Lewis, R., Butters, N., and Goodglass, H. (1973). Role of verbal mediation in performance of motor tasks by Korsakoff patients. *Perceptual and Motor Skills* **37**, 259–62.

——, Butters, N., and Moreines, J. (1974). Some analyses of the verbal encoding deficit of alcoholic Korsakoff patients. *Brain and Language* **1**, 141–50.

Cohen, N.J. and Squire, L.R. (1980). Preserved learning and retention of pattern analyzing skills in amnesia: Dissociation of knowing how and knowing that. *Science* **210**, 207–10.

Corkin, S. (1968). Acquisition of motor skill after bilateral medial temporal-lobe excision. *Neuropsychologia* **6**, 255–65.

Cummings, J.L. and Benson, D.F. (1984). Subcortical dementia: Review of an emerging concept. *Archives of Neurology* **41**, 874–9.

Eslinger, P.J. and Damasio, A.R. (1986). Preserved motor learning in Alzheimer's Disease: Implications for anatomy and behaviour. *Journal of Neuroscience* **6**, 3006-9.

Folstein, M.F., Folstein, S.E., and McHugh, P.R. (1975). 'Mini-Mental State': A practical method for grading the cognitive state of patients for the clinician. *Journal of Psychiatric Research* **12**, 189–98.

Graf, P., Squire, L.R., and Mandler, G. (1984). The information that amnesic patients do not forget. *Journal of Experimental Psychology: Learning, Memory, and Cognition* **10**, 164–78.

Heindel, W.C., Butters, N., and Salmon, D.P. (in press). Impaired learning of a motor skill in patients with Huntington's Disease. *Behavioral Neuroscience*.

Josiassen, R.C., Curry L.M., and Mancall, E.L. (1983). Development of neuropsychological deficits in Huntington's Disease. *Archives of Neurology* **40**, 791–6.

Lezak, M.D. (1983). *Neuropsychological Assessment* (2nd edition). Oxford University Press, New York.

Martin, A. and Fedio, P. (1983). Word production and comprehension in Alzheimer's Disease: The breakdown of semantic knowledge. *Brain and Language* **19**, 124–41.

Martone, M., Butters, N., Payne, M., Becker, J., and Sax, D.S. (1984). Dissociations between skill learning and verbal recognition in amnesia and dementia. *Archives of Neurology* **41**, 965–70.

Mattis, S. (1976). Mental status examination for organic mental syndrome in the elderly patient. In *Geriatric Psychiatry* (ed. L. Bellack and T.B. Karasu), pp. 77–121. Grune & Stratton, New York.

McKann, G., Drachman, D., Folstein, M., Katzman, R., Price, D., and Stadlan, E.M. (1984). Clinical diagnosis of Alzheimer's Disease: Report of the NINCDS-ADRDA Work Group under the auspicies of Department of Health and Human Services Task Force on Alzheimer's Disease. *Neurology* **34**, 939–44.

Moss, M.B., Albert, M.S., Butters, N., and Payne, M. (1986). Differential patterns of memory loss among patients with Alzheimer's Disease, Huntington's Disease and alcoholic Korsakoff's syndrome. *Archives of Neurology* **43**, 239–46.

Shimamura, A.P., Salmon, D.P., Squire, L.R., and Butters, N. (in press). Memory dysfunction and word priming in dementia and amnesia. *Behavioral Neuroscience*.

Squire, L.R. (1982). The neuropsychology of human memory. *Annual Review of Neuroscience* **5**, 241–73.

—— (1984). The neuropsychology of memory. In *The Biology of Learning* (ed. P. Marler and H.S. Terrace), pp. 667–85. Springer-Verlag, Berlin.

—— (1986). Mechanisms of memory. *Science* **232**, 1612–19.

Talland, G.A. (1965). *Deranged Memory*. Academic Press, New York.

Terry, R.D. and Katzman, R. (1983). Senile dementia of the Alzheimer type. *Annals of Neurology* **14**, 497–506.

Victor, M., Adams, R.D., and Collins, G.H. (1971). *The Wernicke–Korsakoff Syndrome*. F.A. Davis, Philadelphia.

Warrington, E.K. and Weiskrantz, L. (1973). An analysis of short-term and long-term memory defects in man. In *The Physiological Basis of Memory* (ed. J.A. Deutsch), pp. 365–96. Academic Press, New York.

2

Computer-aided assessment of dementia: comparative studies of neuropsychological deficits in Alzheimer-type dementia and Parkinson's disease

ROBIN G. MORRIS, JOHN L. EVENDEN,
BARBARA J. SAHAKIAN, AND TREVOR W. ROBBINS

INTRODUCTION

This chapter illustrates how recent developments in computing technology can be exploited for the neuropsychological assessment of dementia. The term 'dementia' covers a large variety of disorders which result in a global impairment in intellectual functioning (Marsden 1978). However, the chapter focuses on a comparative study of Alzheimer-type dementia (AD) and Parkinson's disease (PD).

At first sight the two disorders contrast in several ways. AD is primarily associated with a progressive and substantial impairment of the patient's mental abilities. The underlying neuropathological changes of Alzheimer's disease include profuse and widespread alterations in the cortex, with the formation of neurofibrillary tangles, senile plaques, and granulo-vacuolar degeneration in conjunction with neuronal loss (Terry and Katzman 1983). In contrast, PD is primarily associated with deficits in motor functioning relating to neurochemical alterations in subcortical areas, the basal ganglia in particular (Rossor 1982).

More detailed investigation of the neurochemical alterations in AD and PD, however, indicate that although the conditions broadly speaking may be characterized as involving cortex on the one hand and basal ganglia on the other, there are some common neurochemical deficiencies in the two disorders (Edwardson et al. 1985). Of particular interest is the loss of acetylcholine in frontal cortex of some PD patients. In this regard it is notable that a significant proportion of PD patients develop dementia (Brown and Marsden 1984) and there is growing evidence that even at the early stages of the disease PD patients suffer from mild cognitive deficits, including impairments in attention and visuospatial processing and memory (Bowen *et al.*

21

1972; Lees and Smith 1983; Boller *et al.* 1984). It is therefore of interest to make careful studies and comparisons of the cognitive impairments in AD and PD.

There are a number of reasons for using microcomputers to study these deficits. First, it is widely recognized that automated testing procedures may provide a means for improving the assessment of dementia in general. For example, in 1981 the Royal College of Physicians produced a report on organic mental impairment in the elderly in which it recommended the rapid development of automated testing procedures. It was suggested that these would be most suited to the serial measurements necessary for clinical trials because of the greater reliability of automated procedures.

About 10 years ago these proposals would have been unrealistic, largely because of the high cost and inflexibility of computer systems. Indeed, the design of automated psychological assessment procedures was confined to a few specialized research departments, often removed from clinical settings (cf. Watts *et al.* 1982). However, with the rapid decrease in the cost of computing technology and the development of the microprocessor it is now possible for most research groups to design, implement and standardise psychological assessment procedures (Skillbeck 1984; Morris 1985).

In fact, comparative psychology has employed automated testing procedures for many years because of the better control they offer over stimulus presentation, the recording of responses and the programming of feedback contingencies. The second rationale for using such procedures with human subjects would be to facilitate a direct comparison of behavioural or cognitive deficit in animals bearing specific neurochemical lesions with the impairments shown by human patients with PD or AD. Such a comparison would help to validate the animal models of the disorder, and establish the neural or neurochemical substrates of deficits in the patients, from the plethora of possibilities apparent from the post-mortem literature. Indeed, this approach has been used to tackle the question of whether specific neural changes might be causal to cognitive impairments or merely correlate with them.

One problem with this approach has been that computerized neuropsychological testing in primates (as distinct from rats and pigeons) has not been developed any better than that for humans. However, Bartus and Johnson (1976) have developed an elegant paradigm which can, for example, be used to study the effects of ageing and drugs in a test of short-term memory (delayed response) in the rhesus monkey. More recently, Gaffan *et al.* (1984) have utilized a touch-sensitive screen which enables experimental primates to touch the discriminative stimuli presented on a VDU screen. Our own research group is testing marmosets with the same type of procedure. Despite the advantages of the automated procedures for primates, there are several handicaps to overcome. For example, unlike the traditional Wisconsin General Testing Apparatus, where the animal can find food in a well beneath the correct object in a discrimination task (e.g. Mishkin and Pribram 1956), stimulus and reinforcer are necessarily separated spatially if a touch-sensitive screen is employed. Thus, considerable care has to be taken in the design of appropriate tests for primates.

The design of psychological tests for demented patients also requires careful consideration (Miller 1980; Morris and Miller 1986). Tests which a normal person would find simple and obvious may present difficulties for the demented patient because of the nature of their intellectual impairment. For example, a normal person can usually assimilate a series of instructions necessary to perform a test. Often, the demented patient cannot assimilate more than one instruction at a time and even this may have to be repeated several times. Computer-aided assessment procedures are not exempt from these problems and a specialist knowledge is essential for the design of appropriate tests.

The aim of this chapter is to outline some of the principles appropriate for the design and implementation of microcomputer aided tests with special reference to testing procedures that have been developed as the Cambridge Neuropsychological Test Automated Battery (CANTAB) by our research group at the University of Cambridge and at the Institute of Psychiatry in London. These procedures can be used to test such functions as memory, attention, and planning, and are specifically designed for use with AD and PD patients (Sahakian *et al*. 1987; Morris *et al*. 1987).

TECHNOLOGICAL DEVELOPMENTS

Choice in the use of microcomputers is considerable, but in relation to neuro-psychological assessment, the most popular types in Britain are the Commodore, Apple, and the Acorn BBC personal computers. These have all been used successfully either experimentally or with patients (French and Beaumont 1984; Morris 1985; Skillbeck 1984). The low cost of such systems means that they are already routinely used for clinical assessment. The Acorn BBC computers are perhaps the most economical and probably at least equal to the other systems in terms of their potential for administering neuropsychological tests. They have the advantage that a large quantity of educational software has been developed for their use which could form the basis for automated psychological tests for demented patients. The BBC computer has already been used in clinical trials with dementia patients by Simpson and Linney (1983).

An important aspect of the new technology in relation to dementia is the availability of peripheral systems which enable the computer to interact with the patient. The computers listed above have relatively versatile graphics systems, stemming from their design for computer games, with fast moving visual displays. However, with a clinical population that is likely to have sensory deficits, this facility is probably best limited to simple visual material which does not change very rapidly over time. The CANTAB series of tests has been designed so that the material is generally restricted to simple shapes or slabs of colours which can be clearly seen by the patient. Because of the links with comparative psychology, the material is designed so that it is difficult to label verbally, thus minimizing verbal mediation in the human subjects.

In the past a limiting factor in designing computerized tests with demented patients has been the response media. The conventional QWERTY keyboard is unsuitable, even if certain keys are 'locked', because it provides a confusing array of response keys which rely on fine motor co-ordination to execute a response. In some systems only the relevant keys are left uncovered, but this still leaves the problem that for a Parkinsonian or demented patient the response keys may be far too small to operate. Frequently, the patient is unable to maintain a single finger over the response key and is distracted from attending to the Visual Display Unit (VDU). One solution is to use a response key, purpose built for each test, which is placed close to the VDU. The CANTAB research group have used large response levers, which have the advantage that they allow the patient to rest more than one finger on top, ready to respond.

A more direct mode of responding such as the light-pen or touch sensitive screen is preferable. The main advantage of such methods is that the stimulus material which guides the decision of the subject is the same as the cues which guide the response, a frequent consideration in comparative neuropsychology. For example, in a matching-to-sample task (described in more detail below) the patient chooses from among a number of alternatives the one which matches a sample stimulus. The response can be made directly by touching one of the choice stimuli. This means that the patient does not have to divide his or her attention between the VDU and the response media, a most important consideration given that demented patients are known to be impaired on tasks that divide their attention (Morris 1986; Baddeley *et al.* 1986).

The light-pen is designed so that the patient makes a response by touching a particular part of the VDU screen. It still has the disadvantage that the patient has to hold the pen, which may present difficulties if the subject has problems with motor co-ordination. With the touch sensitive screen the patient simply points with a finger to a particular area on the VDU screen. The applicability of the touch sensitive screen for clinical testing has been researched by French and Beaumont (1984), and by Carr *et al.* (1986) specifically with elderly psychiatric patients. Until recently it was either very expensive or relatively unreliable. However, cheaper reliable systems are now available, including the Microvitec Touch Screen which is proving highly successful in the CANTAB studies of AD and PD.

THE DESIGN OF COMPUTER-AIDED TESTS

Making full use of the new technology is largely dependent on the design of computerized tests. One advantage of designing an automated test is the precision, speed, and reliability of the microcomputer. For example, it is possible to control with precision the length of time that material is presented and the delays between successive trials in a test. It is also possible to measure response latencies to millisecond accuracy, necessary for sophisticated reaction time studies. This is a useful feature for studies of information processing capacity in dementia where it may be necessary to have simultaneous information about the speed and accuracy of the patient.

The computer can also be used to give feedback to the subject in a systematic and objective fashion. In many 'paper and pencil' tests, feedback is discouraged because this can bias the performance of the patient, depending on how the feedback is given. In other words, it is not generally possible to standardize the provision of feedback. For example, in the Wechsler Adult Intelligence Scale (WAIS), the examiner is instructed to refuse to inform patients about the accuracy of their responses (Wechsler 1955), impassively deflecting their questions and so placing at risk the motivation of the patient.

In contrast, the computer can provide feedback which motivates the patient, but in a manner consistent for all subjects. Feedback can be given at different levels to guide and facilitate the patient's interaction with the computer, particularly when using the touch-sensitive screen. For example, in many of the CANTAB tests, the computer emits a short tone whenever the patient touches an appropriate part of the screen. This provides immediate feedback which encourages the patient to respond in the correct fashion. The frequency of the tone can be altered according to whether the patient has made a correct or incorrect response. Another form of feedback is to change the colour of the portions of the visual display, using red for errors and green for correct responses. To supplement this procedure words or short phases can be displayed, such as 'correct', 'very good', 'wrong' or 'try again'. These serve to give further information which will guide the patient's responses on subsequent trials.

Feedback is most effective in the context of a tailored test which systematically prevents the patient from experiencing a high degree of failure. Again, the computer has the advantage over conventional 'paper and pencil' tests in that quite complex contingencies can be implemented using software which monitors the performance of the patient online and branches when appropriate (Vale 1981). For example, the computer is able to keep an accurate running score as a test progresses and to terminate the test if the patient falls below a certain level of performance. Complex contingencies that determine decisions when to terminate can be rigorously applied for each patient. If these rules are carefully constructed they can trim the length of testing sessions by cutting out superfluous testing trials.

Another related feature is that the patient can be systematically trained to do a task before proceeding to the test phase. Many of the protocols for this type of training already exist in comparative psychology where computers have long been used to control testing sessions. Thus, the patient can be systematically and gradually introduced to the components of a task. The patient is not confronted with a long list of instructions and successful performance on the earlier stages of the task demonstrates that the requirements of the task have been understood. This may, of course, be especially useful in patients whose receptive or expressive speech has been affected.

THE CAMBRIDGE NEUROPSYCHOLOGICAL TEST AUTOMATED BATTERY

Many of the general aspects of design referred to above are included in the CANTAB (Cambridge Neuropsychological Test Automated Battery). This includes a series of tests currently under development which measure different aspects of intellectual functioning such as visuo-spatial memory, learning, attention, and planning. These tests incorporate the following main additional features.

1. They are designed to test different aspects of mental functioning so that a profile of performance can be constructed for a particular patient group.
2. Where appropriate they are graded in difficulty to ensure that the subject understands the principles of the test and to assess a broad range of cognitive ability. This ensures that they can be used to compare the performance of demented patients with a normal control sample.
3. They employ non-verbal stimuli and require non-verbal responses. The structures of the tests are designed to be largely self evident to the patient, minimizing the number of verbal prompts needed for guidance. Many are designed to parallel those used in studying animals, such as the delayed-matching-to-sample, the delayed response and recognition memory tests (Bartus and Johnson 1976; Gaffan *et al.* 1984). We have also been comparing discrimination learning and reversal in marmosets and man, together with tests of selective attention to particular dimensions using extra-and intra-dimensional shifts (Roberts *et al.* 1987).
4. They are designed to be visually attractive and interesting, using positive feedback which will maintain motivation. Some of them have a 'game-like' quality, and care was taken to ensure that they do not appear threatening to the patient.

A selection of tests which we have developed to compare the performance of AD and PD patients is described below. This includes tests of visuo-spatial memory, learning, and planning which have been used to study a sample of 15 AD patients taken from those who attend the Memory Clinic at the Maudsley Hospital in London and are under the care of Professor Raymond Levy. An equivalent number of Parkinson's disease patients has been tested, mainly in collaboration with Dr Andrew Heald at Addenbrooke's Hospital in Cambridge. Each group has also been compared with a control sample of healthy subjects who are matched individually for age and verbal intelligence. The results given below represent the preliminary findings of studies that are to be described in more detail elsewhere (Sahakian *et al*. 1987; Morris *et al*. 1987).

The motor screening test

Each session starts with a simple test of pointing which trains the patient how to respond to the touch-sensitive screen and also gives a brief assessment of their manual motor abilities. The patient is required to touch the centre of a flashing cross which appears on the VDU screen. An auditory warning tone begins whenever the patient's finger deviates from the centre of the cross. The computer monitors any impairment in maintaining an accurate pointing response (caused for example by tremor). The cross moves to a new location after a short period and the subject has to rapidly follow suit. The test can be used as a simple test of tremor and response speed. Not surprisingly, PD patients are impaired on both of these measures, but AD patients are only impaired on their speed of movement.

Delayed matching-to-sample

This test has been used successfully to elucidate the characteristics of short-term memory in animals (D'Amato 1973; Roberts and Grant 1976; Mahut *et al*. 1982) and more recently adapted for use with human subjects (Perez 1980; Albert and Moss 1985). In our version of the test the subject is shown an abstract pattern, the target stimulus, which disappears and is followed by four abstract patterns, the choice stimuli. The patient has to touch the choice stimulus which matches the target stimulus and a correct match is greeted by a red tick accompanied by a satisfying 'bleep' from the computer. Initially there is no memory load, for the sample and choice stimuli are presented together. This condition is used to assess whether the patient can successfully attend and do the basic feature matching. Then the choice stimuli are shown following the removal of the sample stimulus and the delay between these two events is increased (see Plates 2.1 and 2.2). By averaging performance for a series of trials, a measure of forgetting over the retention interval can be obtained. A special feature of our test is that the abstract stimuli differ in

terms of colour and shape, thus requiring the patients to attend to two stimulus dimensions, which increases the attentional demands of the test
(Oscar-Berman and Bonner 1985).

Performance of the patients on the delayed matching-to-sample test is
expressed in terms of the percentage of items correctly recognized at different
delays between presentation of the sample and choice stimuli. In Fig. 2.1, the
performance of the AD patients is compared to that of their matched controls. This shows that the controls do not show significant forgetting over the
retention interval. In contrast, the AD patients start at the same level of performance but then show increasing forgetting with the longer delays until
their performance is approaching chance after 16 seconds. The rapid forgetting can be interpreted as a deficit in retention, that is the decay or increasing inaccessibility of the memory trace with time (Huppert and Piercy 1978).

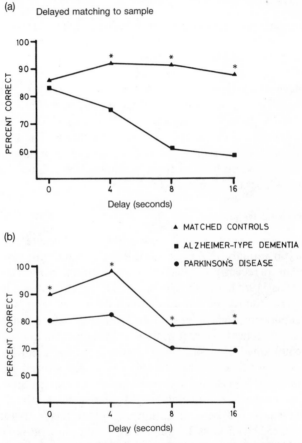

FIG. 2.1.   Performance of Alzheimer-type dementia patients, Parkinson's disease
patients and matched controls on delayed matching-to-sample.

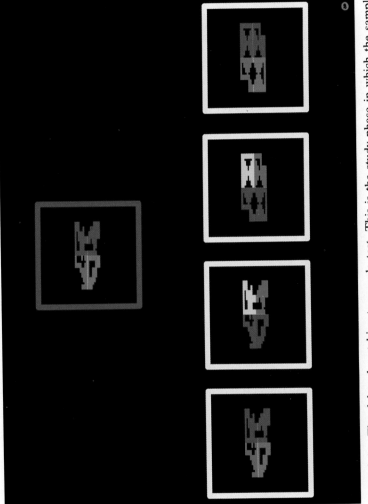

PLATE 2.1. The delayed-matching-to-sample test. This is the study phase in which the sample stimulus is shown in the red box in the centre of the screen. (Drawn from photograph of display.)

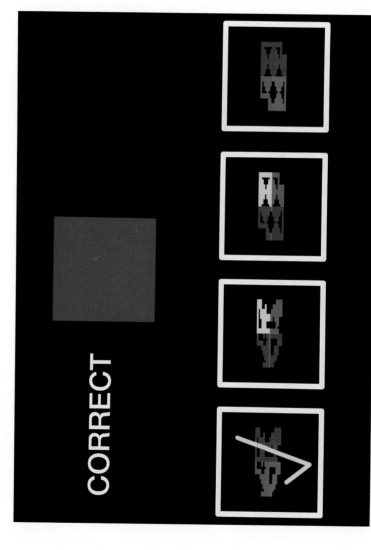

PLATE 2.2. The delayed-matching-to-sample test following the retention interval. The subject has been shown the comparison stimuli in the four white boxes and has made the correct response. (Drawn from photograph of display.)

PLATE 2.3. The delayed response test. In the study phase, the stimuli appear in the peripheral boxes one at a time. This photograph shows the appearance of one of the stimuli. (Drawn from photograph of display.)

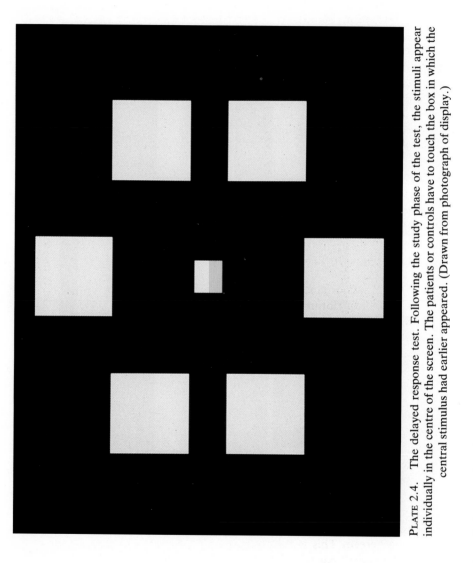

PLATE 2.4.   The delayed response test. Following the study phase of the test, the stimuli appear individually in the centre of the screen. The patients or controls have to touch the box in which the central stimulus had earlier appeared. (Drawn from photograph of display.)

There is a different pattern of results for the PD patients. Here the performance of their control group is more variable, but essentially there is no significant forgetting over the delay interval. The PD patients are significantly impaired even at zero delay, but do not have a greater rate of forgetting as the delay interval increases. This parallel decline in performance does not suggest a deficit in retention, but rather in the encoding of the sample stimulus or feature matching at the recognition stage. This deficit may relate to an inability to attend sufficiently to the features of the target and choice stimuli.

To date, there have been no directly comparable studies using delayed-matching-sample on animals with lesions which might match the pathology of AD or PD. However, similar paradigms have been used which suggest some interesting parallels. For example, Dunnett (1985) has used a simple matching-to-position procedure for rats showing that lesion of the nucleus basalis of Meynert produces effects on performance which are independent of delay. In contrast, rats with fornix lesions (that largely isolate the hippocampus) exhibit delay-dependent effects. Mahut *et al.* (1982) report that monkeys with hippocampal lesions (alone or in combination with removal of the amygdala) show delay-dependent effects in the delayed non-matching-to-sample task, a near equivalent of delayed matching-to-sample.

Delayed response

This paradigm was originally developed to investigate memory in monkeys (e.g. Mishkin and Pribram 1956) and is now automated for this use (Bartus and Johnson 1976). Human analogues have been developed more recently and used clinically, showing that patients with right hippocampal lesions are particularly impaired (Smith and Milner 1981). In the animal version the subject has to remember the spatial location of a hidden object and it is thus a test of visuo-spatial memory.

As a prelude to our version of the delayed response test, there are separate measures of both visual pattern and spatial recognition memory. These were constructed to assess independently whether a patient was impaired either in recognizing the appearance of a pattern or in the ability to remember spatial locations. In the pattern recognition memory test the patient is shown a set of abstract patterns, presented individually. At the end of presentation, pairs of patterns are presented, one of which is the same as one from the previous set. The patient must select and touch the one that has been seen before. In the spatial 'equivalent' the patients are shown a square which moves to five locations on the screen. This is followed by a series of five displays with two squares. The patient has to touch the square which has a location matching one of the five previously shown locations.

In both the pattern and spatial recognition memory tests, the AD patients were impaired relative to control subjects. A similar result occurred for the

PD patients. This latter result is consistent with the suggestion that PD patients generally show greater impairments on tasks requiring the visual processing of information (Boller *et al.* 1984).

In our version of the delayed response task patients are required to remember the locations of abstract visual stimuli (patterns), thus combining the 'pattern' and 'spatial' demands of the recognition memory tests. Six boxes appear on the screen which open up in turn to reveal what is inside (Plate 2.3). The abstract visual stimuli are hidden in the boxes. Following presentation, these stimuli appear in the centre of the boxes and the patient has to point to the boxes where they are located (see Plate 2.4). At the lowest level, there is only one stimulus hidden in one of the boxes. At this stage the task is formally similar to the procedure used by Bartus and Johnson (1976) for delayed response in rhesus monkeys.

The number of stimuli, however, then increases to two, three, and then six stimuli, filling all the boxes. There is a more difficult version at the end of the test with eight boxes and corresponding stimuli. At each level, if the patient makes at least one error, all the boxes are re-opened and the patient has to try again. This continues until either the patient has learnt where all the stimuli are placed or has made ten consecutive unsuccessful attempts to do so. Thus, the test provides a measure of visuo-spatial memory from the number correctly placed at the first attempt and a test of visual-spatial learning from how

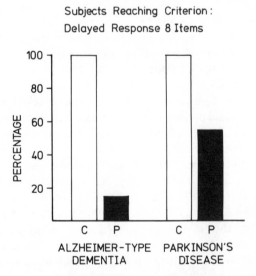

FIG. 2.2. Percentage of Alzheimer-type dementia patients, Parkinson's disease patients and appropriate matched control subjects reaching the criterion of learning with eight items on the delayed response test. C = control; P = patient.

quickly the patient learns to associate the stimuli with the correct boxes.

Performance on this task can be expressed in terms of the number of patients reaching the criterion of learning the spatial location of the patterns within ten attempts. Figure 2.2 shows the proportion of patients reaching the end of the test, that is reaching criterion with eight items. It should be noted that there was successful performance in all groups with just two items suggesting that all the subjects understood the procedure of the task. The results showed that virtually all of the controls were capable of reaching the eight-item test and learning to criterion. There was a striking deficit in the AD patients with many of them unable to learn even the three-item problem and only 2 out of 16 were able to reach criterion with eight items. The deficit in the PD patients was milder, but still substantial, with a proportion of patients either failing to reach the eight-item problem or to learn it within ten attempts.

The computerized Tower of London test

This test is a measure of planning ability or the organization of goal-directed behaviour. It is based on the test developed by Shallice (1982) to assess the effects of frontal lobe damage on planning. In the original version, a set of coloured beads is threaded into three upright sticks of unequal length. The beads have to be moved one at a time to create a different arrangement, provided by the experimenter. The task requires forward planning in order to rearrange the beads within a restricted number of moves. By varying the number of moves needed to rearrange the beads, the test can be graded in difficulty for the patient.

So far, we have used a computerized version to test PD patients, because it has been suggested that this group has possible 'frontal lobe'-type deficits in mental functioning (Bowen *et al.* 1975; Lees and Smith 1983). The patient moves around three coloured rectangles into a new arrangement shown in the upper portion of the screen. Moves are made by first touching a rectangle and then touching where it should be placed. The computer can be used to obtain a more detailed analysis of performance than is possible with the original test. It records the exact sequence of moves and the time taken to initiate and execute each move. If the patient is consistently making too many moves on each problem the computer terminates the test (cf. Morris *et al.* 1987).

There is also a measure of 'thinking time', the delay before initiating the first move. Because of the slowing of response speed in AD and PD, irrespective of planning ability, a 'motor' control test was included in which the computer 'leads' the patient through a sequence of moves one by one. The computer makes changes by one move at a time, with the patient following suit, until the goal is reached. Since the sequence of moves is given by the computer, there is no planning involved. This gives a control measure of the

TABLE 2.1. *Performance of the 18 Parkinson's disease patients and matched controls on the Tower of London test. The total number of moves for 12 problems is presented, as well as the mean thinking, initiation, and execution times*

	Parkinson's disease patients		Control subjects	
	Mean	SE	Mean	SE
Total number of moves	66.7	2.0	66.1	2.9
Mean latencies Thinking time				
Test	17.3	2.7	12.5	0.9
Control	4.3	2.7	3.1	0.7
Initiation time				
Test	6.3	0.9	4.9	2.2
Control	2.3	0.4	1.6	0.2
Execution time				
Test	3.2	0.4	1.9	0.2
Control	2.1	0.1	1.4	0.1

speed with which the patients can make the moves. A control of this type would be difficult, if not impossible, to implement with the more conventional 3-D version of the task.

The principal measure is the number of moves, totalled for the twelve problems (Table 2.1). This shows that, on average, the PD patients are able to solve the problems in the same number of moves as the controls. The response times indicate that the PD patients are slower on all measures including mean thinking, initiation and execution times.

The procedures described above illustrate the principles of test design that we have used with the CANTAB series, with feedback, training on simple versions, graded difficulty, and the link with animal research. These tests also show how a profile of test performance can be constructed for each patient group. On the delayed-matching-to-sample test, the AD patients show a greater rate of forgetting than the PD patients, whose performance essentially runs parallel to the controls with increasing delay. This suggests that the AD patients have a deficit in retention whilst the PD patients are impaired in feature matching and/or attention. Both groups of patients are impaired on visual and spatial recognition, but the PD patients are more impaired on the visual task, as predicted from previous studies indicating deficits in spatial processing (Bowen *et al.* 1972; Boller *et al.* 1984). The delayed response test, which combines both spatial and visual memory, is particularly sensitive to the memory impairments in AD. For the PD patients the lack of deficit on some aspects of the planning task is interesting because it shows that not all aspects of visuo-spatial processing are impaired.

DISCUSSION

The results of the initial studies are promising in that they show that computerized tests can be successfully used for the experimental study of demented patients, at least at the early stages of the disorder. They demonstrate that carefully designed tests, using a touch-sensitive screen as the response mode, are suitable for AD patients—a group known to be difficult to investigate because of their mental impairments (Miller 1980).

It might be argued by some that computerized tests are not suitable for an elderly brain-damaged population, because the novelty of the technology might appear threatening. However, the facility for systematically grading the tests in terms of difficulty, and using tailoring to avoid a high level of subjective failure, combined with the ease of response, outweighs this possible disadvantage. The majority of patients from the CANTAB group reacted more positively to the computerized tests than conventional psychometric tests, including the WAIS. Furthermore, they seemed to derive satisfaction from the sensation of being in 'control' of the testing session by interacting with the computer rather than the psychologist. The presence of the psychologist, however, is still important in order to maintain the motivation of the patient and to supervise the testing session. In this respect, the computer-aided tests enable the psychologist to adopt a more personal role in the testing session and to devote more attention to clinical observation which can be used to supplement the information given by the tests.

The results of the preliminary studies described above indicate qualitative differences in the mental deficits associated with AD and PD. This comparison has recently been strengthened by obtaining a sample of PD patients who are not on medication and who essentially show a similar pattern of deficits, although in a more mild form (Sahakian *et al.* 1987). Thus, the attentional and memory deficits in the PD group cannot be solely attributed to their anti-Parkinsonian medication, although much further research is needed to elucidate the relative contributions of the disease and different forms of treatment on mental functioning (Brown *et al.* 1984).

In conclusion, this chapter has shown how computer-aided tests have been used to forge a link between different areas in psychology. The link between neurochemistry and human neuropsychology is aided by developing tests for animals and patients which have a high degree of procedural similarity. This development is greatly enhanced by the microcomputer, which can be used to implement tests for demented patients which are interactive and visual in nature. The CANTAB series is being further developed to include tests of Working Memory and Attention, and parallels are being drawn with behavioural research into the neurochemistry of AD and PD, and models of these conditions.

ACKNOWLEDGEMENTS

This work was supported by a Major Award from the Wellcome Trust to TWR, B.J. Everitt, and S.B. Dunnett, and a Nuffield Foundation grant and Eleanor Peel Lectureship to BJS. We are grateful to Drs N. Legg, J.A. Morgan-Hughes, J. Pilling, R. Smith, and Professor C.D. Marsden for enabling us to see patients under their care.

REFERENCES

Albert, M. and Moss, M. (1985). The assessment of memory disorders in patients with Alzheimer Disease. In *The Neuropsychology of Memory* (ed. L.R. Squires and N. Butters), pp. 236–46. The Guilford Press, New York.

Baddeley, A.D., Bressi, S., Logie, R., Della Sala, S., and Spinnler, H. (1986). Dementia in working memory. *Quarterly Journal of Experimental Psychology* **38A**, 603–18.

Bartus, R.T. and Johnson, H.K. (1976). Short-term memory in the rhesus monkey—disruption from the anti-cholinergic scopolamine. *Pharmacology, Biochemistry and Behaviour* **5**, 39–46.

Boller, F., Passafiume, D., Keefe, N.C., Rogers, K., Morrow, L., and Kim, Y. (1984). Visuospatial impairment in Parkinson's disease. *Archives of Neurology* **41**, 485–90.

Bowen, F.P., Hoehn, M.M., and Yahr, M.D. (1972). Parkinsonism: alterations in spatial orientation as determined by a route-walking test. *Neuropsychologia* **10**, 335–61.

——, Kamienny, R.S., Burns, M.M., and Yahr, M.D. (1975). Parkinsonism: effects of levo-dopa treatment on concept formation. *Neurology* **25**, 701–4.

Brown, R.G. and Marsden, C.D. (1984). How common is dementia in Parkinson's disease? *Lancet* **ii**, 1263–5.

——, Marsden, C.D., Quinn, N., and Wyke, M.A. (1984). Alterations in cognitive performance and affect-arousal state during fluctuations in motor function in Parkinson's disease. *Journal of Neurology, Neurosurgery and Psychiatry* **47**, 454–65.

Carr, A.C., Woods, R.T., and Moore, B.J. (1986). Automated cognitive assessment of elderly patients: A comparison of two types of response device. *British Journal of Clinical Psychology* **25**, 305–6.

D'Amato, M.R. (1973). Delayed matching and short-term memory in monkeys. In *The Psychology of Learning and Motivation: Advances in Research and Theory* Vol. 7 (ed. G.H. Bower), pp. 227–69. Academic Press, New York.

Dunnett, S.B. (1985). Comparative effects of cholinergic drugs and lesions of nucleus basalis or fimbria-fornix on delayed matching in rats. *Psychopharmacology* **87**, 357–63.

Edwardson, J.A., Bloxham, C.A., Candy, J.M., Oakley, A.E., Perry, R.H., and Perry, E.K. (1985). Alzheimer's disease and Parkinson's disease: pathological and biochemical changes associated with dementia. In *Psychopharmacology. Recent Advances and Future Prospects* (ed. S.D. Iversen), pp. 131–45. Oxford University Press, Oxford.

French, C.C. and Beaumont, J.G. (1984). The Leicester-DHSS project on micro-computer aided assessment. Paper presented at the Symposium on automated Testing, Royal Hospital and Home for Incurables, London.

Gaffan, D., Saunders, R.C., Gaffan, E., Harrison, S., Shields, C., and Owen, M.J. (1984). Effects of fornix transection upon associative memory in monkeys: role of hippocampus in learned action. *Quarterly Journal of Experimental Psychology* **36B**, 173–221.

Huppert, F.A. and Piercy, M. (1978). Dissociation between learning and remembering in organic amnesia. *Nature* **275**, 317–18.

Lees, A.J. and Smith, E. (1983). Cognitive deficits in the early stages of Parkinson's disease. *Brain* **106**, 257–70.

Mahut, H., Zola-Morgan, S., and Moss, M. (1982). Hippocampal resections impair associative learning and recognition memory in the monkey. *Journal of Neuroscience* **2**, 1214–29.

Marsden, C.D. (1978). The diagnosis of dementia. In *Studies of Geriatric Psychiatry* (ed. A.D. Isaacs and F. Post), pp. 95–118. John Wiley, Chichester.

Miller, E. (1980). Cognitive assessment of the older adult. In *Handbook of Mental Health and Ageing* (ed. J.E. Birren and R.B. Sloane), pp. 520–36. Prentice Hall Inc., Engelwood Cliffs, New Jersey.

Mishkin, M. and Pribram, K.H. (1956). Analysis of the effects of frontal lesions in monkeys: II. Variations in delayed response. *Journal of Comparative and Physiological Psychology* **49**, 36–40.

Morris, R.G. (1985). Automated Clinical Assessment. In *New Directions in Clinical Psychology* (ed. F. Watts), 121–38. John Wiley, Chichester, UK.

—— (1986). Short-term forgetting in senile dementia of the Alzheimer's type. *Cognitive Neuropsychology* **3**, 77–97.

—— and Miller, E. (1987). On the psychological assessment of dementia. (In preparation.)

——, Downes, J., Sahakian, B.J., Evenden, J., Heald, A., and Robbins, T.W., (1987). Working memory and planning functioning in Parkinson's Disease. (In preparation.)

Oscar-Berman, M. and Bonner, R.T. (1985). Matching-and Delayed Matching-to-Sample performance as measures of visual processing, selective attention, and memory in aging and alcoholic individuals. *Neuropsychologia* **23**, 639–51.

Perez, F.I. (1980). Behavioural studies of dementia: Methods of investigation and analysis. In *Psychopathology in the Aged* (ed. J.O. Cole and J.E. Barrett), pp. 81–95. Raven Press, New York.

Roberts, A., Everitt, B.J., and Robbins, T.W. (1987). Extra and intra dimensional shifts in man and marmoset: a comparative study. (In preparation.)

Roberts, W.A. and Grant, D.S. (1976). Studies of short-term memory in the pigeon using delayed matching-to-sample procedure. In *Processes of Animal Memory* (ed. R.T. Davis, D.L. Medin, and W.A. Roberts), pp. 79–112. Lawrence Erlbaum Associates, Hillsdale, New Jersey.

Rossor, M.N. (1982). Neurotransmitters and central nervous system disease. *Lancet* **ii**, 1200–4.

Royal College of Physicians (1981). Organic mental impairment in the elderly: Implications for research, education and the provision of services. *Journal of the Royal College of Physicians of London* **15**, 3–38.

Sahakian, B.J., Morris, R.G., Evenden, J., Robbins, T.W., Heald, A., and Levy, R. (1987). A comparative study of visuo-spatial memory and learning deficits in Alzheimer-type dementia and Parkinson's disease. (Submitted.)

Shallice, T. (1982). Specific impairments in planning. *Philosophical Transactions of the Royal Society* **B298**, 199–209.

Skillbeck, C. (1984). Computer assistance in the management of memory and cognitive impairments. In *Clinical Management of Memory Problems* (ed. B. Wilson and N. Moffat), pp. 112–31. Churchill Livingstone.

Simpson, J.M. and Linney, A. (1983). Computer controlled assessment of cognitive change in organically mentally-impaired old people. Paper presented at the Symposium on Automated Testing, Royal Hospital and Home for Incurables, London.

Smith, M.L., and Milner, B. (1981). The role of the right hippocampus in the recall of spatial location. *Neuropsychologia* **19**, 781–93.

Terry, R.D. and Katzman, R. (1983). Senile dementia of the Alzheimer type. *Annals of Neurology* **14**, 497–506.

Vale, C.D. (1981). Design and implementation of a microcomputer-based adaptive testing system. *Behaviour Research and Instrumentation* **13**, 399–406.

Watts, K., Baddeley, A.D., and Williams, M. (1982). Automated tailored testing using Raven's Matrices and Mill Hill Vocabulary Tests: A comparison with manual administration. *International Journal of Man-Machine Studies* **17**, 331–44.

Wechsler, D. (1955). *Manual for the Wechsler Adult Intelligence Scale*. The Psychological Corporation, New York.

3

Modelling cognitive function in dementia

ELAINE FUNNELL

The purpose of this chapter is to consider the differing approaches currently applied to the psychological study of cognitive breakdown. In particular, the aims and methodology which guide investigations into the nature and extent of intellectual impairments in dementia will be discussed. These will be compared with the aims and methodology used in the information-processing approach successfully applied recently to cognitive impairments arising principally from cerebral vascular accident (CVA). A strong case will be made in favour of applying a similar approach to the study of dementia.

COGNITIVE IMPAIRMENTS IN DEMENTIA

At present, the cognitive impairments associated with dementia are poorly described and poorly understood. This has far-reaching consequences for diagnosis and treatment. The gross level at which cognitive impairments are described and tested makes it difficult to define sub-types of dementia. Furthermore, possible beneficial effects of treatment on specific systems within the brain may be too limited in their effect to be quantified with these methods and therefore not recognized.

The diagnosis of dementia depends crucially upon identifying the cognitive features of the disorder. However, despite the fact that cognitive impairments are the central sign of dementia, there are no cognitive tests available at present which alone provide a reliable diagnosis of primary dementia syndromes, as opposed to other cognitive disorders, such as amnesia and aphasia, or from dementing symptoms secondary to other disorders, such as severe depression (Pearce 1984a). There are also no cognitive tests which will distinguish reliably among the differing underlying causes of dementia (Miller 1984; Huppert and Tym 1986).

The failure so far to find distinctive cognitive features associated with a particular diagnosis does not mean that no such features exist. It could be argued that the gross level of the descriptions of cognitive function, together with the methodology used would be insufficiently sensitive to detect such features, even if they were present.

The cognitive impairments observed in dementia are frequently referred to in general terms, such as deficits in memory, intellect, and language (e.g. Pearce 1984*b*). While these are meaningful descriptions of the real-life areas of difficulty experienced by a patient, they are too general to be useful as psychological descriptions of impaired cognitive processing (Huppert and Tym 1986). Memory, intellect, and language are global functions which embrace a variety of processes which act upon a range of different types of material. Only some aspects of these functions may be impaired in particular patients (see Butters, this volume). The cognitive deficits of a dementia patient who could define correctly the meaning of uncommon abstract words, for example: *supplication* as 'making a serious request for help', but could not retrieve the meaning of uncommon concrete words, saying 'I've forgotten' to the word *hay* (Warrington 1975), cannot be explained in terms of a general deficit in memory. The reason why the deficit affects some classes of words and not others must be addressed. Similarly, the problem of the aphasic patient who could write the names of pictures, but not speak them (even though articulation itself was unaffected) (Patterson 1986), or of the dementia patient who could understand the syntactic construction of a sentence, but not the lexical content (Schwartz *et al.* 1979), cannot be accounted for simply in terms of problems in language. Again, it must be explained why one patient could produce written, but not spoken words, and why the other could understand syntax, but not word meanings.

The tests used to assess cognitive function in patients reflect the global nature of the descriptions of cognitive function used. Although these tests include a range of sub-tests which tap a variety of different skills (e.g. Pearce 1984*b*), these tend not to be analysed separately. Instead, an overall test score is obtained which provides only a gross-level description. Hidden beneath this score, there may be informative patterns of scores on the sub-tests; two patients sharing the same overall score may have quite different patterns of impairment.

Different patterns of impairments on particular sub-tests may indicate different sub-types of dementia. Evidence in support of this comes from Marin (1987) who reports that while a perceptive agnosia is a frequent early sign occurring in Creutzfeldt–Jakob disease, it occurs only rarely in the early stages of Dementia of the Alzheimer type (DAT). It is possible then that the nature of the *first* signs of dementia, before the disease has become too severe, may be a promising place to look for clinically distinguishing features displayed in differing patterns of impairment.

A major problem in the current searches for differences between diagnostic types arises from comparisons made between groups of patients. As Marin notes, the cognitive impairments of individuals within a diagnostic category tend to vary—no two individuals being exactly alike. Such variance in performance within a group, makes it unlikely that clear distinctions *between* diagnostic groups can be obtained; and unless clear distinctions can be drawn between *all* members of the different groups a reliable diagnostic test is clearly unobtainable. As Miller (1984) observes, although average differences between groups of patients may be highly significant statistically, there may be considerable overlap between the groups when individual performance is examined.

An additional problem for studies of patients with a progressive disorder (such as is found in most types of dementia) is that two patients with differing patterns of cognitive deficits may represent different degrees of severity of an identical type of dementia. For reasons outlined below, matching patients for degree of severity is not a solution to the problem.

Just as the insensitivity of the tests and methods used may be detrimental to the diagnosis and assessment of dementia, so also may they be detrimental to the assessment of the effects of intervention. Changes in behaviour at the level of a specific circumscribed cognitive function are unlikely to be detected by tests sampling behaviour at a gross level.

In sum, our understanding of the cognitive impairments in dementia is likely to be hindered by an approach which operates at too gross a level for any possible sub-sets of dementia to be exposed, or for treatment effects to be adequately assessed. For a sensitive assessment of the cognitive deficits in dementia individual patients need to be studied in depth; detailed descriptions are needed of the range and nature of the cognitive deficits which are found; and the cognitive processes which underlie these deficits need to be identified.

COGNITIVE IMPAIRMENTS AFTER CEREBROVASCULAR ACCIDENT (STROKE): AN INFORMATION PROCESSING APPROACH

Until recently, the study of language disorders after focal brain damage centred upon a similar approach. Assessment techniques used (and still use) batteries of tests, grouped according to general levels of description, such as production and comprehension. Sub-types of aphasia were (and still are !) associated with particular areas of damage to the left cerebral hemisphere and investigated further in order to determine the complex of symptoms associated with a particular area of the brain.

A new approach to the study of cognitive deficits in brain-damaged patients appeared when Marshall and Newcombe (1973) applied information processing techniques to cognitive impairments in language and changed the nature of the inquiry. Instead of trying to relate performance deficits to site of lesion, performance deficits were related to models of cognitive function, with no reference to neurology.

Information processing models are composed of functionally independent

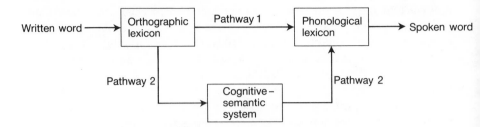

FIG. 3.1.   A model of the lexical processes involved in oral reading.

knowledge structures connected in a particular organization. It has been argued by Marr (1976) that a modular organization of specialized sub-systems is the optimal arrangement for the representation of information. Figure 3.1 is an extremely simple model which describes the processes and information required in order to be able to read words aloud. Although it is simple, and is only part of the full reading system (for example, the processes involved with the ability to read unfamiliar words, or non-words, have been excluded for the sake of simplicity); nevertheless, the model has the power to account for a number of different dissociations in performance.

The model explains word reading in the following way. A written word is first analysed visually and is recognized as a familiar written word form if it is represented within the orthographic lexicon (this is a store of knowledge about written word forms). From this point, the phonological word form may be accessed directly in the phonological lexicon by the use of pathway 1. The meaning of the word may be addressed in the cognitive-semantic system, from where the phonological lexicon may be accessed (pathway 2). Thus, there are two lexical routes for obtaining a spoken word form from print. Once the phonological word form has been activated, it can be pronounced by means of a set of articulatory processes (not included in the model).

The study of acquired dyslexia in patients with cognitive impairments has helped to develop models of normal processing, such as that illustrated in Fig. 3.1. Dissociations in performance have enabled impaired processes responsible for performance deficits to be distinguished from processes responsible for those parts of a task that can still be performed. For example, Marshall and Newcombe (1973) described two patients who could read aloud meaningful words but who frequently made semantic errors: reading *antique* as 'vase' and *nephew* as 'uncle'. From this they hypothesized that the patients were reading on the basis of a pathway to phonology via the meaning of the word (i.e. pathway 2 in Fig. 3.1). The patient appeared to have lost the use of possible further pathways that normally would enable words with little meaning to be read orally. In contrast, Schwartz *et al.* (1980) describe a demented patient who could read irregular words aloud (that could not be worked out by the use of letter to sound rules) yet could not decide whether the word signified an animal, colour or body part. 'Hyena . . . hyena . . . what in the heck is that?' she asked. On the basis of this dissociation, the direct lexical pathway 1, by-passing the cognitive-semantic system was hypothesized to exist; the lexical-semantic route (pathway 2) appeared to be impaired.

Using this approach, a variety of types of acquired dyslexia have been identified whose characteristic impairments can be pin-pointed on a more fully elaborated model (e.g. Patterson 1986; Howard and Franklin 1987; Coltheart and Funnell 1987). While syndrome names were originally given to each symptom complex, for example, *deep dyslexia* (Coltheart *et al*. 1980), *surface dyslexia* (Patterson *et al*. 1985), *phonological dyslexia* (Derouesne and Beauvois 1979), it has become clear that such classification is not useful, since patients who share many features may diverge on others. It is now beginning to be appreciated that it is more meaningful to describe patients in terms of specific impairments within a model of normal processing (Funnell 1983; Coltheart 1984; Ellis 1987). This point is referred to again when the classification of patients with dementia is discussed.

It might be supposed that the cognitive organization of individuals may be sufficiently unique, or the effects of brain damage be so bizarre, that a different model of cognitive processing would be required to explain the performance of each new patient examined. So far this has not been so. New patients do appear with performance deficits which question the model and the model is then developed to account for the new data. However, this is the strength of the discipline of cognitive neuropsychology. In fact, the patterns of cognitive impairments have produced a surprising degree of conformity across individuals (see, for example, the patients described within Coltheart *et al*. 1980; Patterson *et al*. 1985), and necessary adjustments to the model have been minor. Such uniformity is reassuring, and it is now generally assumed that the processes exposed after brain damage reflect the partial operation of the normal system (Saffran 1982).

While discussion here has been limited to a simplified model of reading performance, models are also being developed to account for other language tasks, such as the processing of sentences (Schwartz *et al*. 1985) and for other areas of cognitive processing, such as recognizing objects (Warrington and Taylor 1978; Ratcliff and Newcombe 1982) and faces (Hay and Young 1982).

MODELLING COGNITIVE FUNCTION IN DEMENTIA

The modular organization of psychological information into specialized sub-systems illustrates how misleading it is to think in terms of global psychological constructs, such as memory, intellect, and language. In just one small part of the full language system, the model of reading in Fig. 3.1 shows how

reading may be further sub-divided into a number of functionally independent parts. For example, the functionally independent orthographic and
phonological lexicons (Morton and Patterson 1980a; Allport and Funnell
1981), which embody written and spoken word forms, respectively, cle: rly
represent separate memory systems, specialized for specific types of material.
It is insufficient, therefore, to note simply that memory or language problems
are associated with dementia. The precise nature of the problem needs to be
specified with respect to the underlying organization of the cognitive system.

It is clear also from the model depicted in Fig. 3.1 why describing impairments in terms of symptoms and grouping patients according to symptoms
may be misleading, since a particular symptom may arise from a variety of
causes. An inability to read words aloud may arise because the phonological
lexicon is impaired, preventing the patient from finding the word for the oral
response; or, the orthographic lexicon may be impaired so that the patient
cannot recognize any written word as familiar; or, both pathways connecting
the orthographic and phonological lexicons may be impaired, preventing
material from being passed between them. Exactly the same arguments apply
to comparisons between patients on the basis of degree of severity: two
patients may be equally severely impaired at performing the same task, but
for quite different reasons.

The use of a model allows the impairments of an individual patient to be
described precisely. Using tests devised to tap each component and pathway
in the model, it is possible to pin-point the locus of the processing impairment. For example, assuming that the visual processes preceding access to the
orthographic lexicon are intact, then the functional level of the orthographic
lexicon itself may be tested by asking a patient to sort written words from
written nonsense words which look similar to words (Patterson 1979). If the
patient succeeds at this task, then the functional level of the semantic system
may be tapped with a test requiring the patient to make decisions about the
meanings of written words (for example to choose the word that 'goes with'
mitten, from the word pair *glove* and *sock* (Funnell 1983).

The model of normal processing enables the cognitive impairments of
individual patients to be compared and contrasted. Individuals who share
particular selective impairments (although differing in others) may then be
selected to form a group for the study of the properties *of that process*. The
benefits of such an approach to the assessment of treatment in the individual
cannot be underestimated. The ability to locate the impaired process, or processes, by the use of specific tests, means that the effect of specific treatments
upon specific cognitive deficits can be evaluated. Any sub-sets of patients
sharing *identical* patterns of impairments on the model should be discernible.
If consistent patterns of impairments emerge, these may indicate the presence
of a particular pathological substrate.

The decline of cognitive function over time should also be able to be mapped in terms of the model. Whether the impairments have a focal starting point with a spread to other processes, or whether low level impairments may be observed to affect a wide range of processes even at early stages, are questions which it should be possible to answer using this technique.

Although dementia is often described as a diffuse cognitive impairment affecting all processing, some processes seem particularly vulnerable, while others particularly resistant. Whitaker (1976) described an echolalic patient who could repeat sentences, correcting the syntactic errors and phonological errors of the experimenter, but not the semantic errors. Schwartz *et al.* (1979) reported a dementing patient who could understand the syntactic construction of the sentence, but could not comprehend the meaning of the referential words. The apparent dissociation between spared syntactic processing and impaired semantic processing observed in each of these dementing patients has been further supported by evidence from Bayles (1982). It is noteworthy that in cases of aphasia the opposite dissociation is more common: that is, a loss of syntactic knowledge, with relatively good retention of semantic knowledge (e.g. Morton and Patterson 1980*b*). These opposite patterns of impairment may be important to the differential diagnosis of aphasia and dementia and to the understanding of the syntactic and semantic functions of language.

MODELS OF COGNITIVE PROCESSING AND BIOLOGICAL STRUCTURE

The development of models of cognitive processing can proceed without reference to neurological or pharmacological location of function. However, the appearance of dissociations in function between discrete cognitive processing modules raises the important question as to whether these discrete cognitive processes map onto discrete biological sub-systems within the brain. Whether or not dissociations in cognitive function reflect the influence of different neurotransmitters (cf. Rossor, this volume) is a further question worthy of attention. When techniques are available which make it possible to investigate the functioning of discrete biological sub-systems within the living human brain, then much will be learned if such systems are studied through tasks whose individual cognitive processes are understood.

In the study of organic brain disease (unlike the study of impairments after vascular accident, where patterns of impairments are likely to be due to the fact that they are coincidentally dependent upon the same part of the vascular system), it is possible that impairments may be confined, particularly in the early stages, to specific biological sub-systems. Schwartz *et al.* (1979) speculate that the perceptual aspects of semantic knowledge (for example, what an object looks like) may be especially vulnerable to the effects of a diffuse cerebral pathology due to representation in a diffusely organized neuronal system. In contrast, the relative sparing of syntactic and phonological knowledge in dementia may reflect a 'tightly wired' lateralized system. The mapping between functional systems and biological systems is an exciting prospect which will apply further constraints to the development of cognitive theory.

The models of cognitive function presently available are generally under-specified. In the case of some cognitive tasks (such as spatial processing) models scarcely exist. The language system has at present the most elaborated models, but even these are incomplete. Unfortunately, therefore, it is not the case that the information processing approach can offer the biological sciences a complete theory of cognitive processing. Just as the study of neuro-chemistry is developing all the time, so is the study of cognitive processing. Certainly, however, both will benefit if their development proceeds in harmony.

REFERENCES

Allport, D.A. and Funnell, E. (1981). Components of the mental lexicon. *Philosophical Transactions of the Royal Society of London* **B295**, 397–410.

Bayles, K.A. (1982). Language function in senile dementia. *Brain and Language* **16**, 265–80.

Coltheart, M. (1984). Acquired dyslexia and normal reading. In *Dyslexia: A Global Issue* (ed. R.N. Malatesha and H. A. Whitaker) pp. 357–73. Martinus Nijhoff Publishers, The Hague, The Netherlands.

—— and Funnell, E. (1987). Reading and writing: one lexicon or two? In *Language Perception and Production: Shared Mechanisms in Listening, Speaking, Reading and Writing* (ed. D.A. Allport *et al.*) (in press). Academic Press, London.

——, Patterson, K.E., and Marshall, J. (ed.) (1980). *Deep Dyslexia*. Routledge and Kegan Paul, London.

Derousesne, J. and Beauvois, M.F. (1979). Phonological processing in reading: data from alexia. *Journal of Neurology, Neurosurgery and Psychiatry* **42**, 1125–32.

Ellis, A.W. (1987). Intimations of modularity or the modelarity of mind: doing cognitive neuropsychology without syndromes. In *The Cognitive Neuropsychology of Language* (ed. M. Coltheart *et al.*) (in press). Lawrence Erlbaum Assoc., London.

Funnell, E. (1983). Phonological processes in reading: new evidence from acquired dyslexia. *British Journal of Psychology* **74**, 159–80.

Hay, D.C. and Young, A.W. (1982). The human face. In *Normality and Pathology in Cognitive Functions* (ed. A.W. Ellis), pp. 173–202. Academic Press, London.

Howard, D. and Franklin, S. (1987). Three ways for understanding written words, and their use in two contrasting cases of surface dyslexia. In *Language Perception and Production: Common Processes in Listening, Speaking, Reading and Writing* (ed. D.A. Allport *et al.*) (in press). Academic Press, Hillsdale, New Jersey.

Huppert, F.A. and Tym, E. (1986). Clinical and neuropsychological assessment of dementia. *British Medical Bulletin* **42**, 11-18.

Marin, O.S.M. (1987). Dementia and visual agnosia. In *Visual Object Processing: A Cognitive Neuropsychological Approach* (ed. G.W. Humphreys and M.J. Riddoch) (in press). Lawrence Erlbaum Assoc., London.

Marr, D. (1976). Early processing of visual information. *Philosophical Transactions of the Royal Society of London* **B942**, 483-519.

Marshall, J.C. and Newcombe, F. (1973). Patterns of paralexia: A psycholinguistic approach. *Journal of Psycholinguistic Research* **2**, 175-99.

Miller, E. (1984). Psychological aspects in dementia. In *Dementia: A Clinical Approach*, (ed. J.M.S. Pearce) pp. 135-53. Blackwell Scientific Publications, Oxford.

Morton, J. and Patterson, K.E. (1980*a*). A new attempt at an interpretation, or, an attempt at a new interpretation. In *Deep Dyslexia* (ed. M. Coltheart *et al.*), pp. 91-118. Routledge and Kegan Paul, London.

—— and —— (1980*b*). 'Little words—No!' In *Deep Dyslexia* (ed. M. Coltheart *et al.*) pp. 270-85. Routledge and Kegan Paul, London.

Patterson, K.E. (1979). What is right with 'deep' dyslexic patients? *Brain and Language* **8**, 111-29.

—— (1986). Lexical but non-semantic spelling? *Cognitive Neuropsychology* **3**, 341-67.

——, Marshall, J.C., and Coltheart, M. (1985). *Surface Dyslexia: Neuropsychological and Cognitive Analysis of Phonological Reading*. Lawrence Erlbaum Associates, London.

Pearce, J.M.S. (1984*a*). Differential diagnosis. In *Dementia: A Clinical Approach* (ed. J.M.S. Pearce), pp. 46-55. Blackwells Scientific Publications, Oxford.

—— (1984*b*) Neurological signs in dementia. In *Dementia: A Clinical Approach* (ed. J.M.S. Pearce), pp. 56-70. Blackwells Scientific Publications, Oxford.

Ratcliff, G. and Newcombe, F. (1982). Object recognition: Some deductions from the clinical evidence. In *Normality and Pathology in Cognitive Functions* (ed. A.W. Ellis), pp. 147-71. Academic Press, London.

Saffran, E.M. (1982). Neuropsychological approaches to the study of language. *British Journal of Psychology* **73**, 317-38.

Schwartz, M.F., Linebarger, M.C., and Saffran, E.M. (1985). The status of the syntactic deficit theory of agrammatism. In *Agrammatism* (ed. M.L. Kean) pp. 83-125. Academic Press, Orlando, Florida.

——, Marin, O.S.M., and Saffran, E.M. (1979). Dissociations of language function in dementia. *Brain and Language* **7**, 277-306.

——, Saffran, E.M., and Marin, O.S.M. (1980). Fractionating the reading process in dementia: Evidence from word-specific print-to-sound associations. In *Deep Dyslexia* (ed. M. Coltheart *et al.*), pp. 259-69. Routledge and Kegan Paul, London.

Warrington, E.K. (1975). The selective impairment of semantic memory. *Quarterly Journal of Experimental Psychology* **27**, 635-57.

—— and Taylor, A.M. (1978). Two categorical stages of object recognition. *Perception* **7**, 695-705.

Whitaker, H. (1976). A case of isolation of the language function. In *Studies in Neurolinguistics*, Vol. 2 (ed. H. Whitaker and H.A. Whitaker), pp. 1-58. Academic Press, New York.

4

Amnesia, personal memory, and the hippocampus: experimental neuropsychological studies in monkeys

DAVID GAFFAN

There are many experimental tasks in which densely amnesic patients can be demonstrated to acquire information, retain it for long periods, retrieve it successfully and utilize it appropriately (for examples, see Warrington and Weiskrantz 1982). However, these patients lack personal recollection, the ability to describe their own recent experience, and they are severely impaired in recognition memory, the ability to discriminate items which they have recently experienced from those which they have not. The contrast between these deficiencies of personal memory and the normal retrieval of non-personal information in amnesia has often been remarked, and several technical terms have been proposed in attempts to explain or label the distinction between the impaired and unimpaired kinds of memory, from Claparède's (1911) 'moïté' to Kinsbourne and Wood's (1975) 'episodic and semantic' memories.

Delay and Brion (1969) were the first to systematically expound the hypothesis that human amnesia is caused by any bilateral interruption of the hippocampal system: the hippocampal formation itself in the medial temporal lobe; the fornix, which is the main output pathway of the hippocampal formation; and the mammillary bodies, which the fornix projects to. This hypothesis rests on an accumulation of neuropathological evidence which began in the last century and is still continuing (Heilman and Sypert 1977; Mair *et al.* 1979). Although the force of this evidence has sometimes been disputed (Horel 1978), the objections to it are not conclusive (Gaffan 1985*b*). However, there are considerable difficulties in drawing firm conclusions from the correlation of impairment with neuropathology in the clinical data. For this reason, memory tests for experimental animals have been developed in order to test the effect of experimentally controlled brain lesions upon aspects of memory performance similar to those in which amnesic patients are impaired.

Experimental studies of lesions of the hippocampal system in monkeys have shown that they produce a number of disorders of learning and memory, which are discussed below. These disorders, I argue, are fundamentally similar to those of human amnesics. Human memory in the strong sense of personal recollection implies knowledge not just of facts and events in the world, but of one's personal history in relation to events; monkeys with lesions of the hippocampal system are normal in tasks which require them to acquire and use information only about the relations of environmental events as such, but are deficient when they need to remember and learn about their own interaction with the environment. The hippocampal system receives information from temporal-lobe association cortex which is responsible for the identification of objects in the environment (D. Gaffan *et al*. 1986; E.A. Gaffan *et al*. 1986); and it also receives information from parietal-lobe association cortex (Fig. 4 of Jones and Powell 1970; Seltzer and Van Hoesen 1979) which is responsible for the orientation and direction of the body in and through personal space (Mountcastle *et al*. 1984) and for mnemonic spatial representations (Bisiach and Luzzatti 1978). Thus, the hippocampal system brings together information about environmental stimuli and information about personal space, and enables animals and people to remember and learn about the interaction of those two types of information.

LEARNING ABOUT THE DIRECTION OF MOVEMENTS

Mahut (1971, 1972) discovered that lesions of the hippocampal system in monkeys, either ablation of the hippocampal system or transection of the fornix, retarded learning rates in a task where the monkeys had to learn whether to go left or right in search of food. In an attempt to expand on these findings and clarify their interpretation, several experiments have recently examined the effects of fornix transection in a variety of tasks which require learning and memory of an animal's own direction of movement in relation to environmental stimuli, and these data can be briefly summarized.

A simple learning task will illustrate one kind of impairment which is produced in monkeys by transection of the fornix. In this task a monkey is presented on some trials with a visual stimulus, A, and on other trials with a different visual stimulus, B. If the monkey approaches and contacts A, a food reward is delivered; but if he avoids contact with A, no food is delivered. On the other trials, those in which B is the stimulus presented, the opposite rules apply: if he avoids contact with B a food reward is delivered, but if he approaches and contacts B no food reward is delivered. The monkey's learning of these relationships between the visual stimuli, his own movements in relation to them, and the delivery of food reward, is reflected in his behaviour by an increasingly reliable tendency to approach the visual stimulus on trials where it is A, and to avoid it on trials where it is B. Fornix-transected monkeys learn tasks of this form much more slowly than normal monkeys do

(Rupniak and Gaffan 1987; Experiment 1 of Gaffan and Harrison 1984).

The task just described requires learning about the animal's own movements in relation to visual stimuli as predictors of the delivery of reward. However, an impairment is also observed in the conceptually simpler task which requires a monkey to remember whether or not he has approached and contacted a visual stimulus. This task takes the form of a series of 'acquisition trials' and 'retention tests' in a memory task. At a pair of acquisition trials the monkey sees one stimulus A and makes manual contact with it, and sees another stimulus B and makes no contact with it. At the subsequent retention test with these two stimuli the monkey is offered a choice between A and B, and is rewarded for choosing B, the stimulus which evokes no memory of approach and contact. A similar procedure, a pair of acquisition trials followed by a retention test at which the monkey is rewarded for a choice of the stimulus not contacted in the acquisition trials, is then performed with C

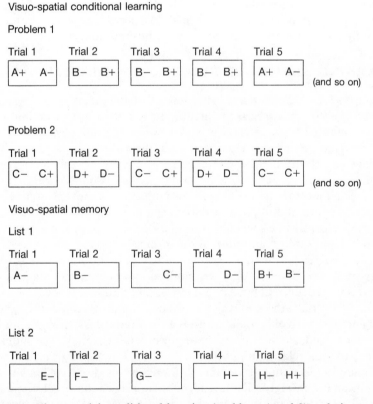

FIG. 4.1. Visuo-spatial conditional learning (problems 1 and 2) and visuo-spatial memory (lists 1 and 2).

and D, and so on with many pairs of new visual stimuli. The monkey rapidly learns to choose at retention tests whichever object he does not remember touching. The ability to make that choice correctly, reflecting the ability to remember which stimulus was approached and contacted, was severely impaired by fornix transection (Experiment 4 of Gaffan *et al*. 1984*b*; Experiment 2 of Gaffan *et al*. 1984*c*).

Similar impairments are observed in learning tasks which require a monkey to learn about his leftward and rightward movements in relation to visual stimuli. The learning task illustrated in the top part of Fig. 4.1 is a task which fornix-transected monkeys learn much more slowly than normal animals (Experiment 5 of Gaffan *et al*. 1984*a*). At each trial the animal must choose whether to reach out and touch the stimulus on his left, or on his right; opposite directions are rewarded for visually different stimuli, as illustrated. This is analogous to the task described above in which monkeys had to learn whether to approach and contact a visual stimulus or, on the contrary, to withdraw from it and avoid contact; the leftward and rightward movements on the part of the monkey determine whether a stimulus is rewarded or not, just as in the other experiments the monkey's approach and avoidance movements did. Similarly, just as in the conceptually simpler task of memory for approach and avoidance, there is also an impairment[1] when the monkey is asked to remember whether a stimulus has appeared on his left or on his right, as illustrated in the lower part of Fig. 4.1.

Although these tasks are labelled 'visuo-spatial' learning and memory tasks in Fig. 4.1, the vague word 'spatial' needs to be used with some caution in this context. O'Keefe and Nadel (1978) performed a great service by clearly distinguishing some of the very many senses in which an animal can be said to discriminate spatially, but even their lengthy treatment does not cover every possible sense. As discussed at length by Rupniak and Gaffan (1987),

---

1. The test of memory for visuo-spatial configurations shown in the lower part of Fig. 4.1 is one which Gaffan and Harrison employed in an unpublished experiment. At every choice trial, trial 5 of each list, the monkey could obtain a reward by touching that member of the displayed pair of identical stimuli which was in the same position as it had occupied during the acquisition trials of the list (trials 1 to 4). Every list had new visual stimuli in it, presented to the monkey on the computer display described by Gaffan *et al*. 1984*a*. Two monkeys were trained to a criterion of 90 per cent correct choices on the task with lists of four stimuli as illustrated, and were then subjected to fornix transection; subsequently, neither could re-attain criterial performance, even when the task was made easier by presenting lists of only two stimuli instead of four. This unpublished result is very similar to results from two previous experiments. Gaffan and Saunders (1985) observed similar effects of fornix transection in a task which similarly required memory of visuo-spatial configurations, the main difference being that in Gaffan and Saunders' experiment the discrimination was a successive (go, no-go) discrimination with single stimuli presented at retention tests, not pairs. Parkinson *et al*. (1987) observed a profound impairment following hippocampectomy in a similar task in the Wisconsin General Test Apparatus, requiring memory of visuo-spatial configurations. The unpublished task illustrated in Fig. 4.1 is similar to these previous experiments, but it has the advantage that its design makes as clear as possible the similarity between the conditional learning task (top) and the memory task (bottom).

'spatial' in the present context does not refer to O'Keefe and Nadel's 'cognitive map' of relationships between environmental stimuli, but rather to the relationships of environmental stimuli to the disposition of the person (limbs, head, and eyes).

These and other impairments of the fornix-transected monkeys in remembering their own orientation and direction of movement in relation to a stimulus, or in learning about the consequences of different possible orientations and movements in relation to a stimulus, contrast with inimpaired performance in many tasks requiring sensory memory only. One example of the latter is memory for the spatial (in a different sense) relation of objects to each other, when an animal has to learn that one object has another specific object under it (Experiment 3 of Gaffan *et al*. 1984*a*). Many other examples could be given; the unimpaired learning ability of fornix-transected monkeys extends, for instance, to auditory-visual conditional discrimination learning (Experiment 7 of Gaffan *et al*. 1984*a*), and this example is important in showing that the learning impairments in the tasks we have discussed above are specific to tasks involving personal orientation, and not representative of a general impairment in cross-modal integration or in learning about conditional relationships. I have reviewed elsewhere (Gaffan 1985*a*) the evidence for this distinction between purely sensory memory and memory about the animal's own movements. The new point I have to make concerns the comparison of these animal experiments to human amnesia. When one designs an experimental task to test some aspect of a monkey's associative memory—memory for what was under an object, for example—one may unofficially think of oneself as asking the monkey a question like 'What colour was the penny you found under this object last time you displaced it?' (This refers to the task of Gaffan and Bolton 1983, and of Experiment 3 of Gaffan *et al*. 1984*a*); but in fact, unless this object is associated with some penny of a different colour in circumstances of a different relation between the environmental stimuli and the monkey—unless, that is, the above question is contrasted with some question like 'What colour was the penny you saw revealed under this object last time you saw it without yourself displacing it?'—then it is a more correct analogy to think of oneself as asking 'Guess what colour is under this'. The importance of these analogies is that an amnesic patient may respond correctly and efficiently in the light of past experience when asked 'Please guess which of these has a penny under it' even though the same patient responds with 'I do not know, I have never seen these things before' to the question 'Which of these did you find a penny under?' (Gaffan 1972, pp. 339–40). However, in the case of the tasks for the monkey which ask questions like 'Was this rewarded when you approached it, or when you did not approach it?', the personal reference is inescapable.

RECOGNITION MEMORY

One of the most powerful forms of human memory is the ability to recognize items which are familiar from past experience, and to tell them apart from novel items. The items may be faces, words, pictures, places, tunes, or anything else. The experience of recognizing such an item is a common and easily understood example of memory as cognition, knowing that something is in one's own personal experience. The impressive power of human recognition memory is demonstrated in formal laboratory experiments in which a subject is shown a number of 'samples' to remember, and is subsequently asked at a retention test to discriminate the previously presented samples from novel 'foils' (Standing 1973; Gaffan 1978). Operationally, such recognition memory tasks can be distinguished from associative memory tasks; in the latter, of which the tasks in Fig. 4.1 are examples, the items at retention tests do not differ systematically in novelty or in recency of presentation. In formal experimental tests, defective recognition memory is observed in amnesic patients (Warrington 1974; Heilman and Sypert 1977). Experiments have also shown that animals, particularly macaque monkeys, possess excellent recognition memory ability, qualitatively comparable to that of people (Gaffan 1974, 1977). Lesions limited to the hippocampal system have often been shown to produce impairments of recognition memory in monkeys (Carr 1982; Gaffan 1974, 1977; Gaffan and Weiskrantz 1980; Mishkin 1978; Owen and Butler 1981; Zola-Morgan and Squire 1986). Thus, it is an attractive hypothesis that monkeys' recognition memory offers an animal model of exactly the 'awareness of the past', as opposed to mere acquisition of information, which is absent in amnesia. For all these reasons, tests of recognition memory have been by far the most commonly used behavioural task in experiments on memory and amnesia in monkeys, over the last 12 years, not only with hippocampal system lesions, but with a variety of other experimental interventions as well; Zola-Morgan and Squire (1986) have said that the recognition memory tasks of delayed matching or non-matching are the 'standard test' of experimental models of human amnesia in the monkey. It is important to realize, therefore, that there remain many important unanswered questions about the mechanisms of recognition memory performance in the monkey, about the categorization of different types of recognition memory task, and about the true analogies to be drawn between monkey and human recognition memory.

The severity of the impairment in recognition memory following damage to the hippocampal system in the monkey has varied widely from experiment to experiment. For example, fornix transection produced a much bigger recognition memory deficit in the experiment of Owen and Butler (1981) than in the experiment of Mahut *et al.* (1982), and ablation of the hippocampal formation in the medial temporal lobe produced a much bigger recognition memory deficit in the experiment of Zola-Morgan and Squire (1986) than in the experiment of Mishkin (1978). Other examples could be given. There are a large number of technical factors which explain at least some of these differences in the severity of impairment. These factors include: 'ceiling effects', that is the masking of an underlying impairment by a task that is so easy, for all animals, that very few errors are made by any animal; the difference between sub-total lesions, total lesions, and super-total lesions; and practice effects, as discussed at length and documented by Gaffan *et al.* 1984*b*. However, all of these technical factors taken together cannot explain all of the differences in existing results. Rather, it appears that the hippocampal system in the monkey makes some specific contribution to some specific aspect of recognition memory ability which some experimental procedures stress more than others. We do not yet know exactly what this contribution is.

One valuable line of evidence has come from asking what brain structures mediate the recognition memory ability which survives hippocampal system lesions. Although Mishkin (1978) observed a statistically significant impairment in recognition memory for lists of samples in his animals with hippocampal lesions alone, he observed a much more severe impairment in animals with lesions of the amygdala added to hippocampal lesions. Since amygdala lesions alone had little effect on recognition memory, he argued that only combined lesions to the amygdala and to the hippocampus produced a recognition memory impairment severe enough to be comparable to that of human amnesics. There are difficulties in accepting the argument that the combination of amygdala lesions with hippocampal system lesions produces a syndrome analogous to that of the human amnesic; amygdala lesions produce gross changes in the reponsiveness of monkeys to food, social signals, threats, and other emotional stimuli, and these changes, first described as an effect of amygdalectomy by Weiskrantz (1956), are not reminiscent of the behaviour of human amnesics. However, an important implication can be drawn from the fact that recognition memory, to the extent that it survives lesions of the hippocampal system, is dependent on the amygdala. It suggests that the specific involvement of the hippocampal system in recognition memory may be complementary to the specific involvement of the amygdala.

We have recently been performing experiments in associative learning and memory in the expectation that information about the specific role of the amygdala and hippocampus in associative memory would suggest what their specific roles in recognition memory might be. Our interpretation of our results from associative learning and memory tasks with lesions of the hippocampal system has been summarized above. With amygdalectomy, we have argued (Gaffan and Harrison 1987; E. A. Gaffan *et al.* 1987) that the amygdala is important for associating stimuli with primary reinforcers, that is, events of intrinsic motivational significance to the animal, such as food; and that it is not involved in associating stimuli with events other than primary reinforcers, such as secondary reinforcers or motivationally neutral stimuli. This conclusion may also have relevance for recognition memory. When a monkey learns to perform delayed non-matching in the Wisconsin General Test Apparatus, the apparatus employed in many of the experiments we have discussed above, the animal is repeatedly faced with two different objects and has to choose which one to displace. It is always the case that one object is novel (has not been seen by the monkey before) while the other is a familiar object (i.e. one of the objects that the monkey has recently seen). The novel object always, on every such choice trial, has a piece of food under it; the familiar object never does. Thus, visual novelty is directly, consistently predictive of the presence of food; visual novelty has come to play, for this monkey, a role similar to the colour and shape of a banana skin. This suggests one possible hypothesis about the role of the amygdala in recognition memory, namely, that when visual novelty or familiarity are themselves differentially associated with food reward (as in delayed non-matching and delayed matching, respectively), the amygdala in interaction with the visual association cortex is sufficient to allow the animal to discriminate visual familiarity from visual novelty, independently of the hippocampal system if necessary. However, conversely, when familiarity and novelty are not differentially predictive of primary reinforcement, we would expect the hippocampal system to be more exclusively important for recognition memory performance.

One result from fornix transection which this hypothesis explains is its abolition of spontaneous non-matching (Experiment 1 of Gaffan *et al.* 1984*b*). In spontaneous non-matching, the task is given to the animal in exactly the same way as delayed matching or non-matching except that both the sample and the foil are baited at each retention test. Thus, in spontaneous non-matching the animal demonstrates recognition memory by spontaneous preference for novel objects, and novelty is not differentially associated with food reward in the way that it is when an animal is trained in ordinary non-matching, with the foils and not the samples baited at retention tests. However, most strikingly, this hypothesis demands further experimental tests of recognition memory outside the popular matching and non-matching paradigms. Human recognition memory is not performed under circumstances where novelty or familiarity themselves directly predict primary reinforcement. Insofar as correct memory performance by people produces primary reinforcement, that reinforcement is not differentially associated with novelty or familiarity, but with both of them so long as they elicit correct conditional responses, such as verbal categorization of novel and familiar items. Thus, to model correctly this aspect of human recognition memory—the role of primary reinforcement—it will be necessary in future research to move away from the matching and non-matching paradigms, and to use instead, for example, some conditional discriminations in which both novel and familiar stimuli can lead to food reward, so long as the animal makes the correct discriminative conditional response to them.

In concentrating on analogies with human memory, however, one should not forget that the simple existing results from spontaneous recognition memory performance in animal exploration (Experiments 1 and 2 of Gaffan 1972; Experiment 1 of Gaffan *et al.* 1984*b*) suggest an important link between the two kinds of learning and memory which the hippocampal system is involved in. In the control of an animal's movement in relation to environmental stimuli, one of the main influences is exploration, and exploration requires an efficient recognition memory (Dember 1956).

CONCLUSION

No memory task should be regarded as a 'standard test' of animal models of human amnesia. Instead, experimental tasks should be subjected to task analysis to determine which of their features are crucial to the underlying psychological processes one is trying to measure, and which of their possible points of analogy with human memory performance are crucial to the adequacy of the model. The present brief review of some recent experimental tests of Delay and Brion's hypothesis of amnesia has shown that some progress has been made in the design of analytic experimental tasks in this area, but that more work remains to be done. However, it is a reasonable expectation that experimental analysis of memory processes in the monkey will soon reach the point where we will be able with confidence either to accept or to reject Delay and Brion's hypothesis as applied to monkeys. If it is accepted—the outcome which I expect, as has been made clear—then the same behavioural analysis of memory processes in the monkey can be applied to experimental tests of neurochemical manipulations designed to alleviate or prevent amnesia.

REFERENCES

Bisiach, E. and Luzzatti, C. (1978). Unilateral neglect of representational space. *Cortex* **14**, 129–33.

Carr, A.C. (1982). Memory deficit after fornix section. *Neuropsychologia* **20**, 95–8.

Claparede, E. (1911). Recognition et moiite. *Archives de Psychologie (Geneve)* **11**, 79–90.

Delay, J. and Brion, S. (1969). *Le syndrome de Korsakoff*. Masson, Paris.

Dember, W.N. (1956). Response by the rat to environmental change. *Journal of Comparative and Physiological Psychology* **49**, 93–5.

Gaffan, D. (1972). Loss of recognition memory in rats with lesions of the fornix. *Neuropsychologia* **10**, 327–41.

—— (1974). Recognition impaired and association intact in the memory of monkeys after transection of the fornix. *Journal of Comparative and Physiological Psychology* **86**, 1100–9.

—— (1977). Monkeys' recognition memory for complex pictures and the effect of fornix transection. *Quarterly Journal of Experimental Psychology* **29**, 505–14.

—— (1978). Measurement of trace strength in memory for pictures. *Quarterly Journal of Experimental Psychology* **30**, 263–81.

—— (1985*a*). Hippocampus: memory, habit and voluntary movement. *Philosophical Transactions of the Royal Society (London)* **B308**, 87–99. (Reprinted in *Animal Intelligence* (ed. L. Weiskrantz), pp. 87–99. Clarendon Press, Oxford.)

—— (1985*b*). Human and animal amnesia. In *Perspectives on Learning and Memory* (ed. L. Nilsson and T. Archer), pp. 279–89. Lawrence Erlbaum Associates, Hillsdale, NJ.

—— and Bolton, J. (1983). Learning of object-object associations by monkeys. *Quarterly Journal of Experimental Psychology* **35B**, 149–55.

—— and Harrison, S. (1984). Reversal learning by fornix-transected monkeys. *Quarterly Journal of Experimental Psychology* **36B**, 223–34.

—— and —— (1987). Amygdalectomy and disconnection in visual learning for auditory secondary reinforcement by monkeys. *Journal of Neuroscience* (in press).

—— and Saunders, R.C. (1985). Running recognition of configural stimuli by fornix-transected monkeys. *Quarterly Journal of Experimental Psychology* **37B**, 61–71.

—— and Weiskrantz, L. (1980). Recency effects and lesion effects in delayed non-matching to randomly baited samples by monkeys. *Brain Research* **196**, 373–86.

——, Saunders, R.C., Gaffan, E.A., Harrison, S., Shields, C., and Owen, M.J. (1984*a*). Effects of fornix transection upon associative memory in monkeys: role of the hippocampus in learned action. *Quarterly Journal of Experimental Psychology* **36B**, 173–221.

——, Gaffan, E.A., and Harrison, S. (1984*b*). Effects of fornix transection on spontaneous and trained non-matching by monkeys. *Quarterly Journal of Experimental Psychology* **36B**, 285–303.

——, Shields, C., and Harrison, S. (1984*c*). Delayed matching by fornix-transected monkeys: the sample, the push and the bait. *Quarterly Journal of Experimental Psychology* **36B**, 305–17.

——, Harrison, S., and Gaffan, E.A. (1986). Visual identification following inferotemporal ablation in the monkey. *Quarterly Journal of Experimental Psychology* **38B**, 5–30.

Gaffan, E.A., Harrison, S., and Gaffan, D. (1986). Single and concurrent discrimination learning by monkeys after lesions of inferotemporal cortex. *Quarterly Journal of Experimental Psychology* **38B**, 31–51.

——, Harrison, S., and Gaffan, D. (1987). Role of the amygdala in discrimination learning for food reward. (In preparation.)

Heilman, K.M. and Sypert, G.W. (1977). Korsakoff's syndrome resulting from bilateral fornix lesions. *Neurology* **27**, 490–3.

Horel, J.A. (1978). The neuroanatomy of amnesia: a critique of the hippocampal memory hypothesis. *Brain* **101**, 403–45.

Jones, E.G. and Powell, T.P.S. (1970). An anatomical study of converging sensory pathways within the cerebral cortex of the monkey. *Brain* **93**, 793–820.

Kinsbourne, M. and Wood, F. (1975). Short-term memory processes and the amnesic syndrome. In *Short-term memory* (ed. D. Deutsch and J.A. Deutsch) pp. 43–63. Academic Press, New York.

Mahut, H. (1971). Spatial and object reversal learning in monkeys with partial temporal lobe ablations. *Neuropsychologia* **9**, 409–24.

—— (1972). A selective spatial deficit in monkeys after transection of the fornix. *Neuropsychologia* **10**, 65–74.

——, Zola-Morgan, S., and Moss, M. (1982). Hippocampal resections impair associative learning and recognition memory in the monkey. *Journal of Neuroscience* **2**, 1214–29.

Mair, W.G.P., Warrington, E.K., and Weiskrantz, L. (1979). Memory disorder in Korsakoff's psychosis: a neuropathological and neuropsychological investigation of two cases. *Brain* **102**, 749–83.

Mishkin, M. (1978). Memory in monkeys severely impaired by combined but not by separate removal of amygdala and hippocampus. *Nature* **273**, 297–8.

Mountcastle, V.B., Motter, B.C., Steinmetz, M.A., and Duffy, C.J. (1984). Looking and seeing: the visual functions of the parietal lobe. In *Dynamic Aspects of Neocortical Function* (ed. G.M. Edelman, W.E. Gall and W.M. Cowan), pp. 159–93. Wiley, New York.

O'Keefe, J. and Nadel, L. (1978). *The Hippocampus as a Cognitive Map*. Oxford University Press, Oxford.

Owen, M.J. and Butler, S.R. (1981). Amnesia after transection of the fornix in monkeys: long-term memory impaired, short-term memory intact. *Behavioural Brain Research* **3**, 115–23.

Parkinson, J.K., Murray, E.A., and Mishkin, M. (1987). Evidence for the participation of the hippocampus in memory for the location of objects. (In preparation.)

Rupniak, N.M.J. and Gaffan, D. (1987). Monkey hippocampus and learning about spatially directed movements. *Journal of Neuroscience* (in press).

Seltzer, B. and Van Hoesen, G.W. (1979). A direct inferior parietal lobule projection to the presubiculum in the Rhesus monkey. *Brain Research* **179**, 157–61.

Standing, L. (1973). Learning 10,000 pictures. *Quarterly Journal of Experimental Psychology* **25**, 207–22.

Warrington, E.K. (1974). Deficient recognition memory in organic amnesia. *Cortex* **10**, 289–91.

—— and Weiskrantz, L. (1982). Amnesia: a disconnection syndrome? *Neuropsychologia* **20**, 233–48.

Weiskrantz, L. (1956). Behavioral changes associated with ablation of the amygdaloid complex in monkeys. *Journal of Comparative and Physiological Psychology* **49**, 381–91.

Zola-Morgan, S. and Squire, L.R. (1986). Memory impairment in monkeys following lesions limited to the hippocampus. *Behavioral Neuroscience* **100**, 155–60.

5

Primate models of senile dementia

N.M.J. RUPNIAK AND S.D. IVERSEN

INTRODUCTION

The complexity of intellectual and neuropathological changes occurring in senile dementia constitutes a daunting challenge for the development of a convincing animal model of the disease. The establishment of such a model would represent a major advance in our understanding of dementia and related conditions, and enable rational approaches for the development of effective drug therapies. The considerable degree of neuroanatomical and behavioural sophistication of non-human primates affords a unique opportunity for the experimental examination of cognitive processes under controlled conditions following systematic manipulation of brain function, and has contributed greatly to our understanding of brain mechanisms and cognition. This chapter will review several different approaches to the development of primate models of dementia and their utility for pharmaceutical research.

TRANSMISSIBLE VIRAL DEMENTIA

Dementia is not a condition *known* to occur spontaneously in any non-human species. In man, infection by the viral agent responsible for Creutzfeldt–Jakob disease causes dementia and neuropathological changes resembling those of Alzheimer's disease. The ability to transmit Creutzfeldt–Jakob disease to primates has raised the possibility that Alzheimer's disease might also be virally transmitted, and that dementia with identical symptomatology and pathology might be induced in infected primates.

Subacute viral encephalopathy has been transmitted to several Old and New World primate species following intracerebral inoculation with brain tissue from patients suffering from Creutzfeldt–Jakob disease (Gajdusek and Gibbs 1975). Neuropathology, characterized predominantly by the slowly progressive induction of status spongiosus, and destruction and loss of neurones, is confined mainly to cortical areas. Although infected primates exhibit 'apparent dementia' the behavioural syndrome has not been evaluated systematically and it is not known whether cognitive impairments comparable to those seen in Alzheimer's disease are present. The utility of this particular primate model of dementia for studies of cognitive ability appears to be severely limited by the extreme severity of illness, which normally results in death less than 12 months after symptoms first develop, and the presence of other neurological signs (such as myoclonus) which might interfere with performance in behavioural tests. The successful transmission of Alzheimer's disease to primates might be expected to induce less severe illness and longer survival times. However, despite numerous experiments involving almost 150 inocula from patients suffering from Alzheimer's disease, encephalopathy has not been reliably induced in any primate species examined (see Goudsmit *et al.* 1980; Brown *et al.* 1982). The most plausible interpretation of this failure is that Alzheimer's disease is not virally trans-mitted, and that any neuropathological and symptomatological similarities with Creutzfeldt–Jakob disease result from neurodegenerative changes of unknown cause.

The possibility of inducing Alzheimer's disease in primates should, how-ever, not be completely dismissed since the failure of transmission experi-ments may not be surprising for several reasons. Unlike Creutzfeldt–Jakob disease, Alzheimer's disease predominantly affects patients in old age. The induction of neuritic plaques in Creutzfeldt–Jakob patients is also seen most frequently in older patients. The formation of plaques (a neuropathological hallmark feature of Alzheimer's disease) in the brains of the young primates inoculated with Alzheimer tissue might, therefore, be expected to follow a greatly protracted time course, and might not emerge at all until the animals were greatly advanced in age. Indeed, it is particularly noteworthy in this res-pect that such young primates which exhibited spongiform encephalopathy after inoculation with Kuru or scrapie-infected tissue were not reported to have developed neuritic plaques, even though plaques are frequently present in these diseases in man and rodents (Gajdusek and Gibbs 1975). Moreover, slow acting viruses appear to undergo transformation on serial passage from host to host which may greatly alter the duration of the incubation period as well as the type of neuropathology (Gajdusek and Gibbs 1975; Bruce 1984). For example, fifteen different strains of scrapie have now been identified in mice with incubation periods ranging from 140 days to the animal's lifespan, depending on the particular murine strain and the route of inoculation. Whereas certain scrapie/mouse strain combinations result in plaque forma-tion, others do not (Bruce 1984). Thus, the problems of successfully trans-mitting Alzheimer's disease to primates using compatible inocula and hosts, even following a suitably prolonged incubation period, might be consider-able. However, if Alzheimer's disease *is* induced by a viral pathogen, the demonstration of transmission to primates could have widespread implica-tions for the control and prevention of the disease, as well as providing an extremely valuable model for the evaluation of therapeutic agents to alleviate the symptoms of dementia.

NEUROTOXIC METALS

Exposure to organic and inorganic metals such as trimethyltin, organoleads (petroleum additives) and aluminium may induce progressive encephalopathy in man with neurological signs including convulsions, hallucinations, insomnia, emotional disturbances, disorientation, and memory deficits (Ross *et al*. 1981; Valpey *et al*. 1978; Goldings and Stewart 1982; Alfrey *et al*. 1976). Neuropathological changes in man and rodents following exposure to these agents are confined predominantly to the limbic cortex, cerebellum, hippocampus, and septum (Dyer *et al*. 1982; Wisniewski and Terry 1970; Valpey *et al*. 1978; Seawright *et al*. 1980). It has been suggested that environmental exposure to toxins such as these might be responsible for certain neurodegenerative changes characteristic of Alzheimer's disease. Rabbits treated with aluminium or lead develop neurofibrillary tangles consisting of bundles of smooth or occasionally twisted tubules (Wisniewski *et al*. 1965; Niklowitz 1975). With the finding of increased aluminium levels in areas showing neuronal degeneration in the brains of Alzheimer patients it was suggested that aluminium might play a role in the pathogenesis of the disease (Crapper *et al*. 1973). This proposal now seems unlikely for the following reasons.

1. Tangles in aluminium-treated rabbits do not resemble those seen in Alzheimer's disease, which consist of abnormal helically wound, paired neurofilaments (Wisniewski *et al*. 1976)

2. There are considerable species variations in the ability of neurotoxic metals to induce neurofibrillary changes. Although tangles have been induced by aluminium or lead treatment in rabbits and dogs, they are not associated with lead or aluminium-induced encephalopathy in rodents or primates including man (see Wisniewski *et al*. 1977; Zook and Paasch 1980)

3. The finding of increased brain aluminium levels in Alzheimer's disease has not been replicated in other studies, although there does appear to be an increase in aluminium concentrations with age which is not associated with dementia (Wisniewski *et al*. 1977). Moreover, there is no evidence of a causal link between high aluminium levels and tangle formation in man or that chronic, low level environmental exposure to such a neurotoxin is capable of inducing any neuropathological change.

Despite these findings, neurotoxic metals provide an extremely interesting method of producing an animal model of cognitive impairment owing to their apparently selective effects on the limbic system. Tin and lead-induced neurotoxicity in rodents is asssociated with impairments in learning and memory tasks, such as maze learning, visual reversal learning, and retention of passive avoidance (see Walsh and Tilson 1984 for review). There are no published studies of the cognitive effects of tin or aluminium exposure in primates, but there have been several reports on the behavioural effects of lead intoxication.

Infant primates are more susceptible to lead intoxication than adults or juveniles, a finding which may be attributable to their greater capacity to absorb lead from the gastrointestinal tract (see Willes *et al.* 1980). Behavioural effects of lead poisoning have, therefore, been evaluated in young macaques (aged 2–3 years) following neonatal ingestion. Such animals are not impaired in learning various tactile or visual discriminations, but are severely impaired in visual discrimination *reversal* learning (see Willes *et al.* 1980). Using a similar paradigm, Zook *et al.* (1980) found no impairment in *tactile* discrimination reversals. It is not clear whether the impairment in reversal learning is modality-specific, or whether the degree of lead exposure in the study by Zook *et al.* was insufficient to induce cognitive impairment since visual discrimination reversal was not examined. It is also not known whether the deficit in visual reversal learning is attributable to a memory impairment or the acquisition of an abnormally strong response bias towards the original positive and negative discriminative stimuli as might occur from an emotional or motivational disturbance. The effects of psychoactive drugs known to alter short-term memory processes have not been examined in this model. More detailed analysis of the nature and degree of cognitive impairment induced by lead in primates using a wider range of task paradigms is required to enable evaluation of this syndrome as a behavioural model of dementia.

AGED PRIMATES

A number of neuropathological changes occurring in Alzheimer's disease are also seen to a lesser degree during normal aging in man and primates. The aged primate is an extremely important model of senile dementia since these alterations are associated with a well characterised decline in cognitive abilities.

The brains of macaques aged 16–30 years are slightly shrunken with enlarged lateral ventricles and contain a low density (< 5 per mm^2) of neuritic plaques rich in amyloid in the amygdala, hippocampus, and cerebral cortex which are not present in the brains of young animals (Wisniewski *et al.* 1973; Struble *et al.* 1982). The density of plaques appears to increase with age, as does amyloid content (Struble *et al.* 1982). More rarely, pairs of helically wound neurofilaments have also been observed (Wisniewski *et al.* 1973).

The deposition of plaques appears to be related to the disruption of activity in a number of neurotransmitter systems since they show abnormal acetycholinesterase activity and immunoreactivity to somatostatin and tyrosine hydroxylase (Struble *et al.* 1984*a*, *b*; Kitt *et al.* 1985). Over the age span of 2–18 years, primate brains exhibit a marked reduction (50 per cent) in endogenous dopamine levels in the frontal and temporal lobes, and reduced synthesis of dopamine and noradrenaline in all sensory and association cortical areas (Goldman-Rakic and Brown 1981). The possible relationship between these neurochemical alterations and cognitive impairment has been examined pharmacologically in aged primates using drugs which alter acetylcholine, catecholamine, and neuropeptide function.

Aged rhesus monkeys manifest a number of cognitive impairments, including poor performance in learning lists of concurrent visual discriminations and visual discrimination reversals (Medin *et al.* 1973; Bartus *et al.* 1979; Table 5.1). These deficits do not appear attributable to perceptual or associative abnormalities since visual discrimination learning is unimpaired (Davis 1978; Bartus *et al.* 1979), and performance is not improved by the presentation of stimuli to be remembered either several times or for longer periods (Bartus *et al.* 1978). Rather, the deficit seems related to an increased sensitivity to proactive and retroactive interference since the inclusion of irrelevant visual distractors during retention of a spatial task drastically impairs the performance of aged, but not young monkeys (Davis 1978; Bartus and Dean 1979). Aged monkeys also show more rapid rates of forgetting over long retention intervals (up to 60 seconds) than young animals in the spatial delayed response task, although their performance at short delays (around 5 seconds) is normal (Medin 1969; Bartus *et al.* 1979; Marriott and Abelson 1980; Arnsten and Goldman-Rakic 1985). In summary, aged primates show specific impairments in visual and spatial short-term memory processes which may be comparable to those observed in senile dementia.

TABLE 5.1. *Cognitive impairment in aged macaques*

Age (years)	Task	Reference
20–21	Concurrent visual discriminations (lists of 20)	Medin *et al.* (1973)
18–23	Visual discrimination reversal	Bartus *et al.* (1979)
18–23	Spatial delayed response	Medin (1969); Bartus *et al.* (1979); Marriott and Abelson (1980); Arnsten and Goldman-Rakic (1985)
18–21	Spatial delayed response with distraction during retention interval	Davis (1978); Bartus and Dean (1979)

TABLE 5.2 *Pharmacological enhancement of memory in aged primates*

Neurotransmitter system	Drug and dose	Improvement	Reference
Acetylcholine	Physostigmine (0.0025–0.1 mg/kg)	Yes	Bartus (1979); Bartus *et al.* (1980)
	Arecoline (0.025–0.1 mg/kg)	Yes	
	Choline (30–400 mg/kg)	No	
Dopamine	Apomorphine (0.025–0.2 mg/kg)	No	Bartus (1981); Arnsten and Goldman-Rakic (1985)
	L-dopa (0.025–0.2 mg/kg)	No	
Noradrenaline	Clonidine (0.0025–0.08 mg/kg)	No Yes	Bartus (1981); Arnsten and Goldman-Rakic (1985)
Neuropeptide	ACTH (100–400 mg/kg)	Yes	Bartus *et al.* (1982)
	Vasopressin (1–8 mg/kg)	Yes	
	Oxytocin (2–128 mg/kg)	No	
	Somatostatin (0.1–100 mg/kg)	Yes	

The effects of psychoactive drugs on memory in aged primates have been examined exclusively in the spatial delayed response paradigm (Table 5.2). The ability of various classes of drugs to improve memory in aged primates correlates well with their clinical effects in senile dementia. Facilitation of memory has been most consistently observed following treatment with the acetylcholinesterase inhibitor physostigmine and the muscarinic agonist arecoline (Bartus 1979; Bartus *et al.* 1980). However, the dose-response curve for facilitation by physostigmine is extremely steep, with improvement at only one or two doses being followed by impairment at higher doses. Considerable individual differences in response to physostigmine occur, with 100-fold variations in best dose. The performance of some aged animals was not improved by physostigmine, and no significant overall group effect was obtained for this drug. The between subject reliability was greater for enhancement with arecoline, but again the effective dose range was very narrow (0.05–0.075 mg/kg). Attempts to stimulate central acetylcholine synthesis following administration of choline (30–400 mg/kg) did not improve performance (Bartus *et al.* 1980).

Drugs stimulating central dopamine function (apomorphine and L-dopa) have not been found to improve memory in old monkeys (Bartus *et al.* 1983; Arnsten and Goldman-Rakic 1985). In contrast, treatment with the α-adrenoreceptor agonist clonidine markedly improved performance in a dose-related manner in one study (Arnsten and Goldman-Rakic 1985), but not in another (Bartus 1981). A possible reason for this discrepancy may relate to the duration of treatment with this drug, which has sedative effects when given acutely.

A number of neuropeptides have been administered to aged primates (Bartus *et al.* 1982). ACTH, vasopressin, and somatostatin all improved performance in some individuals. Between subject variability was extremely high (one out of six monkeys responded positively to somatostatin and approximately half of the group to ACTH or vasopressin); beneficial effects were typically seen at a single dose only. Administration of oxytocin did not improve performance in any animal and impaired that of three out of six monkeys.

Whilst these findings do not demonstrate a clear-cut effect of any treatment on cognition in aged monkeys, they do indicate a contribution of cholinergic, monoamine, and peptide systems in age-related cognitive decline. It is not known why consistent dose-related improvements are generally not observed, but the following possibilities are suggested:

(i) a modulatory, rather than a direct mediational role for these neurotransmitters in short-term memory; or

(ii) aberrant neuronal transmission in the aged brain, as is suggested by the greater sensitivity and more consistent facilitation of performance by physostigmine in young monkeys (Bartus 1979).

It would be extremely interesting to determine whether the degree of cognitive impairment in any aged individual determines or influences the ability of these agents to enhance performance, and to examine any possible interactions between agents acting on different neurotransmitter systems in improving performance in individual animals.

Finally, a group of compounds which do not act selectively on any single neurotransmitter system have been examined in this primate model. Drugs belonging to this category are the so-called nootropics ('metabolic enhancers') or central nervous system stimulants whose putative therapeutic efficacy in Alzheimer's disease remains controversial. Some evidence for improved performance was obtained in a proportion of individuals treated chronically, but not acutely, with the nootropics piracetam, vincamine, centrophenoxine, and dihydroergotoxine (Bartus *et al.* 1983). Four CNS stimulants (methylphenidate, magnesium penoline, caffeine, and a pentylenetetrazole/niacin mixture) did not improve and often impaired memory in aged as well as young primates (Bartus 1979).

DRUG-INDUCED AMNESIA

The induction of a temporary amnesic syndrome in monkeys following systemic drug administration has certain advantages over other models of dementia. In screening for agents to facilitate memory, this technique enables precise experimental control over the degree of cognitive impairment induced and, because the effect is short-lasting, the effects of agents which reverse drug-induced amnesia can be compared with each individual's normal performance in an undrugged state.

Psychoactive drugs can affect complex behaviours in a number of ways which are not directly involved in memory processes (attentional, perceptional, motivational, emotional, and motor effects). In order to determine whether drugs specifically influence retention of new information, the testing procedure should enable examination of the interaction of dose with forgetting over retention intervals of different duration. Drugs which selectively disrupt performance over long, but not short retention intervals can be argued to act selectively upon memory mechanisms. The paradigm should be designed so as to vary the length of the retention interval from trial to trial in a random counterbalanced manner so that performance at any given retention interval is not confounded by the effects of rising or falling drug levels. Remarkably few drugs have been specifically tested in such paradigms and it is entirely possible that many classes of compounds other than those reviewed here may be capable of selective disruption of memory processes in primates.

Scopolamine

In man, the muscarinic antagonist scopolamine is known to impair delayed (but not immediate) recall of information such as a string of digits (Drachman and Leavitt 1974). A similar effect on recent memory is also induced in primates by scopolamine. Choice of appropriate tasks in which to reveal a selective disruption of memory is complicated by the induction of dose-related perceptual impairments in visual discrimination tasks in monkeys (Evans 1975; Bartus and Johnson 1976). This effect probably accounts for the failure to demonstrate an interaction between dose and performance over short to long retention intervals in delayed matching-to-sample tasks using complex visual stimuli (Bohdanecky *et al.* 1967; Robustelli *et al.* 1969). Although scopolamine induces a dose-related impairment in non-matching-to-sample using similar discriminative stimuli (Aigner *et al.* 1984; Aigner and Mishkin 1986) it was not stated whether this effect is greater after long than short retention intervals in this paradigm. If the discriminative stimuli are altered so that the cue to problem solution is colour rather than any other visual attribute (delayed colour matching-to-sample), a dose related impairment, increasing in magnitude at longer (over 30 sec) retention intervals and absent at short (0 seconds) delays, has been reported after treatment of monkeys with either scopolamine (0.025–0.1 mg/kg i.m.; Glick and Jarvik 1970) or atropine (0.014–0.44 mg/kg i.m.; Penetar and McDonough 1983). A similar specific disruption of short-term memory after long retention intervals by scopolamine (0.01–0.03 mg/kg i.m.), resembling that seen spontaneously in aged monkeys, has been observed in the spatial delayed response task (Bartus and Johnson 1976). Using a single dose of scopolamine (0.06 mg/kg i.m.) administered either just prior to or immediately following visual discrimination learning, or just before a reversal or retention task 24 hours later, Ridley *et al.* (1984) have suggested that scopolamine impairs new learning, impairs the consolidation of new information into long-term memory and may induce a retrieval deficit in marmosets. The ability of scopolamine to induce amnesia in both visual recognition and spatial memory paradigms suggests the involvement of the cholinergic system in similar age-related cognitive impairments.

Scopolamine-induced (0.01–0.02 mg/kg amnesia in the spatial delayed response task can be partially, but consistently reversed by concurrent treatment with physostigmine (0.02–0.03 mg/kg i.m.), presumably as a consequence of elevated synaptic acetylcholine concentration (Bartus 1978). It is not known whether physostigmine is also capable of reversing the scopolamine deficit on delayed colour matching. It is also not known whether the scopolamine deficit can be reversed by co-administration of any other direct or indirectly acting muscarinic agonist in either paradigm, or by clonidine or the nootropic drugs claimed to enhance the performance of aged monkeys. Until the effects of these agents are examined it is not possible to assess fully the utility of the scopolamine-induced impairment in young animals as a model of age-related cognitive decline or its potential as a screening test for novel therapeutic agents. There are no published studies demonstrating reversal of the scopolamine deficit by any drugs in the colour matching task.

Amphetamine

Treatment of primates with amphetamine or methylphenidate is capable of inducing amnesia bearing a striking similarity to that caused by scopolamine. This finding is rather paradoxical since in man it is well-known that scopolamine impairs short-term memory, but there is no evidence for a similar effect of amphetamine or methylphenidate. On the contrary, both of these agents are believed to improve memory in normal subjects and hyperactive children (see Evans *et al.* 1986).

In primates amphetamine treatment (0.1–0.56 mg/kg i.m.) impairs visual discrimination reversal learning (Ridley *et al.* 1980, 1981) and delayed colour matching-to-sample (Glick and Jarvik 1970; Bauer and Fuster 1978). The latter deficit is dose-related and of greatest magnitude at long retention intervals, suggesting a specific disruption of short-term memory. Amphetamine has not been examined in the spatial delayed response paradigm, but methylphenidate (0.1–0.8 mg/kg i.m.) caused dose-dependent impairment of performance in this task (Bartus 1979). The author does not state whether there was an interaction between dose and duration of the retention interval.

There is some evidence for an interaction between central cholinergic and dopaminergic systems in mnemonic processes since co-administration of scopolamine with amphetamine (0.2 mg/kg i.m.) in the colour matching task, or with methylphenidate (0.0125 mg/kg i.m.) in spatial delayed response exacerbates the impairment induced by either drug alone (Glick and Jarvik 1969; Bartus 1978). However, it is not known whether amphetamine-induced amnesia can be reversed by physostigmine, or any other drug showing activity in the aged primate model.

Tetrahydrocannabinol

The active constituent of marihuana, delta-9-tetrahydrocannabinol (THC) interferes with immediate memory in man (Tinklenberg *et al*. 1970) and, like scopolamine and amphetamine, appears to selectively disrupt recent memory in primates. Rhesus monkeys treated with THC (0.5–2 mg/kg p.o.) showed dose-dependent impairment of performance on a colour non-matching-to-sample task, with lower doses affecting performance after long delays (32 seconds) and the highest dose disrupting performance across all retention intervals (0–32 seconds; Zimmerberg *et al*. 1971). The effects of other drugs on THC-induced amnesia in monkeys have not been investigated.

Yohimbine

Performance of young monkeys on the spatial delayed response task with variable retention intervals (0–32 seconds) is dose-dependently impaired following administration of the α_2-adrenergic receptor antagonist yohimbine (0.5–1.5 mg/kg i.m.) (Arnsten and Goldman-Rakic 1985). Unfortunately, the authors do not specify whether there was any interaction between dose and retention interval. However, the finding is of great significance since aged monkeys were most susceptible to the effects of yohimbine in the lower dose range of 0.1–0.5 mg/kg, suggesting a link between α_2-adrenoreceptor function and age-related memory deficits in this task. Neither prazosin (an α_1 selective antagonist) nor propranolol (a β-antagonist) induced impairments in either age group. Deficits induced by yohimbine were reversed by concurrent administration of clonidine (0.02 mg/kg). Further characterization of the yohimbine deficit in other (non-spatial) tasks and its possible reversal by other drugs would be of considerable value.

LESION MODELS

Amnesia in Alzheimer's disease is believed to result from widespread neuro-degenerative changes in temporal and frontal lobe structures. Damage to these areas in man and animals causes specific disruption of recent memory and might provide a useful primate model in which to detect memory facilitation by therapeutic agents.

Temporal lobectomy in man (involving damage to the hippocampus and amygdala) causes global amnesia which has been attributed to rapid forgetting of new spatio-temporal information (Milner 1978). Hippocampal damage in primates (following direct hippocampal ablation or transection of the fornix) was originally believed to cause selective impairments in spatial, but not object, reversal learning (Mahut 1972; Jones and Mishkin 1972) and in spatial, but not nonspatial, delayed alternation (Mahut 1971) suggesting a special role of the hippocampus in spatial learning. It was only following combined lesions of both the hippocampus and amygdala (not either structure alone) that profound memory loss after short retention intervals (10–120 seconds) was observed in a visual non-matching-to-sample task (Mishkin 1978). However, Gaffan (1974) found that fornix transection alone induced a retention deficit over delays of 70 and 130 seconds in the matching-to-sample paradigm. The failure to observe such an impairment in Mishkin's nonmatching task might reflect the spontaneous tendency of both normal and fornix-transected monkeys to choose novel stimuli (Gaffan *et al.* 1984). Attempts to reverse memory loss pharmacologically in this model have not yet been undertaken and would be of great interest.

Damage to the prefrontal cortex in man causes spatial disorientation and disorders in temporal sequencing of recent information (Milner 1971). The nature of cognitive impairment observed in aged primates bears a striking similarity to deficits induced by frontal lobe lesions. Monkeys with frontal lobe lesions have severe impairments in delayed matching, reversal and alternation tasks which require memory of previous events to solve spatial and non-spatial problems (Mishkin *et al.* 1969; Goldman and Rosvold 1970). Damage to the dorsolateral prefrontal cortex following local injection of the neurotoxin 6-hydroxydopamine induces a similar marked impairment in spatial delayed response performance which appears attributable predominantly to the regional depletion of dopamine. The impairment could be reversed by administration of the dopamine agonists L-dopa or apomorphine within a restricted dose range (Brozoski *et al.* 1979). However, the role of dopamine in memory is far from clear, since neither agonist is capable of improving the performance of surgically lesioned or aged primates (Brozoski *et al.* 1979; Bartus 1981; Arnsten and Goldman-Rakic 1985). Moreover, administration of amphetamine typically *impairs* delayed matching in normal animals, and further exacerbated the impairment induced by cooling of the prefrontal cortex (Bauer and Fuster 1976). These conflicting findings may be attributable to differences in neurotransmitter alterations in the different models, or to the nature of the dose-response curve to dopamine agonists (facilitation occurring within a narrow range and disruption by higher doses).

A third focus for examination of lesion-induced memory impairments in primates is the nucleus basalis of Meynert, the major source of neocortical cholinergic projections which is thought to be involved in cognitive impairments seen in Alzheimer's disease. The neuroanatomical distribution of this nucleus makes it an extremely difficult target to lesion completely without considerable damage to surrounding structures. Possibly for this reason, attempts to establish memory impairments in basalis-lesioned monkeys have been disappointing, although such animals show increased susceptibility to disruption by scopolamine in delayed nonmatching-to-sample (Aigner *et al.* 1984). The effects of combined lesions to this area and temporal or frontal lobe structures have not yet been evaluated in primates.

CONCLUSIONS

The studies described in this chapter offer a number of exciting possibilities for the preclinical evaluation of drug effects on cognition. The development of a primate model of dementia has been approached from a wide range of perspectives which together indicate clear directions for future research. The potential therapeutic benefits of further development of these models is particularly well illustrated by the remarkably high degree of correlation between drug-induced memory facilitation in aged primates and clinical findings with the same agents in Alzheimer's disease. The knowledge gained from continued research in all these models could have far-reaching consequences for the treatment of many illnesses which affect cognitive function, including Huntington's chorea, schizophrenia, multiple infarct dementia, toxic confusional states, organic amnesic syndromes, and infectious diseases.

REFERENCES

Aigner, T.G. and Mishkin, M. (1986). The effects of physostigmine and scopolamine on recognition memory in monkeys. *Behaviour Neurology and Biology* **45**, 81–7.

Aigner, T.; Mitchell, S., Aggleton, J., Delong, M., Struble, R., Price, D., and Mishkin, M. (1984). Effects of scopolamine and physostigmine on recognition memory in monkeys after ibotenic acid injections into the area of the nucleus basalis of Meynert. In *Alzheimer's Disease: Advances in Basic Research and Therapies* (ed. R.J. Wurtman, S.H. Corkin, and J.H. Growdon), p. 429. Raven Press, New York.

Alfrey A.C., LeGendre, G.R., and Kaehny, W.D. (1976). The dialysis encephalopathy syndrome. Possible aluminium intoxication. *New England Journal of Medicine* **294**, 184–8.

Arnsten, A.F.T. and Goldman-Rakic, P.S. (1985). Catecholamines and cognitive decline in aged nonhuman primates. *Annals of the New York Academy of Sciences* **444**, 218–34.

Bartus, R.T. (1978). Evidence for a direct cholinergic involvement in the scopolamine-induced amnesia in monkeys: Effects of concurrent administration of physostigmine and methylphenidate with scopolamine. *Pharmacology, Biochemistry and Behaviour* **9**, 353–7.

—— (1979). Physostigmine and recent memory: effects in young and aged nonhuman primates. *Science* **206**, 1087–9.

—— (1981). Age-related memory loss and cholinergic dysfunction: Possible directions based on animal models. In *Strategies for the Development of an Effective Treatment for Senile Dementia* (ed. T. Crook and S. Gershon), pp. 71–89. Mark Powley Assoc. Inc., Connecticut.

—— and Dean, R.L. (1979). Recent memory in aged non-human primates: hypersensitivity to visual interference during retention. *Experimental Ageing Research* **5**, 385–400.

——, and Johnson, H.R. (1976). Short-term memory in the rhesus monkey: disruption from the anti-cholinergic scopolamine. *Pharmacology, Biochemistry and Behaviour* **5**, 39–46.

——, Fleming, D.L., and Johnson, H.R. (1978). Ageing in the rhesus monkeys: debilitating effects on short-term memory. *Journal of Gerontology* **33**, 858–71.

——, Dean, R.L., and Fleming, D.L. (1979). Ageing in the rhesus monkey: effects on visual discriminaion learning and reversal learning. *Journal of Gerontology* **34**, 209–19.

——, Dean R.L., and Beer B. (1980). Memory deficits in aged cebus monkeys and facilitation with central cholinanimetics. *Neurobiology of Ageing*, **1**, 145–52.

——, Dean, R.L., and Beer, B. (1982). Neuropeptide effects on memory in aged monkeys. *Neurobiology of Ageing* **3**, 61–8.

——, ——, and —— (1983). An evaluation of drugs for improving memory in aged monkeys: implications for clinical trials in human. *Psychopharmacology Bulletin* **19**, 168–84.

Bauer, R.H. and Fuster, J.M. (1978). Effects of d-amphetamine and prefrontal cortical cooling on delayed matching-to-sample behaviour. *Pharmacology, Biochemistry and Behaviour* **8**, 243–9.

Bohdanecky, Z., Jarvick, M.E., and Carley, J.L. (1967). Differential impairment of delayed matching in monkeys by scopolamine and scopolamine methylbromide. *Psychopharmacologia* **11**, 293–9.

Brown, P., Salazar, A.M., Gibbs, C.J., and Gajdusek, D.C. (1982). Alzheimer's disease and transmissible virus dementia (Creutzfeldt–Jakob disease). *Annals of the New York Academy of Sciences* **396**, 131–43.

Brozoski, T.J., Brown, R.M., Rosvold, H.E., and Goldman, P.S. (1979). Cognitive deficit caused by regional depletion of dopamine in perfrontal cortex of rhesus monkeys. *Science* **205**, 929–31.

Bruce, M.E. (1984). Scrapie and Alzheimer's Disease. *Psychological Medicine* **14**, 497–500.

Crapper, D.R. *et al.* (1973). Brain aluminium distribution in Alzheimer's disease and experimental neurofibrillary degeneration. *Science* **180**, 511–3.

Davis, R.T. (1978). Old monkey behaviour. *Experimental Gerontology* **13**, 237–50.

Drachman, D.A. and Leavitt, J. (1974). Human memory and the cholinergic system.

Archives of Neurology **30**, 113-21.

Dyer, R.S., Deshields, T.L., and Wonderlin, W.F. (1982). Trimethyltin-induced changes in gross morphology of the hippocampus. *Neurobehavioural Toxicology and Teratology* **4**, 141-7.

Evans, H.L. (1975). Scopolamine effects on visual discrimination: modifications related to stimulus control. *Journal of Pharmacy and Therapy* **195**, 105-13.

Evans, R.W., Gualtieri, T., and Amara, I. (1986). Methylphenidate and memory: dissociated effects in hyperactive children. *Psychopharmacology* **90**, 211-6.

Gaffan D. (1974). Recognition impaired and association intact in the memory of monkeys after transection of the fornix. *Journal of Comparative Physiology and Psychology* **86**, 1100-9.

——, Gaffan, E.A., and Harrison, S. (1984). Effects of fornix transection on spontaneous and trained non-matching by monkeys. *Quarterly Journal of Experimental Psychology* **36B**, 285-303.

Gajdusek, D.C. and Gibbs, C.J. (1975). Familial and sporadic chronic neurological degenerative disorders transmitted from man to primates. In *Advances in Neurology*, Vol. 10 (ed. B.S. Meldrum and C.D. Marsden), pp. 291-316. Raven Press, New York.

Glick, S.D. and Jarvik, M.E. (1969). Amphetamine, scopolamine and chlorpromazine interactions on delayed matching performance in monkeys. *Psychopharmacologia* **16**, 147-55.

—— and —— (1970). Differential effects of amphetamine and scopolamine on matching performance of monkeys with lateral frontal lesions. *Journal of Comparative Physiology and Psychology* **73**, 307-13.

Goldings, A.S. and Stewart, R.M. (1982). Organic lead encephalopathy: behavioural change and movement disorder following gasoline inhalation. *Journal of Clinical Psychiatry* **43**, 70-2.

Goldman, P.S. and Rosvold, H.E. (1970). Localization of function within the dorsolateral prefrontal cortex of the rhesus monkey. *Experimental Neurology* **27**, 291-304.

Goldman-Rakic, P.S. and Brown, R.M. (1981). Regional changes of monoamines in cerebral cortex and subcortical structures of ageing rhesus monkeys. *Neuroscience* **6**, 177-87.

Goudsmit, J., Morrow, C.H., Ashep, D.M., Yanagihara, R.T., Masters, C.T., Gibbs, C.J., and Gajdusek, D.C. (1980). Evidence for and against the transmissibility of Alzheimer disease. *Neurology* **30**, 945-50.

Jones B. and Mishkin M. (1972). Limbic lesions and the problem of stimulus-reinforcement associations. *Experimental Neurology* **36**, 362-77.

Kitt C.A. *et al.* (1985). Catecholaminergic neurites in senile plaques in prefrontal cortex of aged nonhuman primates. *Neuroscience* **16**, 691-9.

Mahut, H. (1971). Spatial and object reversal learning in monkeys with partial temporal lobe ablations. *Neuropsychologia* **9**, 409-24.

—— (1972). A selective spatial deficit in monkeys after transection of the fornix. *Neuropsychologia* **10**, 65-74.

Marriott, J.G. and Abelson, J.S. (1980). Age differences in short-term memory of test-sophisticated rhesus monkeys. *Age* **3**, 7-9.

Medin, D.L. (1969). Form perception and pattern reproduction in monkeys. *Journal of Comparative Physiology and Psychology* **68**, 412-9.

——, O'Neil, P., Smeltz, E., and Davis, R.T. (1973). Age differences in retention of concurrent discrimination problems in monkeys. *Journal of Gerontology* **28**, 63-7.

Milner, B. (1971). Interhemispheric differences in the localization of psychological processes in man. *British Medical Bulletin* **27**, 272-7.

—— (1978). Clues to the cerebral organization of memory. In *Cerebral Correlates of Conscious Experience* (ed. P.A. Buser and A. Rougeul-Buser), pp. 139-53. Elsevier, Amsterdam.

Mishkin, M. (1978). Memory in monkeys severely impaired by combined but not by separate removal of amygdala and hippocampus. *Nature* **273**, 297-8.

——, Vest, B., Waxler, M., and Rosvold, H.E. (1969). A re-examination of the effects of frontal lesions on object alternation. *Neuropsychologia* **7**, 357-63.

Niklowitz, A.J. (1975). Neurofibrillary changes after acute experimental lead poisoning. *Neurology* **25**, 927-34.

Penetar, D.M. and McDonough, J.M. (1983). Effects of cholinergic drugs on delayed match-to-sample performance of rhesus monkeys. *Pharmacology, Biochemistry and Behaviour* **19**, 963-7.

Ridley, R.M., Haystead, T.A.J., and Baker, H.F. (1980). An analysis of visual object reversal learning in the marmoset after amphetamine and haloperidol. *Pharmacology, Biochemistry and Behaviour* **14**, 345-51.

——, Baker, H.F., and Haystead, T.A.J. (1981). Perseverative behaviour after amphetamine; dissociations of response tendency from reward association. *Psychopharmacology* **75**, 283-6.

——, Bowes, P.M., Baker, H.F., and Crow, T.J. (1984). An involvement of acetylcholine in object discrimination learning and memory in the marmoset. *Neuropsychologia* **22**, 253-63.

Robustelli, F., Glick, S.D., Goldfarb, T.L., Geller, A., and Jarvik, M.E. (1969). A further analysis of scopolamine impairment of delayed matching with monkeys. *Communications in Behavioural Biology* **3**, 101-9.

Ross, W.D., Emmett, E.A., Steiner, J., and Tween, R. (1981). Neurotoxic effects of occupational exposure to organotins. *American Journal of Psychiatry* **138**, 1092-5.

Seawright, A.A., Brown, A.W., Aldridge, W.N., Verschoyle, R.D., and Street, B.W. (1980). Neuropathological changes caused by trialky lead compounds in the rat. In *Mechanisms of Toxicity and Hazard Evaluation*, (ed. B. Holmstedt, R. Lauwerys, M. Mercier, and M. Roberfroid) pp. 71-4. Elsevier North Holland, Amsterdam.

Struble, R.G., Cork, L.C., Whitehouse, P.J., and Price, D.L. (1982). Cholinergic innervation in neuritic plaques. *Science* **216**, 413-15.

——, Hedrean, J.C., Cork, L.C., and Price, D.L. (1984*a*). Acetylcholinesterase activity in senile plaques of aged macaques. *Neurobiology of Ageing* **5**, 191-8.

——, Kitt, C.A., Walker, L.C., Cork, L.C., and Price, D.L. (1984*b*). Somatostatinergic neurites in senile plaques of aged non-human primates. *Brain Research* **324**, 394-6.

Tinklenberg, J.R., Melges, F.T., Hollister, L.E., and Gillespie, H.K. (1970). Marijuana and immediate memory. *Nature* **226**, 1171-2.

Valpey, R., Sumi, S.M., Dopass, M.K., and Goble, G.J. (1978). Acute and chronic progressive encephalopathy due to gasoline sniffing. *Neurology* **52**, 345-50.

Walsh, T.J. and Tilson, H.A. (1984). Neurobehavioural toxicology of the organoleads. *Neurotoxicology* **5**, 67-86.

Willes, R.F., Rice, D.C., and Truelove, J.F. (1980). Chronic effects of lead in non-human primates. In *Lead Toxicity* (ed. L. Singhal and J.A. Thomas), pp. 213–40. Urban and Schwarzenberg Inc., Baltimore.

Wisniewski, H. and Terry, R.D. (1970). An experimental approach to the morphogenesis of neurofibrillary degeneration and the argyrophilic plaque. In *Alzheimer's Disease and Related Conditions* (ed. G.E.W. Wolstenholme and M. O'Connor), pp. 223–40. Ciba Foundation Symposium.

——, Terry, R.D., Pena, C., Streicher, E., and Klatzo, I. (1965). Experimental production of neurofibrillary degeneration. *Journal of Neuropathology and Experimental Neurology* **24**, 139.

Wisniewski, H.M., Ghetti, B., and Terry, R.D. (1973). Neuritic (senile) plaques and filamentous changes in aged rhesus monkeys. *Journal of Neuropathy and Experimental Neurology* **32**, 566–84.

——, Narang, H.K., and Terry, R.D. (1976). Neurofibrillary tangles of paired helical filaments. *Journal of Neurological Science* **27**, 173–81.

——, Korthals, J.K., Kopeloff, L.M., Ferszt, R., Chusid, J.C., and Terry, R.D. (1977). Neurotoxicity of aluminium. In *Neurotoxicology* (ed. L. Roizin, H. Shirak, and N. Grcevic), pp. 313–15. Raven Press, New York.

Zimmerberg, B., Glick, S.D., and Jarvik, M.E. (1971). Impairment of recent memory by marihuana and THC in rhesus monkeys. *Nature* **233**, 343–5.

Zook, B.C. and Paasch, L.H. (1980). Lead poisoning in zoo primates: environmental sources and neuropathologic findings. In *The Comparative Pathology of Zoo Animals* (ed. R.J. Montali and K. Migaki), pp. 143–52. Smithsonian Institution Press, Washington.

——, London, W.T., DiMaggio, J.F., Rothblat, L.A., Sauer, R.M., and Sever, J.L. (1980). Experimental lead paint poisoning in nonhuman primates. II Clinical pathologic findings and behavioural effects. *Journal of Medical Primatology* **9**, 286–303.

6

Do hippocampal lesions produce amnesia in animals?

J.N.P. RAWLINS

Humans who have undergone bilateral resections of the mesial temporal lobe, including the hippocampus and amygdala, suffer from profound and enduring anterograde amnesia, although their general intellectual function is spared (Scoville and Milner 1957). This discovery has led to major research efforts concerned with establishing in experimental animals the critical neuropathology underlying this memory deficit. A further spur to research has resulted from concern about the psychological problems faced by the steadily increasing section of the population at risk from Alzheimer's disease; one very serious component of this condition is a severe, and so far irreversible, memory loss. In this paper, I shall consider whether hippocampal damage leads to amnesia in humans, and review evidence from animal experiments concerned with manipulation of the neural systems believed to be implicated in human anterograde amnesia. Finally, I shall consider whether the psychological consequences of hippocampal damage in animals can reasonably be said to constitute amnesia, or even a constituent element of amnesia.

AMNESIA AND THE TEMPORAL LOBES

Undoubtedly, the most significant case study of amnesia resulting from temporal lobe damage is that of H.M. (Scoville & Milner 1957). This patient was admitted for surgery in order to treat his severe temporal lobe epilepsy. The areas removed almost certainly included the amygdala, the hippocampal gyrus, the anterior two-thirds of the hippocampus itself, and the uncus. However, the resulting amnesia was generally attributed to the hippocampal damage.

Early attempts to demonstrate that bilateral hippocampal damage produced an equivalent amnesia in animals met with failure; animals with substantial lesions could clearly acquire and perform a variety of new behavioural skills, even when all the training was carried out post-operatively. The animals' greatest difficulty appeared to be in learning to

73

inhibit the recently acquired responses if conditions changed so that behaviour was no longer rewarded, or was simultaneously both rewarded and punished (Douglas 1967; Gray 1982*a*, *b*). More recently, the kinds of behavioural tasks used to assess the consequences of damage to the hippocampus and closely related structures have been significantly refined, and clear deficits in learning and performance are now reliably obtained. Studies in non-human primates have demonstrated significant deficits after lesions that include the hippocampus or fornix on a variety of tasks, for example assessing memory for trial-unique 'junk' objects, (e.g. Gaffan 1977; Mishkin 1978) or in spatial, but not object, reversal tasks (Jones and Mishkin 1972).

Studies in non-primates—most frequently rats—have demonstrated clear impairments in a variety of spatial learning tasks (e.g. review by O'Keefe and Nadel 1978; see also Morris *et al.* 1982), particularly those including an element of working memory (Olton *et al.* 1979; Rawlins and Olton 1982). However, the underlying psychological changes, that are presumed to result from damage to the hippocampus or to areas that are anatomically closely related to it, vary widely from theory to theory (e.g. O'Keefe and Nadel 1978; Olton *et al.* 1979; Gray 1982*a*, *b*; Rawlins 1985). These theories differ considerably in the extent to which they are concerned with memory processes. Nonetheless, if for the moment we simply accept that there are tasks that are highly sensitive to damage that includes the hippocampus and/or its major projections, we can ask what is necessary or sufficient to produce a marked impairment in the performance of these tasks.

CRITICAL ELEMENTS IN MESIAL TEMPORAL LOBE DAMAGE RESULTING IN AMNESIA

Primate studies

Perhaps the most radical interpretation of the human lesion data is that the crucial element in H.M.'s lesion is the inclusion of white matter in the temporal stem (Horel 1978). This hypothesis has been tested by making lesions restricted to the temporal stem in rhesus monkeys, and determining the consequences this has for performance of a delayed non-matching to sample task (Zola-Morgan *et al.* 1982). On such tasks the animal is presented with a sample stimulus, then after a delay, is required to choose between the sample and a novel stimulus, and is rewarded for selecting the novel item. The temporal stem lesion did not affect performance of this task, while a combined lesion of amydala plus hippocampus produced a major impairment. It seems implausible that an amnesic monkey *could* perform this task, so unless we want to attribute very different functions to homologous structures in man and other primates we should accept that damage to the fibres in the temporal stem is unlikely to be the cause of H.M.'s amnesia.

Instead, it appears that damage to the hippocampus and amygdala produces behavioural changes consistent with impairment of memory. However, if we attempt more precisely to identify whether it is damage to the hippocampus or damage to the amygdala that is responsible for the profound impairment that results from combined damage to both structures, we find that damage to neither alone seems sufficient to produce a robust memory impairment, but that a combined lesion is necessary (Mishkin 1978). An intermediate behavioural impairment is seen both when bilateral hippocampal damage is combined with unilateral amygdalar damage, and when bilateral amygdalar damage is combined with unilateral hippocampal damage (Saunders *et al.* 1984).

Rat studies

A similar pattern of results is emerging from recent behavioural studies in which rats were trained to perform a task that conforms quite closely to those typically used to assess memory in primates (Aggleton *et al.* 1986). The rats were required to learn a 'non-matching' rule. Each trial consisted of a choice between two distinctively different goal boxes. One of these was identical to the goal box the rat had chosen on the previous trial, and the rat was rewarded only if it chose the alternative offered. There were 50 distinctive pairs of goalboxes, each of which was used only once in each day's testing. This large set of discriminable stimuli corresponds to the 'junk' objects typically used to provide trial-unique information in experiments using primates, and as such is a novel technique in rat testing.

Rats can learn this version of an 'object non-matching' task fairly readily (Aggleton 1985). Bilateral hippocampal aspiration lesions did not reduce choice accuracy relative to controls, even when an inter-response interval of up to 1 minute was used, at which stage choice accuracy was significantly reduced in all the animals (Aggleton *et al.* 1986). Amygdalar lesions also appear to leave choice accuracy intact, but our pilot observations strongly suggest that lesions combining hippocampal and amygdalar damage produce a profound impairment. These results clearly match Mishkin's (1978) data rather closely. However, before we conclude that hippocampal damage alone has relatively trivial consequences, it is important to note that the same hippocampectomized rats that could choose accurately on the delayed object non-matching task showed a profound and enduring impairment on a discrete-trial working-memory alternation task on an elevated T-maze, a test of spatial working memory (Aggleton *et al.* 1986). Thus, hippocampal lesions reliably reduced performance to chance levels of accuracy on one working memory task, while leaving performance on the other quite unaffected. The dissociation between the sensitivity of the two tasks is striking, and the possible explanations for it illustrate clearly some of the key differences between competing theories of hippocampal function.

These explanations may implicitly suggest what the psychological contribution of the amygdalar damage is in those tasks in which only combined lesions are effective, but they do not actually explicitly address this important question. Of course, a radical answer to the question might be that we are wrong to be looking for general memory systems at all, and different systems determine memory for different tasks.

SOME DIFFERENCES BETWEEN SPATIAL AND NON-SPATIAL WORKING MEMORY

The T-maze task, but not the goal box non-matching task, is a spatial task that might be presumed to require a cognitive map for its solution (O'Keefe and Nadel, 1978), though this presumption has been challenged on empirical grounds (Rawlins and Olton 1982). Alternatively, the T-maze task, but not the non-matching task, might require the rat to use a memory of its own response to guide its next response (see Gaffan *et al.* 1984). While both tasks may require the rat to inhibit responses to stimuli that have recently been paired with reward, the use of a large number of goal boxes in the non-matching task might be expected to reduce interference. By contrast, on the T-maze the rats must repeatedly switch their responses to the same two stimuli: this might be expected to lead to a high level of interference (Gray 1982*a*, *b*). This latter interpretation has been challenged (Rawlins 1982), and is perhaps further weakened by the observation that normal rats apparently show high levels of interference if the same goal boxes are used more often than once a day in the object non-matching task (Aggleton *et al.* 1986), but do not show an analogous, increasing error rate in the T-maze as the day's trials progress (Rawlins and Olton 1982, p. 345). This contrast suggests that the difference between the two tasks lies in more than the differential need to vary responses in the presence of the same stimuli.

Neither the working-memory theory (Olton *et al.* 1979) nor the suggestion that the hippocampus is a temporary memory store (Rawlins 1985) can so deftly accommodate this dissociation between tasks. The most obvious move for both is probably to point to the apparent conflict with the results reported by Olton and Feustle (1981), who used a superficially similar task, but note that their experiment used different lesions—fornix lesions—as well as a quite different form of goal-arm non-matching task. In Olton and Feustle's task rats were prevented from using cognitive mapping, but were forced instead to rely on working memory. Although the topological relationship of the arms of their radial arm maze was changed from trial to trial, each contained a distinctive set of discriminative stimuli by which it could be identified. Lesioned animals performed at chance levels. One might then suggest that Aggleton's task would also be sensitive to hippocampal lesion effects if the rats had to remember a 'list' of distinctive goal boxes (as they had to in Olton and Feustle's experiment) to guide choice responding, rather than only ever needing to remember one goal box—the one most recently visited. We are currently assessing this possibility. Such a suggestion is incomplete, however, without the related suggestion that rats normally learn the relations between locations in spatial arrays by attending to, and storing, information about a wide range of interoceptive and exteroceptive stimuli experienced while in transit (Rawlins 1985; Rawlins in press). Without such a store available, by the time the rats arrive at their goal, they are likely to have forgotten how they got there (see Morris *et al.* 1986). Hence, the spatially extended T-maze problem is insoluble for rats with hippocampal lesions, because in order to remember a response that involves moving through space, a large temporary memory store is required, while an equivalent non-spatial two-lever problem in an operant chamber is soluble, so long as the inter-response interval is kept short (Rawlins and Tsaltas 1983). Even if there is no agreement about which working memory tasks are sensitive to hippocampal dysfunction, one can probably conclude that the hippocampus is a significant element in a memory system that also includes the amygdala, and that both hippocampus and amygdala independently contribute some functional element to the normal working of the whole system.

CORTICAL ROLE IN MEMORY TASKS

Mishkin (1978) originally demonstrated that some tasks required combined lesions of hippocampus and amygdala in order to induce a major behavioural impairment. Subsequent work revealed (Murray and Mishkin 1986) that the contribution he had originally attributed simply to hippocampal damage actually should have been attributed to a combination of hippocampal plus rhinal cortical damage, because the rhinal cortex was damaged during the surgical approach to perform the hippocampectomy. The subsequent study (Murray and Mishkin 1986) demonstrated that lesions which combined damage to the rhinal cortex and the amygdala were also comparable in their consequences to lesions of these regions plus hippocampus. However, if the rhinal cortex was spared, and hippocampus and amygdala were removed, then only a relatively mild behavioural impairment resulted. Does this imply that the memory function previously attributed to the hippocampus should correctly be attributed to the rhinal cortex? There are sufficient examples in the rat literature of clear behavioural dissociations between lesions including hippocampus and some overlying occipital cortex, and controls with the same cortical damage to conclude that at least in some tasks hippocampal damage is the critical element, and the associated cortical damage seems unimportant (e.g. Morris *et al.* 1982;

Aggleton *et al*. 1986). However, the degree of significance of damage confined to cells in the hippocampus proper has been questioned.

SELECTIVE NEUROTOXIC LESIONS: NO ROLE FOR THE HIPPOCAMPUS?

One line of research which suggests that the role of the hippocampus in a variety of tasks may have been overestimated, arises from the use of excitatory neurotoxins to induce highly selective lesions within the hippocampal formation. The aim is to destroy neurones whose cell bodies are in the region of toxin injection, while sparing neurones whose axons project to, or pass through, the region. Conventional lesions of course cannot destroy a structure without at the same time severing fibres merely *en passage* and disrupting the blood vessels in the region; hence, the effects of the surgical lesion may well include a contribution over and above that made by destroying the target structure itself. The first excitotoxin to be used widely was kainic acid. Hippocampal lesions made in rats by kainic acid injection can produce behavioural impairments comparable to those resulting from conventional hippocampal aspiration lesions (Jarrard 1983), encouraging the view that the conventional lesions modify behaviour because they destroy neurones located in the hippocampus. However, kainic acid produces marked extrahippocampal damage at relatively distal sites, if administered by intraventricular injection, and even injections of kainic acid directly into subiculum (just caudal from the hippocampal CA1 cell field) produce a rather similar pattern of extrahippocampal damage, perhaps as a result of diffusion back into the adjacent lateral ventricle (Jarrard 1983). The spread of the damage can be reduced by treating the animal with diazepam at the time of the kainate injection (Jarrard 1983, 1986). This increases the selectivity of kainic acid lesions. When these more selective lesions do not include cells in the subiculum, the animals are apparently quite unimpaired on at least some tasks that are profoundly affected by complete hippocampal aspiration lesions, or more generalized neurotoxin lesions (Jarrard 1986). Hence, there can be a marked loss of hippocampal neurones with only a minimal behavioural effect, again suggesting that the importance of the hippocampus may be overestimated in conventional lesion studies.

Even though diazepam is effective in limiting the spread of damage caused by kainic acid injection, an alternative excitatory neurotoxin, ibotenic acid, seems to be preferable if really selective lesions are to be made (Kohler and Schwarcz 1983; Jarrard 1986). Ibotenate-induced lesions can be restricted successfully to selected cell fields within the hippocampus. As the results from the diazepam-treated kainic acid lesions would suggest, the consequences of substantial cell loss restricted to the hippocampal cell fields are much less striking than those seen when large, conventional hippocampal lesions are studied (Jarrard 1986). Although the neurotoxic lesions can have effects similar to those of conventional lesions, even then the consequences seem to be less marked (Sinden *et al.* 1986). Does this, then, confirm that the hippocampus itself does not really play a part in generating normal performance on most of those tasks that are reliably affected by conventional hippocampal lesions, and that we should turn our attention to other areas, perhaps like the rhinal cortex? I suspect not. Several lines of evidence suggest that the hippocampus is indeed concerned in those functions implicated in the conventional lesion studies.

THE CONTRIBUTION OF SUBICULAR DAMAGE

First, neurotoxic lesions have much more dramatic (and 'traditional') consequences when they include the subiculum (e.g. Jarrard *et al.* 1986; Jarrard 1986). This cell field is not only the main target for output from the CA1 cell field in hippocampus proper, but it is also the source of a pathway to the mammillary bodies and anterior thalamic nuclei, which was until recently believed to originate in the hippocampus itself (Swanson 1979). So it makes some sense to engage in some restricted gerrymandering, and think of the hippocampus as including the subiculum. The justification for this would be primarily hodological (hodology being the study of neuronal pathways and connections) since so much of the hippocampal output can only reach other areas of brain via subiculum: indeed, the hippocampus proper has otherwise only one really major extrinsic projection, and that is to the lateral septal nuclei (Swanson 1979). Most large, conventional hippocampal lesions have included at least some part of the subiculum, or will have destroyed subicular axons coursing rostrally. The neurotoxic lesions have now positively identified the subicular output as particularly significant in what had been regarded as the hippocampal contribution to normal behaviour. The interesting question thus becomes whether this is by virtue of the intimate relations between hippocampus proper and subiculum, or by virtue of the relationship between subiculum and extra-hippocampal sites, several of which project directly to the subiculum (e.g. Swanson 1979).

Secondly, there are good reasons why neurotoxic lesions may be less effective than conventional ones, even if the effects of both lesions are mediated entirely through damage to the target structure (see discussion in Sinden *et al.* 1986, p. 327). Even so, it is not essential to include subicular damage to show a significant behavioural effect of 'complete' hippocampal lesions made with ibotenic acid (Sinden *et al.* 1986), though including subiculum seems to make such lesions still more effective (Jarrard 1986).

ADVANTAGES OF NEUROTOXINS

While the results from the selective neurotoxic lesions have clearly led us to modify our views about critical areas within the hippocampal formation, their contribution need not be restricted simply to producing finer and finer disconnections. Their modus operandi makes them well suited for producing the kinds of neuropathology that appear characteristic in some human clinical conditions (see Schwarcz *et al.* 1984). In this respect, I wonder whether kainic acid has been prematurely abandoned as a research tool. The extrahippocampal sites that appear damaged following intrahippocampal kainic acid injections (e.g. amygdala and the midline thalamic nuclei) have themselves been identified as contributing to normal memory function in man, and may even be best considered as constituents of a single memory system (but for an opposed view see Squire 1981). It is thus possible that the tendency of kainic acid to cause distal lesions may be turned to advantage, if we wish to model the rather diffuse neurological changes that underlie some human amnesias. Such an approach would perhaps require elaborate histological procedures, but might offer a more appealing experimental lesion technique than is provided by the more restricted complete lesions. For example, if one wished to assess the potential of pharmacological agents in alleviating human memory failure, in those classes of human patients (like Alzheimer patients, or perhaps those who suffer from Korsakoff's psychosis) in which the amnesia does not necessarily result from some highly selective intervention, such lesions would be valuable.

SELECTIVE HIPPOCAMPAL DAMAGE IN MAN

Human clinical observations provide additional data that have been taken to indicate a role for the hippocampus and subiculum in normal memory. First, there is a report that bilateral damage confined to the hippocampus was the only change detectable—by pneumoencephalography—in an amnesic patient whose CAT scan appeared normal (Muramoto *et al.* 1979). The possibility of subicular damage could not be critically assessed by this technique, but it seems implausible that the damage stopped neatly at the CA1/subicular border. Secondly, there is postmortem histology on an amnesic patient with a large, left hemisphere infarct in the medial-temporal-occipital region, but with a right hemisphere lesion confined to the posterior two-thirds of the hippocampus and extending into the subiculum (Woods *et al.* 1982). This patient showed complete demyelination of the left fornix and partial (largely lateral) demyelination of the right. There is also a much more striking, recent report of an amnesic patient whose lesion was almost entirely restricted to area CA1, which was apparently eliminated bilaterally, throughout its proximal-distal extent (Zola-Morgan *et al.* 1986). This patient had, in addition, two small unilateral focal lesions, one in the right striatum, and one in the left somatosensory cortex. The mammillary bodies, amygdala, and dorsomedial thalamic nucleus were specifically described as being normal; the subiculum also appeared intact, suggesting that loss of the output from CA1 to subiculum is sufficient to produce a severe memory impairment, so long as the CA1 lesion is sufficiently complete. This last case suggests that bilateral damage to the hippocampus can be sufficient to produce amnesia in man. The first two cases do not permit such a strong conclusion, in the first because the neuropathology must be regarded as interesting, but not yet definitive, and in the second because although the only bilateral damage was in the hippocampus, we have already seen that a combination of unilateral amygdalar damage with such a lesion adds significantly to the impairment observed (Saunders *et al.* 1984).

AMNESIA WITHOUT HIPPOCAMPAL DAMAGE

It is, however, clear that hippocampal damage is not a necessary component of all anterograde human amnesias. Neither two amnesic patients with Korsakoff's psychosis, whose post-mortem pathology was described by Mair *et al.* (1979), nor the amnesic patient N.A., studied by Squire and Moore (1979), and whose lesion has been assessed by CAT scan, appear to have sustained any hippocampal damage (nor, for that matter, damage to the white matter in the temporal stem). Both of the former patients had marked bilateral gliosis, shrinkage, and discolouration in the medial nuclei of the mammillary bodies, and a thin band of gliosis in the dorsal thalamus, medial to the medial dorsal nucleus, and extending anterior to it. The latter patient is believed to have a lesion in the dorsomedial thalamic nucleus itself (Squire and Moore 1979), though there is a strong case for suspecting that he also has damage in, or close to, the mammillary bodies which may have severed the mammillothalamic tract if it has spared the mammillary bodies themselves (Weiskrantz 1985). There is disagreement as to whether Korsakoff patients and N.A. should be considered as having essentially the same amnesia as temporal lobe patients like H.M. (Squire 1981; Weiskrantz 1985).

However, since the mammillary bodies and anterior thalamic nuclei are themselves important targets for the efferent projection from the subiculum, it has at least the virtue of parsimony to assume that this relationship underlies the amnesias seen following either hippocampal/subicular damage or mammillary body and thalamic lesions. According to such a view, these lesions would affect a multicomponent memory system. Lesions at different sites in such a network might well be expected to have discriminably different, though related, consequences. If the lesions had precisely the same effects, we would be left to accept that there is duplication of function, and that there are no differences between the system's information processing at different stages. Given the multiplicity of potentially independent inputs and outputs at each stage of the system, it seems highly implausible that disconnections at different levels should all be equivalent. Hence, we should not necessarily assume that we are dealing with functionally quite independent systems if we find that lesions at different locations have different effects.

AMNESIA AND ALZHEIMER'S DISEASE

A still more complicated neuropathology is associated with the amnesia found in Alzheimer's patients. There is a marked loss of noradrenergic cells in locus coeruleus (Tomlinson *et al*. 1981), loss of cells in the nucleus basalis and a marked drop in cortical and hippocampal choline acetyltransferase levels (see review by Rossor 1982), and loss of cells in the subiculum and those layers of entorhinal cortex from which the descending efferent connections arise (Hyman *et al*. 1984; Braak and Braak 1985). While this is by no means an exhaustive list of the pathology associated with Alzheimer's disease (for a general review see, for example, Katzman 1986), it is worth noting that some particularly marked changes do occur in the hippocampus and in some of the areas most intimately connected with it; and this damage may correlate with the memory loss in Alzheimer's patients (Hyman *et al*. 1985). Alzheimer's patients have memory problems that are in some respects qualitatively different from those affecting H.M or Korsakoff's patients. In particular, the latter patients have spared short-term memory (e.g. Baddeley and Warrington 1970) while the former do not. However, although the Alzheimer patients, in line with their typically more widespread neuropathology, have far more general psychological problems (including general intellectual deterioration) than the relatively pure amnesics, we might nonetheless attribute the psychological problems that the different groups share to the pathology they have in common. Thus, an understanding of the role of the hippocampal system in human amnesias represents one possible way to understand, and perhaps eventually treat, an element of the symptom complex that constitutes Alzheimer's disease. However, it is questionable whether such selective treatment would be of much clinical value, given the potentially devastating and varied additional problems that these patients have.

SUMMARY OF HUMAN NEUROPATHOLOGY

The human data so far considered thus suggest that general amnesias of varying degress of purity can result from a rather variable pathology. However, good support exists for the view that there is a common involvement of regions related to the hippocampus, and perhaps in particular the subiculum and the areas with which it is closely connected. There is also growing evidence of a significant role for the descending projections from the entorhinal cortex (Hyman *et al.* 1984; Braak and Braak 1985). One notable exception is that there is no clinical evidence of a role in memory for the lateral septal nuclei, which receive the major extrinsic projection from area CA3 via the pre-commissural fornix. Overall then, in man the descending afferents from cortex to hippocampus, the hippocampal efferents carried in the post-commissural fornix (the role of the precommissural fornix being uncertain), the target areas for this projection and perhaps the ascending cholinergic afferents to the hippocampus all play a part in normal memory function. Sufficient damage to any part of this system seems to cause a roughly equivalent amnesia, and it is clearly unnecessary for there to be direct damage to the hippocampus itself, to the amygdala or to the rhinal cortex for amnesia to develop. The roles of the ascending noradrenergic and serotonergic innervation seem very much less certain, although they do project to the regions considered critical for memory functions, they do modify electrophysiological activity in at least some of these sites, and they are certainly seriously disrupted in Alzheimer's disease.

The answer to whether the various regions whose dysfunction can give rise to anterograde amnesia share this property because of their anatomical inter-connections must probably await more sophisticated manipulative experiments that can either selectively disrupt particular connections, while leaving the regions and their other connections intact, or more sophisticated analyses of the way activity in one region modifies or determines the activity in the others.

COMPARISON OF CRITICAL NEUROPATHOLOGIES IN MAN AND ANIMALS

There is fairly close correspondence between those neuropathologies which are associated with anterograde amnesias in man, and the experimentally-induced neuropathologies which impair performance on tasks sensitive to hippocampal dysfunction in animals. I have already indicated that damage to the subiculum appears particularly important in producing the behavioural changes that result from large, non-selective hippocampal lesions in animals. Although the mammillary bodies and medial thalamic nuclei are relatively inaccessible to experimenters wishing to manipulate them (and we would assuredly not be able to reproduce accurately the exact thalamic patterns of gliosis reported by Mair *et al.* 1979), some experiments have already been conducted on animals with lesions in these sites, though using conventional lesion techniques. The results suggest that mammillary body lesions and post-commissural fornix lesions have behavioural consequences that at least in a general way resemble those of hippocampal lesions (Rosenstock *et al.* 1977;

Henderson and Greene 1977). Medial thalamic lesions also appear to produce memory impairments (e.g. Aggleton and Mishkin 1983), though these latter effects seem discriminable from those seen after hippocampal lesions (Aggleton and Mishkin 1983; Greene and Naranjo 1986). This perhaps lends some support to the notion that the diencephalic and temporal lobe amnesias may differ (Squire 1981) although I am not convinced that we are yet in a position to talk with certainty of two independent memory systems subserving different functions. Thus, there is good correspondence between sites related to memory processing in man, and sites disrupting tasks that are sensitive to hippocampal lesions in animals.

A ROLE OUTSIDE MEMORY FOR HIPPOCAMPAL CONNECTIONS

Many of the tasks that are sensitive to damage within the system so far considered are also sensitive to septal lesions (Gray and McNaughton 1983); should the entire septum therefore be considered as part of a memory system? My tentative suggestion is that the medial septum, from which the presumed cholinergic input to the hippocampus and entorhinal cortex is derived (Mitchell *et al.* 1982) may indeed contribute significantly to a memory system, but that the dorsolateral septal nuclei, to which the major projection from CA3 is directed, may need to be excluded. This suggestion is based partly on the lack of effect that neurotoxic lesions restricted to CA3 have in spatial working memory tasks (Jarrard 1983, 1986) and partly on pilot data of our own. The lack of positive evidence from human clinical studies is relevant, but less compelling.

Electrolytic lesions of the dorso-lateral septal nuclei produce impaired performance on a discrete-trial working-memory alternation task (Rawlins and Olton 1982). However, when ibotenic acid was used to damage an equivalent (or larger) area of dorsolateral septum, we found that there was no effect on the T-maze task (P. Coffey, unpublished observations). This result suggests the possibility that the conventional lesion may have worked, at least in part by damaging fibres *en passage*. Since the conventional lesions typically include the posterior areas of the septum, the intriguing possibility arises that the electrolytic lesion interrupts the post-commissural projections descending to the mammillary bodies and the thalamus. We are currently investigating this possibility.

Although the ibotenic acid lesion spared performance on the T-maze task, the same subjects had shown a pronounced effect on resistance to extinction in the straight alley. If this behavioural dissociation is still respected in full scale behavioural experiments, then we may be able to start to separate the neurologies underlying the behavioural inhibition emphasized in some theories of hippocampal function (e.g. Gray 1982a, b) from the memory deficits emphasized by others (e.g. Olton *et al.* 1979; Gaffan *et al.* 1984; Rawlins 1985). Hitherto, such developments have been discouraged by the apparent functional equivalence of lesions at quite different sites (e.g. fornix-fimbria transection, hippocampal aspiration lesions, medial septal lesions and lateral septal lesions) all of which have broadly similar effects on spatial working memory tasks in rats.

CONCLUSIONS

I suggest that there is sufficient evidence to encourage us to consider the hippocampus and several of the areas intimately connected with it as critically concerned with normal memory function in man, and as constituting a system concerned with performance of a common set of tasks in animals. But can we relate the underlying psychological demands of these tasks to the clinical observations made in man? In some sense, the answer to this question must be negative: the overwhelming impression gained from studying amnesic patients is that they can remember almost nothing; the impression gained from animals with apparently equivalent neuropathology is that they can remember a good deal. However, a closer scrutiny of the data demonstrates that human amnesics can in fact retain and use a good deal of new information (for review see Weiskrantz 1985), while we have already seen that animals with equivalent damage fail to solve a variety of memory problems. There have been several theoretical attempts to capture the essentials of both the human and the animal memory data. One has been to emphasize the need for cognitive mediation in tasks sensitive to hippocampal dysfunction and in human amnesias (Warrington and Weiskrantz 1982); another has been to emphasize the parallel distinctions between procedural and declarative memory in humans, and reference and working memory in animals (e.g. Olton 1983). I have proposed a way in which the lack of a temporary memory store located in the hippocampus might lead to a failure to associate new items of information with the contexts in which they occurred (Rawlins 1985, pp. 496–7). Thus, although there is still a fairly clear gulf between the clinical observations and the kinds of theories that have been developed in order to explain the effects of hippocampal system lesions in animals, recent descriptions of hippocampal function derived from animal studies do suggest ways in which hippocampal damage could at least contribute to the amnesia seen in humans with damage to related areas.

Further progress probably depends as much on the development of comparable psychological tests in both animals and man, as on the development of more sophisticated neuropsychological manipulations in animals, and better non-invasive scanning techniques in man. It would be encouraging to demonstrate that priming could improve the performance that animals with hippocampal lesions show on memory tasks, and it would be helpful to have more evidence as to whether cueing responses could improve performance in animals as it does in human amnesics (for some possibly relevant evidence see Morris *et al.* 1986). Conversely, it would be valuable to have more evidence as to whether human amnesics show performance deficits on tests directly comparable to those that are sensitive to hippocampal lesions in animals: at least there is some evidence that they do (e.g. Sidman *et al.* 1968; Oscar-Berman *et al.* 1982). The further elaboration of such parallel testing in animals and man seems critical if treatments for at least some human amnesic patients are to be developed in experimental animals.

ACKNOWLEDGEMENT

The research reported in this chapter was supported by the Medical Research Council.

REFERENCES

Aggleton, J.P. (1985). One trial object recognition by rats. *Quarterly Journal of Experimental Psychology* **37B**, 279–94.

—— and Mishkin, M. (1983). Visual recognition impairment following medial thalamic lesions in monkeys. *Neuropsychologia* **21**, 189–97.

——, Hunt, P.R., and Rawlins, J.N.P. (1986). The effects of hippocampal lesions upon spatial and non-spatial tests of working memory. *Behavioural Brain Research* **19**, 133–46.

Baddeley, A.D. and Warrington, E.A. (1970). Amnesia and the distinction between long- and short-term memory. *Journal of Verbal Learning and Verbal Behaviour* **9**, 176–89.

Braak, H. and Braak, E. (1985). On areas of transition between entorhinal allocortex and temporal isocortex in the human brain. Normal morphology and lamina-specific pathology in Alzheimer's disease. *Acta Neuropathologica (Berlin)* **68**, 325–32.

Douglas, R.J. (1967). The hippocampus and behavior. *Psychological Bulletin* **67**, 416–42.

Gaffan, D. (1977). Recognition memory after short retention intervals in fornix-transected monkeys. *Quarterly Journal of Experimental Psychology* **29**, 577–88.

—— (1985). Hippocampus: memory, habit and voluntary movement. *Philosophical Transactions of the Royal Society, Section* **B308**, 87–99.

——, Saunders, R.C., Gaffan, E.A., Harrison, S., Shields, C., and Owen, M.J. (1984). Effects of fornix transection upon associative memory in monkeys: role of the hippocampus in learned action. *Quarterly Journal of Experimental Psychology* **36B**, 173–221.

Gray, J.A. (1982a) *The Neuropsychology of Anxiety: an Enquiry into the Functions of the Septo-hippocampal System.* Oxford University Press, Oxford.

—— (1982b). Multiple book review of The neuropsychology of anxiety: an enquiry into the functions of the septo-hippocampal system. *Behavioral and Brain Sciences* **5**, 469-534.

—— and McNaughton, N. (1983). Comparison between the behavioural effects of septal and hippocampal lesions: a review. *Neuroscience and Biobehavioural Reviews* **7**, 119-88.

Greene, E. and Naranjo, J.N. (1986). Thalamic role in spatial memory. *Behavioural Brain Research* **19**, 123-31.

Henderson, J. and Greene, E. (1977). Behavioural effects of lesions of precommissural and post-commissural fornix. *Brain Research Bulletin* **2**, 123-9.

Horel, J.A. (1978). The neuroanatomy of amnesia. *Brain* **101**, 403-5.

Hyman, B.T., van Hoesen, G.W., Damasio, A.R., and Barnes, C.L. (1984). Alzheimer's disease: cell-specific pathology isolates the hippocampal formation. *Science* **225**, 1168-70.

——, ——, Kromer, L.J., and Damasio, A.R. (1985). The subicular cortices in Alzheimer's disease: neuroanatomical relationships and the memory impairment. *Society for Neuroscience Abstracts* **11**, 458.

Jarrard, L.E. (1983). Selective hippocampal lesions and behavior: effects of kainic acid lesions on performance of place and cue tasks. *Behavioral Neuroscience* **97**, 873-89.

—— (1986). Selective hippocampal lesions and behavior: implications for current research and theorizing. In *The Hippocampus*, Vol. IV (ed. R.L. Isaacson and K. H. Pribram), pp. 93-126. Plenum, New York.

——, Feldon, J., Rawlins, J.N.P., Sinden, J.D., and Gray, J.A. (1986). The effects of intrahippocampal ibotenate on resistance to extinction after continuous or partial reinforcement. *Experimental Brain Research* **61**, 519-30.

Jones, B. and Mishkin, M. (1972). Limbic lesions and the problem of stimulus-reinforcement associations. *Experimental Neurology* **36**, 362-77.

Katzman, R. (1986). Alzheimer's disease. *New England Journal of Medicine* **314**, 964-73.

Kohler, C. and Schwarcz, R. (1983). Comparison of ibotenate and kainate neurotoxicity in the rat brain: a histological study. *Neuroscience* **8**, 819-35.

Mair, W.G.P., Warrington, E.K., and Weiskrantz, L. (1979). Memory disorder in Korsakoff's psychosis: a neuropathological and neuropsychological investigation of two cases. *Brain* **102**, 749-83.

Mishkin, M. (1978). Memory in monkeys severely impaired by combined but not separate removal of amygdala and hippocampus. *Nature* **273**, 297-8.

Mitchell, S.J., Rawlins, J.N.P., Steward, O., and Olton, D.S. (1982). Medial septal lesions disrupt theta rhythm and cholinergic staining in medial entorhinal cortex and produce impaired radial arm maze behavior in rats. *Journal of Neuroscience* **2**, 292-302.

Morris, R.G.M., Garrud, P., Rawlins, J.N.P., and O'Keefe, J. (1982). Place navigation impaired in rats with hippocampal lesions. *Nature* **297**, 681-3.

——, Hagan, J.J., Rawlins, J.N.P. (1986). Allocentric spatial learning by hippocampectomised rats: a further test of the 'spatial-mapping' and 'working-memory' theories of hippocampal function. *Quarterly Journal of Experimental Psychology* **38B**, 365-95.

Muramoto, O., Kuru, Y., Sugishita, M., and Toyokura, Y. (1979). Pure memory loss with hippocampal lesions. *Archives of Neurology* **36**, 54-6.

Murray, E.A. and Mishkin, M. (1986). Visual recognition in monkeys following rhinal cortical ablations combined with either amygdalectomy or hippocampectomy. *Journal of Neuroscience* 6, 1991-2003.

O'Keefe, J. and Nadel, L. (1978). *The Hippocampus as a Cognitive Map*. Oxford University Press, Oxford.

Olton, D.S. (1983). Memory functions and the hippocampus. In *Neurobiology of the Hippocampus* (ed. W. Seifert), pp.335-73. Academic Press, London.

—— and Feustle, W.A. (1981). Hippocampal function required for non-spatial working memory. *Experimental Brain Research* 41, 380-9.

——, Becker, J.T., and Handelman, G.E. (1979). Hippocampus, space, and memory. *Behavioural and Brain Sciences* 2, 315-65.

Oscar-Berman, M., Zola-Morgan, S.M., Oberg, R.G.E., and Bonner, R.T. (1982). Comparative neuropsychology and Korsakoff's syndrome. III—Delayed response, delayed alternation and DRL performance. *Neuropsychologia* 20, 187-202.

Rawlins, J.N.P. (1982). The relationship between memory and anxiety. *Behavioural and Brain Sciences* 5, 498-9.

—— (1985). Associations across time: the hippocampus as a temporary memory store. *Behavioural and Brain Sciences* 8, 479-528.

—— (in press). Time to close the store? *Behavioral and Brain Sciences* (in press).

—— and Olton, D.S. (1982). The septo-hippocampal system and cognitive mapping. *Behavioural Brain Research* 5, 331-58.

—— and Tsaltas, E. (1983). The hippocampus, time and working memory. *Behavioural Brain Research* 10, 233-62.

Rosenstock, J., Field, T.D., and Greene, E. (1977). The role of mammillary bodies in spatial memory. *Experimental Neurology* 55, 340-52.

Rossor, M.N. (1982). Neurotransmitters and CNS disease: Dementia. *Lancet* ii, 1200-3.

Saunders, R.C., Murray, E.A., and Mishkin, M. (1984). Further evidence that amygdala and hippocampus contribute equally to recognition memory. *Neuropsychologia* 22, 785-96.

Schwarcz, R., Foster, A.C., French, E.D., Whetsell, W.O. Jr., and Kohler, C. (1984). Excitotoxic models for neurodegenerative disorders. *Life Sciences* 35, 19-32.

Scoville, W.B. and Milner, B. (1957). Loss of recent memory after bilateral hippocampal lesions. *Journal of Neurology, Neurosurgery and Psychiatry* 20, 11-21.

Sidman, M., Stoddard, L.T., and Mohr, J.P. (1968). Some additional quantitative observations of immediate memory in a patient with bilateral hippocampal lesions. *Neuropsychologia* 6, 245-54.

Sinden, J.D., Rawlins, J.N.P., Gray, J.A., and Jarrard, L.E. (1986). Selective cytotoxic lesions of the hippocampal formation and DRL performance in rats. *Behavioural Neuroscience* 100, 320-9.

Squire, L.R. (1981). Two forms of human amnesia: an analysis of forgetting. *Journal of Neuroscience* 1, 635-40.

Squire, L.R. and Moore, R.Y. (1979). Dorsal thalamic lesion in a noted case of chronic memory dysfunction. *Annals of Neurology* 6, 503-6.

Swanson, L.W. (1979). The hippocampus—new anatomical insights. *Trends in Neuroscience,* 2, 9-12.

Tomlinson, B.E., Irving, D., and Blessed, G. (1981). Cell loss in the locus coeruleus in senile dementia of Alzheimer type. *Journal of Neurological Sciences* **49**, 419–28.

Warrington, E.K. and Weiskrantz, L. (1982). Amnesia: a disconnection syndrome? *Neuropsychologia* **20**, 233–48.

Weiskrantz, L. (1985). On issues and theories of the human amnesic syndrome. In *Memory systems of the brain: animal and human cognitive processes* (ed. N.M. Weinberger, J.L. McGaugh, and G. Lynch), pp. 380–415. Guildford, New York.

Whitehouse, P.J., Price, D.L., Struble, R.G., Clark, A.W., Coyle, J.T., and Delong, M.R. (1982). Alzheimer's disease and senile dementia: loss of neurones in the basal forebrain. *Science* **215**, 1237–9.

Woods, B.T., Schoene, W., and Kneisley, L. (1982). Are hippocampal lesions sufficient to cause lasting amnesia? *Journal of Neurology, Neurosurgery and Psychiatry* **45**, 243–7.

Zola-Morgan, S., Squire, L.R., and Mishkin, M. (1982). The neuroanatomy of amnesia; amygdala-hippocampus vs. temporal stem. *Science* **218**, 1337–9.

——, ——, and Amaral, D.G. (1986). Human amnesia and the medial temporal region: memory impairment following a bilateral lesion limited to field CA1 of the hippocampus. *Journal of Neuroscience* **6**, 2950–67.

7

Circuit basis of a cognitive function in non-human primates

PATRICIA S. GOLDMAN-RAKIC

INTRODUCTION

It is widely recognized that understanding and treating the disorders of cognitive function will require detailed knowledge of the development, circuitry, and neurotransmitters of the association cortex. Such knowledge can be supplied in large part from appropriate anatomical, physiological, and biochemical studies that can be conducted only in non-human animal models, among which, the macaque monkey seems unexcelled for structural-functional analysis of the higher cortical functions. In this chapter I will review a number of studies conducted in my laboratory over the past few years that have yielded new facts and insights into the anatomical and functional organization of the prefrontal areas of the frontal lobe. We have been particularly interested in the prefrontal cortex because its pre-eminent role in cognitive functions makes it a part of the brain with special significance for human society and culture.

A major goal of our studies has been to construct a neurobiological foundation for particular mnemonic functions carried out by prefrontal cortex and use this knowledge to approach issues of development and disease. Progress in this endeavour has been made possible by major technical and conceptual advances in the anatomical and functional disciplines. The circuitry of prefrontal cortex has become understood as never before with the introduction of neuroanatomical tracing methods which allow transport of macromolecules from cell body to terminal and the reverse without damage to the system under investigation. Because the integrity of the tissue is preserved, we are able to determine precisely the cells of origin and their position within the cortical layers of the various classes of cortical efferents as well as to precisely define the laminar targets of afferent projections to the cortex. In addition, the newer methods of anterograde and retrograde transport provide a more reliable picture, approaching, in some instances, the quantitative level, so as to re-evaluate the importance of connections as formerly understood. With combinations of methods, it has become possible to examine the interrelationships of two or more pathways in the same animal by double-labelling with both anterograde (Schwartz and Goldman-Rakic 1982; Selemon and Goldman-Rakic 1985*a*) and retrograde (Kuypers *et al.* 1979; Schwartz and Goldman-Rakic 1984) markers. Finally, the functional contribution of identified prefrontal circuits to behaviour can now be examined with techniques such as Sokoloff's 2-deoxyglucose metabolic mapping in behaving animals (Bugbee and Goldman-Rakic 1981; Friedman and Goldman-Rakic 1985) and with refinement of behavioural paradigms for the correlation of neuronal activity and task parameters in awake behaving monkeys (Funahashi *et al.* 1986).

The major focus of this chapter will be the principal sulcal subdivision of the monkey's prefrontal cortex on the dorsolateral convexity. This area, corresponding to Brodmann's area 49, has long been associated with the spatial cognitive deficits produced by prefrontal lesions. The principal sulcus is particularly important for the guidance of behaviour by stored (representational) information (for review see Goldman-Rakic 1987). We believe that structure-function analysis of this area of prefrontal cortex may provide a model for other prefrontal centres, and that study of its organization and function will hold the key to prefrontal function in general.

PRINCIPAL SULCUS AND REPRESENTATIONAL MEMORY

Our understanding of principal sulcus involvement in representational memory in primates is based on a solid neuropsychological foundation: (i) ablation of the principal sulcus in macaque monkeys results in a profound loss of spatial delayed response ability (e.g. Butters *et al.* 1972; Goldman and Rosvold 1970; Goldman 1971); (ii) a temporal gap of at least 1–2 seconds between stimulus and response is an essential requirement for the deficit to appear (e.g. Goldman and Rosvold 1970); (iii) the deficit is specific to spatial delayed responses and does not extend to short-term memory for the features of objects (Passingham 1975; Mishkin and Manning 1978); (iv) only lesions of the principal sulcus result in this syndrome of selective dissociation of spatial and non-spatial short-term memory. The role of the principal sulcus in the neural mechanisms underlying delayed-response has been amply supported by findings from an impressive range of approaches in addition to ablation studies including comparative analyses of behavioural competence in a variety of species (e.g. Warren *et al.* 1972; Kolb 1984), microstimulation (Stamm 1969; Stamm and Rosen 1973) and unit recording studies in trained monkeys (e.g. Kubota and Niki 1971; Fuster 1973; Kojima and Goldman-Rakic 1982, 1984), cryogenic depression (Alexander and Goldman 1978), and psychopharmacological investigations (Brozoski *et al.* 1979; Arnsten and Goldman-Rakic 1985) and, most recently, metabolic imaging using Sokoloff's 2-deoxyglucose technique (Bugbee and Goldman-Rakic 1981;

Friedman and Goldman-Rakic 1985). The conclusion that can be drawn from all of these approaches is that the cortex of the principal sulcus is essential when behaviour must be guided by the *representation* of a stimulus rather than by the stimulus itself; without the principal sulcus, laboratory animals are unable to use stored information to 'inform' a *correct* choice. Although they are perfectly able to respond, their behaviour is excessively influenced by external stimuli and they exhibit a strong tendency to repeat or perseverate a previously rewarded response, suggesting that the mechanisms for what is usually termed *associative* memory are not those that are lost after prefrontal lesions. In fact, the preservation of associative memory can account for many aspects of the symptoms expressed by human and non-human primates with prefrontal lesions (see Goldman-Rakic 1987 for details).

The ability to guide behaviour by representations of discriminative stimuli rather than by the stimuli themselves may be considered an important achievement of cortical evolution and one of primary significance for human cognition. Study of the neural basis of delayed-response in non-human primates thus provides an opportunity to uncover the circuit and cellular basis of representational guidance of behaviour, and the process of short-term memory that is crucial to that process. I have proposed (Goldman-Rakic 1987) that the ability to guide behaviour by representations requires a combination of neural mechanisms for: (i) selecting relevant information, e.g. attending to the relevant visuo-spatial co-ordinates in a delayed response task; (ii) holding that information 'on line' for a temporal interval, i.e. remembering the relevant information before a response is permitted; and (iii) executing the appropriate motor commands. In addition, mechanisms exist for modulating and enhancing information processing on a moment-to-moment basis. In the succeeding pages, I will describe our recent anatomical findings which reveal that the principal sulcus is part of a wider neural network that can support each of the processes enumerated above. Needless to say, the circuit basis of representational processes here proposed will need to be amended as more is learned about this circuitry at both the light and electronmicroscopic levels, and as physiological studies are carried out to reveal the functional properties of circuit components.

CORTICAL-CORTICAL CONNECTIONS ·

As emphasized by Geschwind (1965), the pathways that connect cortical areas within a hemisphere (associational) or between hemispheres (callosal) can be considered particularly relevant to cognitive processing. Previous studies of the cortico-cortical connections using silver degeneration techniques and large lesions of cortical areas (Jones and Powell 1970; Pandya and Kuypers 1969) established that a network of projections originating from the primary sensory cortices, ultimately reaches the prefrontal cortex. However, the exact nature of visual information flow to prefrontal areas is only recently becoming known through the use of anterograde and retrograde tracing methods. From such studies we have learned that the caudal two-thirds of the principal sulcus and adjacent anterior arcuate cortex (the latter associated with the frontal eyefields) are the major targets of projections from the posterior parietal cortex. Moreover, these major associative centres are

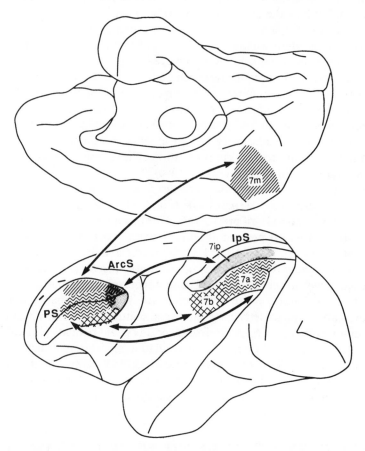

FIG. 7.1. Diagram illustrating subarea to subarea interconnections between the posterior parietal and prefrontal cortex that may be essential for spatial memory. Medial parietal cortex (area 7 m) is interconnected with the dorsal bank and rim of the caudal principal sulcus (PS); lateral parietal cortex (area 7a) is interconnected with the fundus and lower portion of each bank; area 7b is interconnected with the ventral rim of the principal sulcus; and area 7ip in the caudal bank of the intraparietal sulcus (IpS) is connected with the anterior bank of the arcuate (ArcS) and caudal tip of the principal sulcus. Note from the arrowheads that all projections are bidirectional as explained in the text (from Goldman-Rakic 1987).

linked by a precise topographically organized network of connections between particular sectors of the parietal cortex and particular areal subdivisions of the principal sulcal cortex (Fig. 7.1). For example, area 7b of parietal cortex (of Vogt and Vogt 1919) projects to the rim of the ventral bank of the principal sulcus; parietal area 7a is connected with the depth and fundus of the principal sulcus, and, area 7m (PGm of Pandya and Seltzer 1982), the parietal cortex on the medial surface of the lobe, is connected with the dorsal bank of the principal sulcus (Fig. 7.1).

The parallel organization of posterior parietal-prefrontal connections raises the question of whether there is a corresponding parallel organization of specific types of information conveyed to prefrontal cortex and concomitant functional specializations. Our anatomical findings indeed indicate that there is at least a rough organization of different sensory modalities and submodalities in the inferior parietal lobule such that areas 7a, 7ip, and 7m are innervated more by visual afferents whereas 7b receives its innervation primarily from somatosensory-related areas. Physiological studies of the posterior parietal cortex support this general plan of organization (Hyvarinen 1981; Mountcastle *et al*. 1984).

The parieto-prefrontal projections described above are reciprocated by strong prefronto-parietal pathways (Fig. 7.1). Parietal axons terminate mainly in layers I and IV of prefrontal cortex (Cavada and Goldman-Rakic in preparation), whereas prefrontal axons in parietal cortex avoid layer IV, but terminate in layers I and VI (M. L. Schwartz and P. S. Goldman-Rakic, unpublished data). It is therefore important to consider the functions of the feedback as well as the feedforward components of cortico-cortical circuitry. For example, based on our anatomical findings, and what is known or surmised about the parietal cortex and spatial attention, we may hypothesize that, among other processes, prefrontal cortex could regulate parietal cortex projections to the thalamus. Cortico-thalamic projections which originate from layer VI, are thought to play a role in gating thalamic input and, if so, prefrontal cortex could influence the parietal lobe's gating of its own thalamic input, thereby influencing attentional processes at the thalamic level.

LIMBIC CONNECTIONS OF THE PREFRONTAL CORTEX

Supporting the evidence for a role of prefrontal cortex in regulating behaviour by stored representations, i.e. by the memory of events rather than by the events themselves, the principal sulcus is also intimately connected with the hippocampal formation directly and indirectly through a multitude of relays including: (i) entorhinal and extensive areas of parahippocampal gyri; (ii) the presubicular and caudomedial cortices; and (iii) the anterior and posterior cingulate cortices (Fig. 7.2). The full extent of these connections

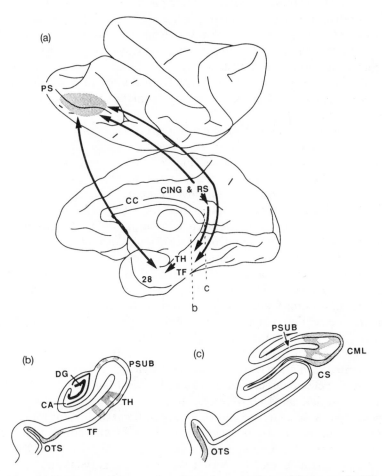

FIG. 7.2. (a) Summary diagram of prefrontal connections with the cortical limbic system, i.e. parahippocampal gyrus (Bonin and Bailey's TH and TF), entorhinal cortex (area 28), presubiculum (PSUB) and caudomedial lobule (CML). (b) and (c) Coronal sections cut through two levels, b and c, showing terminal fields of axons traced (in HRP studies) from the area of prefrontal cortex shown in light stipple. Other abbreviations: CA, Ammon's horn; Cing, cingulate gyrus; CS, collateral sulcus; DG, dentate gyrus; OTS, occipital-temporal sulcus; RS, retrosplenial cortex (from Goldman-Rakic 1987).

has only recently been elaborated in detail and a brief review may assist in discussing the proposed role of prefrontal cortex in representational memory.

According to recent findings, cortical information converges upon the entorhinal cortex via synaptic relays in the posterior parahippocampal areas,

TH and TF (Van Hoesen 1985). The principal sulcus and surrounding dorso-lateral prefrontal cortex, like many other cortical areas, project to the para-hippocampal gyrus, one step removed from the entorhinal cortex. The principal sulcus also projects directly, without relays, to the entorhinal cortex via the fronto-occipital fasciculus (Fig. 7.2). Similar direct projections to entorhinal cortex have also been noted from the orbital prefrontal cortex (Van Hoesen *et al*. 1972) and more recently from the superior temporal gyrus (Amaral *et al*. 1983).

The principal sulcus and surrounding prefrontal cortex project to the presubiculum and extensively to an adjacent area of transitional cortex designated the caudomedial lobule (Fig. 7.2, Goldman-Rakic *et al*. 1984). Since the presubiculum itself projects to entorhinal cortex (Shipley 1974), this may provide another relay to entorhinal cortex. However, the presubiculum represents primarily a major output of the hippocampus to other cortical and subcortical structures (e.g. Meibach and Siegal 1979; Rosene and Van Hoesen 1977; Swanson and Cowan 1975). Accordingly, the prefrontal terminals in this area are in position to gate the output of the hippocampal formation, perhaps in conjunction with recently described projections to the presubicular cortex from the inferior parietal lobule (Seltzer and Van Hoesen 1979). An interesting feature of the prefrontal projections is that they exhibit a columnar distribution in parahippocampal and particularly in presubicular cortices (Fig. 7.2; Goldman-Rakic *et al*. 1984).

By far the most extensive connections of the principal sulcus with the hippocampal formation travel through multisynaptic relays with the anterior and posterior cingulate cortices (e.g. Baleydier and Maugiere 1980; Pandya *et al*. 1971). As indicated above, these areas of cortex, project to the para-hippocampal, subicular, and entorhinal areas. Remarkably, the very areas of entorhinal, parahippocampal, cingulate, and subicular cortex that receive prefrontal input also are the exact targets of parietal projections (Selemon and Goldman-Rakic 1985*b*). Within these targets, parietal and prefrontal axons terminate either in adjacent columns or within different layers of the same cortical columns. We can presume, therefore, that the prefrontal and parietal cortex communicate through 'third party' limbic targets as well as by direct reciprocal interactions (Fig. 7.2).

Moreover, these areas are unified by their thalamic input from the medial pulvinar which projects to the posterior cingulate, retrosplenial and caudo-medial lobule cortex—the limbic areas—as well as to the principal sulcus and parietal area PG (Baleydier and Maugiere 1985; M. Giguere and P.S. Goldman-Rakic, unpublished data). Thus, the medial pulvinar which is prominent only in primates is in position to recruit an entire neural system defined by cortico-cortical connectivity and possibly by common dedication to the complex function of being orientated in time and space.

MOTOR CONTROL CIRCUITS

The anatomy of motor control by which principal sulcus neurons participate in the selection or inhibition of responses has also been examined and worked out in some detail (Fig. 7.3). This anatomy includes projections from the principal sulcus to the caudate nucleus (Goldman and Nauta 1977a; Selemon

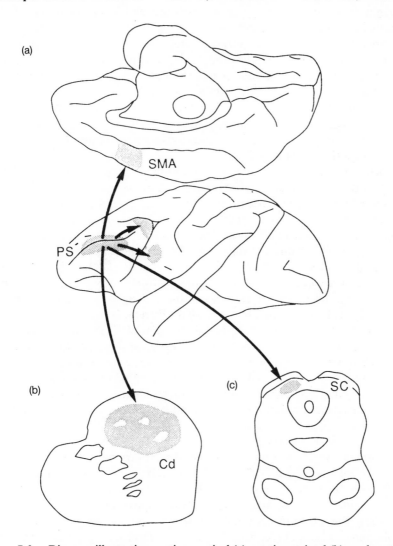

FIG. 7.3. Diagram illustrating cortico-cortical (a), cortico-striatal (b), and cortico-tectal (c) efferent pathways concerned with motor control. Abbreviations: SMA, supplementary motor area; PS, principal sulcus; Cd, caudate nucleus; SC, superior colliculus (from Goldman-Rakic 1987).

and Goldman-Rakic 1985*a*; Yeterian and Van Hoesen 1978); connections with the motor thalamus (Ilinsky *et al*. 1985; Goldman-Rakic and Porrino 1985); and with the deep 'motor' layers of the superior colliculus (Goldman and Nauta 1976; Fries 1984).

The projections from the neocortex to the caudate nucleus and putamen form a massive system of cortical efferents, the cortico-striatal system, long thought to be the major pathway by which associational and sensory cortical areas exert control over motor function (Fig. 7.3). Relevant to the present discussion is the fact that the prefrontal cortex, including the principal sulcus, projects throughout the head, body, and tail of the caudate nucleus as well as to the rostral putamen (Goldman and Nauta 1977*b*; Selemon and Goldman-Rakic 1985*a*). Other areas of cerebral cortex project to the neostriatum and reconstruction of the topography of these multiple cortical inputs indicates that, by and large, each cytoarchitectonic subdivision cortex terminates in a specific region within the neostriatum (Selemon and Goldman-Rakic 1985*a*).

Corticostriatal fibres terminate on the medium spiny neurons of the neostriatum which in turn project to the substantia nigra (pars reticulata) and globus pallidus. The evidence suggests that clusters of cells in the neostriatum that receive specific cortical inputs likewise have specific outputs, and that a 'labelled line' may extend from cortex through neostriatum to the substantia nigra or globus pallidus and then back through the thalamus to cortex (for general discussion, see Delong and Georgopoulus 1981). Presently, while connections between all of the aforementioned structures have been documented, the nature and existence of the 'throughway' originating at one cortical locus and returning to the same (or different) cortical locus remains to be worked out in detail. Although the complex circuitry of the basal ganglia is far from understood, it seems inescapable that the neostriatal complex stands at the headway of a multisynaptic circuit that functions to inhibit most behaviour in order to allow the execution of a single appropriate act. Basal ganglia damage in man disrupts this balance and results in uncontrolled movements ranging from involuntary tremor to ballistic thrusts as well as slowness and/or confusion in motor responses and, often, in thought processes as well.

Another connection from the prefrontal cortex to 'motor' structures is the corticotectal pathway (Fig. 7.3). Long thought to arise in the frontal lobe exclusively from the frontal eye fields, several studies have shown that its cortical innervation actually involves many parts of the cortical mantle including the posterior parietal cortex as well as the principal sulcus (Goldman and Nauta 1976; Fries 1984). These latter areas project to the intermediate and deep layers of the colliculus, which contain multisensory cells (e.g. Meredith and Stein, 1983, 1985) as well as cells that discharge in relation to eye-movements (Mays and Sparks 1980; Mohler and Wurtz 1976; Wurtz and Goldberg 1972). The intermediate and deeper layers of the colliculus could be subcortical centres where principal sulcal innervation facilitates or inhibits head and eye movements, particularly those governed by stored information. Guitton *et al.* (1985) have recently described the difficulties that frontal lobe patients have in inhibiting eye movements to specified visual targets. They proposed that this problem stems from a failure of a mechanism designed to cancel the incorrect response before it has occurred. If we accept that the principal sulcus contains the mechanism for guiding specific responses on the basis of representational knowledge (e.g. in the form of an instruction, 'look to the opposite locations of the target'), then it recommends itself as the structure capable of issuing the cancellation signal in the 'anti-saccade' task described above. Cancellation commands could arise at a number of different levels of the neuroaxis: in the deep layers of the superior colliculus itself, in the substantia nigra (Ilinsky *et al.* 1985), or at the cortex. Regarding the latter, an interaction between the principal sulcus and frontal eye fields is likely because the two cortical areas are interconnected (Pandya and Vignolo 1971; Barbus and Mesulam 1981, 1985). Perhaps prefrontal eyefield neurons are involved in computing or accessing the co-ordinates for an intended eye movement (Bruce and Goldberg 1985) whereas principal sulcus neurons may be concerned more with whether or not to move the eyes to a particular target.

Finally, the principal sulcus, particularly its caudal portion, projects to premotor cortical areas including the anterior supplementary motor cortex (SMA) which, in turn, projects to the upper body and head representation of the primary motor cortex (Fig. 7.3; M. Jouandet and P.S. Goldman-Rakic unpublished observations). Another portion, the ventral rim of the principal sulcus, projects to the mouth area of the lateral premotor cortex which, in turn, projects to the mouth area of the motor cortex (Matelli *et al.* 1986). Thus, neurons in the caudal portion of the principal sulcus, via either SMA or ventral premotor cortex, are only one synapse removed from primary motor cortex. Since the SMA and premotor cortices generally are major targets of the pallido-thalamo-cortical feedback loop (Ilinsky *et al.* 1985; Schell and Strick 1984), the prefronto-premotor connections could interact with and perhaps regulate (suppress or enhance) cortical activation from the motor thalamus. Thus, all circuitry exists in the principal sulcus for influencing motor commands.

BRAIN STEM PROJECTIONS TO PREFRONTAL CORTEX

Systematic biochemical data on the monoaminergic innervation of the cerebral cortex in monkeys have revealed marked regional differences in the endogenous concentrations and turnover rates of dopamine, norepinephrine, and serotonin in different cytoarchitectonic areas of the monkey cerebral cortex (Brown *et al.* 1979). Dopamine and norepinephrine are found in highest concentrations in the prefrontal (and anterior temporal) cortex; with a few exceptions these neurotransmitter substances decrease along the fronto-occipital axis and are present in considerably lower concentration in the primary visual cortex. Synthesis rates for the catecholamines are also especially high in various areas of association cortex, including prefrontal cortex (Brown *et al.* 1979). This basic regional pattern of cortical innervation is supported by evidence from anatomical tracing (Porrino and Goldman-Rakic 1982), histofluorescence (Levitt *et al.* 1984) and immuno-cytochemical (Morrison *et al.* 1982) investigations. In addition, studies employing receptor autoradiography have begun to provide quantitative information on the density and distribution of various classes of neurotransmitter receptors in various cortical areas including prefrontal cortex (Gallager and Goldman-Rakic 1986; Rakic *et al.* 1987).

The findings described above encouraged the idea that endogenous deficits or fluctuations in catecholamines might have a particularly strong influence on the associative functions of prefrontal cortex. Using intracerebral injections of neurotoxins to selectively deplete catecholamines in the principal sulcus of young adult monkeys, we were able to induce impairments of delayed response performance and further to show that the deficits could be partly or wholly reversed by systemic injections of L-dopa, apomorphine, and amphetamine (Brozoski *et al.* 1979). This study provided the first direct evidence for the involvement of catecholamines in a specific cortical function and the possibility that 'replacement' therapy with catecholamine agonists could be used to ameliorate cognitive deficits. We next extended this principal to the aged monkey in which endogenous losses in concentrations and synthesis of dopamine and norepinephrine had been demonstrated (Goldman-Rakic and Brown 1981). Our studies have provided strong evidence that catecholamine agonists, in particular, alpha-2 agonists, may serve as effective treatment of memory decline in aged monkeys (Arnsten and Goldman-Rakic 1985). These studies offer encouragement and clues for the treatment of short-term memory deficits in the human elderly; adrenergic compounds have already proven beneficial in the treatment of memory deficits in Wernicke–Korsakoff's Disease (McEntee and Mair 1978).

SUMMARY OF CIRCUIT BASIS FOR DELAYED RESPONSE

The involvement of the principal sulcus in delayed-response implies access to visuo-spatial co-ordinate systems that register and/or compute the location of objects in space. We have learned from our studies that the main cortical innervation of the caudal principal sulcus, the cortical centre for delayed response, comes from the posterior parietal cortex, the cortical center that is essential for visuo-spatial representations (Mountcastle *et al.* 1984). It is difficult to escape the conclusion that parieto-prefrontal projections are the primary path by which prefrontal cortex gains access to visuo-spatial co-ordinates. Likewise, it seems a reasonable inference that the extensive multiple connections between the principal sulcus and the hippocampal formation that we have elucidated play an important role in the storage and retrieval of data in long-term memory that may be relevant to the task at hand. Finally, neurons lying in the caudal principal sulcus innervate many cortical and subcortical structures that influence motor action. Indeed some of these connections, e.g. those to the SMA, are but one synapse removed from the primary motor cortex. Thus, while probably not able to inde-pendently generate a single motor act, the neurons in the principal sulcus may nevertheless influence motor output by initiating, facilitating, and cancelling commands to structures more directly involved in the programming, computational, and performance aspects of specific motor acts. The brain stem's monoamine projections to the principal sulcus play a modulatory role and normally may enhance or facilitate processing carried out by the information-containing circuitry.

The functional significance of the connections of the principal sulcus are not assumed; rather the contribution of the various components to behaviour can be dissected, tested, and examined by a variety of functional approaches. For one, the use of the 2-deoxyglucose method in monkeys that have been trained to perform a variety of short-term and associative memory tests reveals that metabolic activity in the principal sulcus is reliably enhanced during performance of delayed response tasks relative to performance of associative memory tasks (Bugbee and Goldman-Rakic 1981). Metabolism can also be shown to be significantly higher in structures such as the hippo-campal formation, with which the principal sulcus is connected during the performance of the delayed response task (Friedman and Goldman-Rakic 1985). Electrophysiological studies have shown that neuronal activity is time-locked to one or another parameter of the delayed response task in various parts of the neural network to which the principal sulcus belongs, including the hippocampal formation (Watanabe and Niki 1985; Wilson *et al.* 1986), the caudate nucleus (Goldman and Rosvold, 1972) and the posterior parietal cortex (Alexander, in Rakic and Goldman-Rakic, 1982). Finally, we have shown that the catecholaminergic component of prefrontal function can be specifically analysed by 'biochemical lesions' and pharmacological mani-pulation (Brozoski *et al.* 1979; Arnsten and Goldman-Rakic 1985).

FUTURE PROSPECTS FOR CELLULAR AND MOLECULAR APPROACHES TO COGNITION

The analysis of structure-function relationships also needs to be taken to microstructural level. A major feature of cortico-cortical projections is that they are not uniformly distributed within their prefrontal targets, but are segregated in the form of vertical columns that alternate in rather precise geometric fashion with columns of callosal afferents (Fig. 7.4) (Goldman and Nauta 1977*b*; Goldman-Rakic and Schwartz 1982). The terminal fields of both fibre systems are distributed in side-by-side bands approximately 0.5 mm wide. Knowledge of how, for example, parietal afferents to the prefrontal cortex interact with prefrontal fibres from the opposite hemisphere is not known, but is essential to a further understanding of interhemispheric integration in the cortex. Parietal associational fibres tend to terminate selectively in layers IV and I, while callosal fibres terminate in layers I, IV, and VI (see Fig. 7.4 for details). Associational and callosal neurons are extensively intermixed in layers III, V, and VI of the principal sulcus in columns innervated by callosal or associational axons (Schwartz and Goldman-Rakic 1984; Fig. 7.4). The light microscopic evidence thus suggests that modular units within prefrontal cortex reciprocally interconnect with the parietal cortex both within and between hemispheres.

Evidence on the physiological correlates of cortical columns is still lacking. However, in studies of the dorsolateral prefrontal cortex in behaving monkeys neurons with the same properties have been observed to occur in clusters or in electrode penetrations orthogonal to the cortex (Fuster 1973; Fuster *et al*. 1982). In research on the frontal eye field region of prefrontal cortex, neurons involved in vertical eye movements have been identified in callosally innervated columnar zones while those involved in horizontal eye movements tend to be found in acallosal columnar zones that receive intrahemispheric innervation (Bruce and Goldman-Rakic 1984). A recent observation is that gamma-amino-butyric acid (GABA) containing interneurons in the principal sulcal cortex exhibit a mosaic distribution that possibly coincides with the periodicity of cortical (associational and callosal) fibre columns (Schwartz *et al*. 1985) and may be the basis of the adjacent high and low metabolic zones observed in this region (Goldman-Rakic 1984*a*). It can be presumed that columns of neurons in prefrontal cortex, like those in sensory regions of the cortex, share local functions and that the nature of these functional units will in time be identified.

The cortex also has a radial or horizontal geometry due to its distinct six-layered organization. The significance of the layering pattern is that cells in the specific layers which are targets of incoming callosal or associational afferents have distinctive projection targets in subcortical and/or other cortical structures. For example, as shown in Fig. 7.3, pyramidal neurons in layer III project primarily to other cortical areas, both within the same (associational neurons) and opposite (callosal neurons) hemispheres. While over 80 per cent of callosal and associational neurons originate in layer III, layer V is the major source of projections to the caudate nucleus and putamen, to the colliculus, and to other subcortical structures; neurons in

PRINCIPAL SULCUS

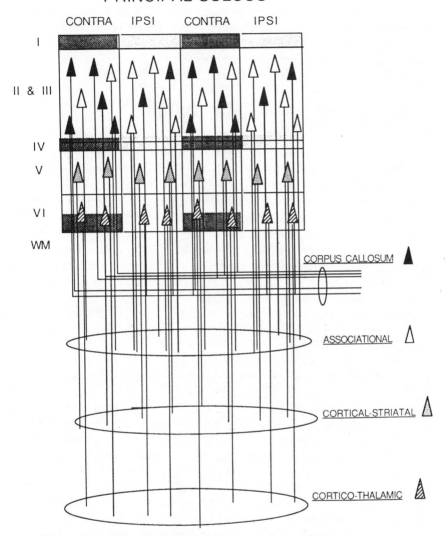

FIG. 7.4. Simplified circuit diagram illustrating the terminal fields of parietal associational fibres (light stipple) and callosal fibres (dark stipple) in layers I and IV of
adjacent columns in the principal sulcus. Also shown are the callosal (dark triangles)
and associational (light triangles) projection neurons that send their axons to the contralateral prefrontal and ipsilateral parietal cortex, respectively. Note that callosal
neurons are more concentrated in columns that receive callosal innervation and associational neurons are present in higher numbers in columns that are innervated by ipsilateral parietal afferents. This organization is indicative of input–output modules.
Also shown are the layer V neurons that project to the caudate nucleus and putamen
and the layer VI cells that target the thalamus (from Goldman-Rakic 1986).

layer VI project selectively to the thalamus (Fig. 7.4). Receptor autoradiographic studies currently underway in our laboratory are linking the neurotransmitter receptors to specific classes of cells in the various layers (Kritzer *et al.* 1987; Gallager and Goldman-Rakic 1986; Rakic *et al.* 1988). One can use these facts to begin to appreciate the synaptic architecture, neurochemistry and wiring diagrams underlying a specific cortical function.

RELEVANCE TO NEUROLOGICAL AND PSYCHIATRIC DISEASE

One or more areas of the prefrontal cortex have been implicated in the cognitive symptoms of a variety of neurological and mental diseases, e.g. schizophrenia (Ingvar 1980; Weinberger *et al.* 1986), Parkinson's disease (Lees and Smith 1983; Taylor *et al.* 1986) and Korsakoff's disease (Oscar-Berman *et al.* 1982; Warrington and Weiskrantz 1982). Although these diseases are very different from one another, they share the common feature of a selective loss in performance on tasks like the Wisconsin Card Sort or other tests that indicate prefrontal cortex involvement, and according to theoretical considerations, reflect disruption of behavioral regulation by representational knowledge (see Goldman-Rakic 1987 for amplification). In schizophrenia, although we cannot pinpoint a lesion site, the thought disorder in this disease must either directly or indirectly involve the cortico-cortical parthways that establish the inner models of reality and adjust them to contemporary demands. It may be this information-processing network that is influenced most severely by hyperactivity in dopamine systems and is mollified by neuroleptic medication. The relatively pure cognitive disorder of Korsakoff's psychosis originates in a lesion of the thalamo-cortical and brain stem pathways that we have argued perform activational and modulatory roles in cortico-cortical processing, while Parkinson's disease which affects the caudate nucleus profoundly disrupts the mechanisms by which prefrontal cortex exerts control over motor acts (see Ilinsky *et al.* 1985; Selemon and Goldman-Rakic 1985*a*). Although in each of these disorders, the lesion is known or can be inferred to occur at a different locus in the circuitry, and thus to compromise a different process, the final outcome may be the same—lack of initiative, compromised affect, disordered thinking and fragmented or incoherent behaviour that might be predicted if behaviour were not regulated by representational knowledge. It is possible that catecholamine deficits—dopamine in schizophrenia and Parkinson's disease, and norepinephrine in other disorders—contribute to this common outcome. Although any formulation concerning the mental diseases is necessarily inadequate, it seems inescapable that future efforts to understand the frontal lobe may pay dividends in contributing and possibly treating a family of mental disorders that reflect breaks in its circuitry.

REFERENCES

Alexander, G.E. and Goldman, P.S. (1978). Functional development of the dorsolateral prefrontal cortex: an analysis utilizing reversible cryogenic depression. *Brain Research* **143**, 233–50.

Amaral, D.G., Insausti, R., and Cowan, W.M. (1983). Evidence for a direct projection from the superior temporal gyrus to the entorhinal cortex in the monkey. *Brain Research* **275**, 263–77.

Arnsten, A.F.T. and Goldman-Rakic, P.S. (1984). Selective prefrontal cortical projections to the region of the locus coeruleus and raphe nuclei in the rhesus monkey. *Brain Research* **306**, 9–18.

—— and —— (1985). Alpha-adrenergic mechanisms in prefrontal cortex associated with cognitive decline in aged nonhuman primates. *Science* **230**, 1273–6.

Baleydier, C. and Mauguiere, F. (1980). The duality of the cingulate gyrus in monkey: Neuroanatomical study and functional hypothesis. *Brain* **103**, 525–54.

—— and —— (1985). Anatomical evidence for medial pulvinar connections with the posterior cingulate cortex, the retrosplenial area, and the posterior parahippocampal gyrus in monkeys. *Journal of Comparative Neurology* **232**, 219–28.

Barbas, H. and Mesulam, M-.M. (1981). Organization of afferent input to subdivisions of area 8 in the rhesus monkey. *Journal of Comparative Neurology* **200**, 407–31.

—— and —— (1985). Cortical afferent input to the principalis region of the rhesus monkey. *Neuroscience* **15**, 619–37.

Brown, R.M., Crane, A.M. and Goldman, P.S. (1979). Regional distribution of monoamines in the cerebral cortex and subcortical structures of the rhesus monkey: concentrations and in vivo synthesis rates. *Brain Research* **168**, 133–50.

Brozoski, T., Brown, R.M., Rosvold, H.E., and Goldman, P.S. (1979). Cognitive deficit caused by depletion of dopamine in prefrontal cortex of rhesus monkey. *Science* **205**, 929–31.

Bruce, C.J. and Goldberg, M.E. (1985). Primate frontal eye fields. I. Single neurons discharging before saccades. *Journal of Neurophysiology* **53**, 603–35.

—— and Goldman-Rakic, P.S. (1984). Columnar organization of callosal connectivity in macaque frontal eye fields and its relation to elicited eye movements. *Society of Neuroscience Abstracts* **10**, 513.

Bugbee, N.M. and Goldman-Rakic, P.S. (1981). Functional 2-deoxyglucose mapping in associated cortex: Prefrontal activation in monkeys performing a cognitive task. *Society of Neuroscience Abstracts* **7**, 416.

Butters, N., Pandya, D., Stein, D., and Rosen, J. (1972). A search for the spatial engram within the frontal lobes of monkeys. *Acta Neurobiologiae Experimentalis* **32**, 305–29.

Cavada, C. and Goldman-Rakic, P.S. Subdivisions of posterior parietal cortex have distinct connection with limbic and sensory cortex. *Journal of Neuroscience* (in preparation).

DeLong, M.R. and Georgopoulos, A.P. (1981). Motor functions of the basal ganglia. In (ed. J.M. Brookhart, V.B. Mountcastle, V.B. Brook, and S.R. Geiger) *Handbook of Physiology*, Section 1, The Nervous System, vol II: Motor Control, Part 2, pp. 1017–61. American Physiological Society. Bethesda, Md.

Friedman, H.R. and Goldman-Rakic, P.S. (1985). Enhancement of trisynaptic pathway in the hippocampus during performance of spatial working memory tasks:

A 2-DG behavioral study in rhesus monkeys. *Society of Neuroscience Abstracts* **11**, 460.

Fries, W. (1984). Cortical projections to the superior colliculus in the macaque monkey. A retrograde study using horseradish peroxidase. *Journal of Comparative Neurology* **230**, 55–76.

Funahashi, S., Bruce, C.J., and Goldman-Rakic, P.S. (1986). Perimetry of spatial memory representation in primate prefrontal cortex: Evidence for a mnemonic hemionopia. *Society of Neuroscience Abstracts* **12**, 554.

Fuster, J.M. (1973). Unit activity in prefrontal cortex during delayed-response performance: Neuronal correlates of transient memory. *Journal of Neurophysiology* **36**, 61–78.

Fuster, M.J., Bauer, R.H., and Jervey, J.P. (1982). Indications of functional relationship between prefrontal and inferotemporal cortex. *Society of Neuroscience Abstracts* **8**, 681.

Gallager, D. and Goldman-Rakic, P.S. (1986). Differential distribution of neurotransmitter receptors in prefrontal cortex of rhesus monkeys revealed by quantitative autoradiography. In *Abstracts of 25th Annual Meeting of the American College of Neuropsychopharmacology*, p. 234.

Geschwind, N. (1965). Disconnection syndromes in animals and man. *Brain* **88**, 237–94.

Goldman, P.S. (1971). Functional development of the prefrontal cortex in early life and the problem of neuronal plasticity. *Experimental Neurology* **32**, 366–87.

—— and Nauta, W.J.H. (1976). Autoradiographic demonstration of a projection from prefrontal association cortex to the superior colliculus in the rhesus monkey. *Brain Research* **116**, 145–9.

—— and —— (1977*a*). Columnar distribution of cortico-cortical fibres in the frontal association, limbic, and motor cortex of the developing rhesus monkey. *Brain Research* **122**, 393–413.

—— and —— (1977*b*). An intricately patterned prefrontocaudate projection in the rhesus monkey. *Journal of Comparative Neurology* **171**, 369–86.

—— and Rosvold, H.E. (1970). Localization of function within the dorsolateral prefrontal cortex of the rhesus monkey. *Experimental Neurology* **27**, 291–304.

—— and —— (1972). The effects of selective caudate lesions in infant and juvenile rhesus monkeys. *Brain Research* **43**, 53–66.

Goldman-Rakic, P.S. (1984). Toward a neurobiology of cognition. In *Neonate Cognition: Beyond the Blooming, Buzzing Confusion* (ed. J. Mehler and R. Fox), pp. 285–6. L. Erlbaum Associates, New Jersey.

—— (1984*a*). Modular organization of prefrontal cortex. *Trends in Neurosciences* **7**, 419–24.

—— (1986). Setting the stage: neural development before birth. In *The Brain, Cognition and Education* (ed. S.L. Friedman, K.A. Klivington, and R.W. Peterson), pp. 265–90. Academic Press, New York.

—— (1987). Circuitry of the prefrontal cortex and the regulation of behaviour by representational knowledge. In *Handbook of Physiology: The Nervous System*, Vol. 5 (ed. F. Plum and V. Mountcastle), pp. 373–417. American Physiological Society, Bethesda, Md (in press).

—— and Brown, R.M. (1981). Regional changes of monoamines in cerebral cortex and subcortical structures of aging rhesus monkeys. *Neuroscience* **6**, 177–87.

—— and Porrino, L.J. (1985). The primate mediodorsal (MD) nucleus and its projections to the frontal lobe. *Journal of Comparative Neurology* **242**, 535–60.

—— and Schwartz, M.W. (1982). Interdigitation of contralateral and ipsilateral columnar projections to frontal association cortex in primates. *Science* **216**, 755–7.

——, Selemon, L.D., and Schwartz, M.L. (1984). Dual pathways connecting the dorsolateral prefrontal cortex with the hippocampal formation and parahippocampal cortex in the rhesus monkey. *Neuroscience* **12**, 719–43.

Guitton, D., Buchtel, H.A., and Douglas, R.M. (1985). Frontal lobe lesions in man cause difficulties in suppressing reflexive glances and in generating goal-directed saccades. *Experimental Brain Research* **58**, 455–72.

Hyvarinen, J. (1981). Regional distribution of functions in parietal association area 7 of the monkey. *Brain Research* **206**, 287–303.

Ilinsky, I.A., Jouandet, M.L., and Goldman-Rakic, P.S. (1985). Organization of the nigrothalamocortical system in the rhesus monkey. *Journal of Comparative Neurology* **236**, 315–30.

Ingvar, D.H. (1980). Abnormal distribution of cerebral activity in chronic schizophrenia: A neurophysiological interpretation. In *Perspectives in Schizophrenia* (ed. C. Baxter and T. Melnechuk), pp. 107–25. Raven Press, New York.

Jones, E.G. and Powell, T.P.S. (1970). An anatomical study of converging sensory pathways within the cerebral cortex of the monkey. *Brain* **93**, 793–820.

Kojima, S. and Goldman-Rakic, P.S. (1984). Functional analysis of spatially discriminative neurons in prefrontal cortex of rhesus monkey. *Brain Research* **291**, 229–40.

——, Kojima, M. and Goldman-Rakic, P.S. (1982). Operant behavioral analysis of delayed-response performance in rhesus monkeys with prefrontal lesions. *Brain Research* **248**, 51–9.

Kritzer, M.F., Innis, R.B., and Goldman-Rakic, P.S. (1987). Autoradiographic localization of cholecystokinin receptors in primate cortex. *Journal of Comparative Neurology* (in press).

Kubota, K. and Niki, H. (1971). Prefrontal cortical unit activity and delayed cortical unit activity and delayed alternation performance in monkeys. *Journal of Neurophysiology* **34**, 337–47.

Kuypers, H.G.J.M., Bentivoglio, M., Van der Kooy, D., Catsman-Berrevoets, C.E., and Bharos, A.T. (1979). Retrograde transport of bisbenzamide and propedium rodide through axons to their parent cell bodies. *Neuroscience* **12**, 1–7.

Lees, A.J. and Smith, E. (1983). Cognitive deficits in the early stages of Parkinson's disease. *Brain* **106**, 257–70.

Levitt, P., Pakic, P., and Goldman-Rakic, P.S. (1984). Region-specific distribution of catecholamine afferents in primate cerebral cortex: A fluorescence histochemical analysis. *Journal of Comparative Neurology* **225**, 1–14.

Matelli, M., Camarda, R., Glickstein, M., and Rezzolatti, G. (1986). Afferent and efferent projections of the inferior area 6 in the macaque monkey. *Journal of Comparative Neurology* **251**, 281–98.

Mays, L.E. and Sparks, D.L. (1980). Dissociation of visual and saccadic-related responses in colliculus neurons. *Journal of Neurophysiology* **43**, 207.

McEntee, W.J. and Mair, R.G. (1978). Memory impairment in Korsakoff's psychosis: a correlation with brain noradrenergic activity. *Science* **202**, 905–7.

Meibach, R.C. and Siegel, A. (1979). Efferent connections of the hippocampal formation in the rat. *Brain Research* **124**, 197–224.

Meredith, M.A. and Stein, B.E. (1983). Interactions among converging sensory inputs in the superior colliculus. *Science* **221**, 389–91.

—— and —— (1985). Descending efferents from the superior colliculus relay integrated multisensory information. *Science* **227**, 657–9.

Mishkin, M. and Manning, F.J. (1978). Non-spatial memory after selective prefrontal lesions in monkeys. *Brain Research* **143**, 313–23.

Mohler, C.W. and Wurtz, R.H. (1976). Organization of monkey superior colliculus: intermediate layer cells discharging before eye movement. *Journal of Neurophysiology* **39**, 722–44.

Morrison, J.H., Foote, S.L., O'Connor, D., and Bloom, F.E. (1982). Laminar, tangential, and regional organization of the noradrenergic innervation of monkey cortex: Dopamine-B-hydroxylase immunohistochemistry. *Brain Research Bulletin* **9**, 309–19.

Mountcastle, V.B., Motter, B.C., Steinmetz, M.A., and Duffy, C.J. (1984). Looking and seeing: the visual functions on the parietal lobe. In *Dynamic Aspects of Neocortical Function*. (ed. G.M. Edelman, W.E. Gall, and W.M. Cowan), pp. 159–93. Wiley & Sons, New York.

Oscar-Berman, M., Zola-Morgan, S.M., Oberg, R.G.E., and Bonner, R.T. (1982). Comparative neuropsychology and Korsakoff's syndrome. III-Delayed response, delayed alternation and DRL performance. *Neuropsychologia* **20**, 187–202.

Pandya, D.N. and Kuypers, H.G.J.M. (1969). Cortico-cortical connections in the rhesus monkey. *Brain Research* **13**, 13–36.

—— and Seltzer, B. (1982). Intrinsic connections and architectonics of posterior parietal cortex in the rhesus monkey. *Journal of Comparative Neurology* **204**, 196–210.

—— and Vignola, L.A. (1971). Intra- and interhemispheric projections of the precentral, premotor and arcuate areas in the rhesus monkey. *Brain Research* **26**, 217–33.

——, Dye, P., and Butters, N. (1971). Efferent cortico-cortical projections of the prefrontal cortex in the rhesus monkey. *Brain Research* **31**, 35–46.

Passingham, R. (1975). Delayed matching after selective prefrontal lesions in monkeys (*Macaca mulatta*). *Brain Research* **92**, 89–102.

Porrino, L. and Goldman-Rakic, P.S. (1982). Brainstem innervation of prefrontal and anterior cingulate cortex in the rhesus monkey revealed by retrograde transport of HRP. *Journal of Comparative Neurology* **205**, 63–76.

Rakic, P. and Goldman-Rakic, P.S. (1982). Development and modifiability of the cerebral cortex. *Neuroscience Research Program Bulletin* **20**, 429–611.

——, ——, and Gallager, D. Areal and laminar distribution of major neurotransmitter receptors in the monkey visual cortex. *Journal of Neuroscience* (submitted).

Rosene, D.L. and Van Hoesen, G.W. (1977). Hippocampal efferents reach widespread areas of cerebral cortex and amygdala in the rhesus monkey. *Science* **198**, 315–17.

Schell, G.R. and Strick, P.L. (1984). Origin of thalamic input to the arcuate premotor and supplementary motor areas. *Journal of Neuroscience* **4**, 539–60.

Schwartz, M.L. and Goldman-Rakic, P.S. (1982). Single cortical neurones have axon collaterals to ipsilateral and contralateral cortex in fetal and adult primates. *Nature* **299**, 154–6.

Schwartz, M.W. and Goldman-Rakic, P.S. (1984). Callosal and intrahemispheric connectivity of the prefrontal association cortex in rhesus monkey: relation between intraparietal and principal sulcal cortex. *Journal of Comparative Neurology* **226**, 403–20.

Schwartz, M.L., Zheng, D.S., and Goldman-Rakic, P.S. (1985). Laminar and tangential variation in the morphology and distribution of GABA-containing neurons in rhesus monkey prefrontal cortex. *Society of Neuroscience Abstracts* **11**, 503.

Selemon, L.D. and Goldman-Rakic, P.S. (1985*a*). Longitudinal topography and interdigitation of corticostriatal projections in the rhesus monkey. *Journal of Neuroscience* **5**, 776–94.

—— and —— (1985*b*). Common cortical and subcortical target areas of the dorsolateral prefrontal and posterior parietal cortices in the rhesus monkey. *Society of Neuroscience Abstracts* **11**, 323.

Seltzer, B. and Van Hoesen, G.W. (1979). A direct inferior parietal lobule projection to the presubiculum in the rhesus monkey. *Brain Research* **179**, 157–61.

Shipley, M.T. (1974). Presubiculum afferents to the entorhinal area and the Papez circuit. *Brain Research* **67**, 162–8.

Stamm, J.S. (1969). Electrical stimulation of monkeys' prefrontal cortex during delayed-response performance. *Journal of Comparative Physiology and Psychology* **67**, 535–46.

—— and Rosen, S.C. (1973). The locus and crucial time of implication of prefrontal cortex in the delayed response task. In *Psychophysiology of the Frontal Lobes* (ed. K.H. Pribram and A.R. Luria), pp. 139–53. Academic Press, New York.

Swanson, L.W. and Cowan, W.M. (1975). Hippocampal-hypothalamic connections: origin in subicular cortex, not Ammon's horn. *Science* **189**, 303–4.

Taylor, A.E., Saint-Cyr, J.A., and Lang, A.E. (1986) Frontal lobe dysfunction in Parkinson's disease: the cortical focus of neostriatal outflow. *Brain.* **109**, 845–83.

Van Hoesen, G. W. (1985). Neural systems of the non-human primate forebrain implicated in memory. *Annals of the New York Academy of Sciences* **444**, 97–112.

——, Pandya, D.N., and Butters, N. (1972). Cortical afferents to the entorhinal cortex of the rhesus monkey. *Science* **175**, 1471–3.

Vogt, C. and O. Vogt (1919). Allgemeinere ergebrisse unserer hirnforschung. *J. F. Psychol. u. Neurol.* **25**, 279–461.

Warren, J.M., Warren, H.B., and Akert, K. (1972). The behaviour of chronic cats with lesions in the frontal association cortex. *Acta Neurobiologica Experimentalis* **32**, 345–92.

Warrington, E.K. and Weiskrantz, L. (1982). Amnesia: a disconnection syndrome? *Neuropsychologia* **20**, 233–48.

Watanabe, T. and Niki, H. (1985). Hippocampal unit activity and delayed response in the monkey. *Brain Research* **325**, 241–54.

Weinberger, D.R., Berman, K.F., and Zec, R.F. (1986). Physiological dysfunction of dorsolateral prefrontal cortex in schizophrenia: 1. Regional cerebral blood flow (rCBF) evidence. *Archives of General Psychology* **43**, 114–25.

Wilson, F.A.W., Brown, M.W., and Riches, I.P. (1986). Sensory-related and motor unit neuronal activity in the monkey hippocampal formation. *Society of Neuroscience Abstracts.* **12**, 556.

Wurtz, R.H. and Goldberg, M.E. (1972). Activity of superior colliculus in behaving monkey. III. Cells discharging before eye movements. *Journal of Neurophysiology* **35**, 575.

Yeterian, E.H. and Van Hoesen, G.W. (1978). Cortico-striate projections in the rhesus monkey: the organization of certain cortico-caudate connections. *Brain Research* **139**, 43–63.

8

Drugs and the processing of information

DAVID M. WARBURTON

This chapter will discuss research that has identified some of the mechanisms that control information processing in the brain. Information processing is conceptualized as a flow of neurally-coded data through the brain systems and which, through a series of transformations, results in complex psychological events such as perception, thought, and memory.

A FUNCTIONAL STATE MODEL

A functional state model considers information processing as the outcome of the states of functioning of brain structures. Here 'state' does not refer to discrete states of the brain but to a continuum of states. It argues that parts of the continuum may be more optimal for some sorts of cognitive operations than other parts. It is important to remember that the brain is not fixed in a 'state', but that the states are varying from moment to moment, and the balance of activity of one particular type will depend on the nature of the requirements of the situation. At one time we may require a state for concentrating hard, interpreting information from the outside world and ignoring distracting external and internal information, like thoughts (Warburton 1979). The information processing performance that is observed is the result of some functional mix of states. A person can control the transitions between different states to meet the demands of the current task. In this chapter, it will be argued that drugs modify information processing by altering the functional state of the brain and so give clues about the biochemistry of behaviour.

One index of the functional state of the brain is the pattern of electrocortical arousal. Psychophysiological studies have revealed that the efficiency of information processing is correlated with the type of electrical activity at the cortex (Lindsley 1952). Cortical desynchronization (beta activity), is correlated with alertness and concentration in the person, while slower, more synchronized cortical activity (alpha activity) is correlated with relaxed wakefullness. Oswald (1962) found that when performing a monitoring task over a considerable period of time, alpha waves on the EEG recording were associated with good target detection scores, whereas theta waves were typical of non-detection.

Similarly, Groll (1966) found that the efficiency of attentional performance was directly related to electrocortical arousal. There were decreases in the percentage of a subject's correct detections in parallel with the average EEG frequency over the session. The average EEG frequency immediately before a missed target was slower than before a detected target. The latency of responses for detected targets was negatively correlated with the EEG frequencies preceding the targets. O'Hanlon and Beatty (1977) found that beta activity was co-related to quicker detection of targets while theta activity was associated with slower detection. Thus, the state of electrocortical arousal will fluctuate to try and match the requirements of the situation; performance will reflect the match between the brain state and the cognitive operations demanded by the task.

It is the thesis of this paper that cholinergic and noradrenergic systems in the brain are involved in the control of state, and thus in different types of information processing. The evidence for this hypothesis has come from studies in which cholinergic and noradrenergic drugs were used as tools to investigate the processes of information processing. The first process to be considered is attention.

ATTENTION

Kahneman, (1973) proposed that attention is the allocation of the available 'mental effort' to particular aspects of information processing, and Norman and Bobrow (1975) suggested that attention is the direction of processing resources to the various potential sources of information input. Both of these approaches consider that the available resources can be allocated to the various stages of processing in a focused attentional task as well as among different concurrent sources of information in divided attention. In addition, Kahneman (1973) noted that physiological arousal and effort co-varied and hypothesized that changes in information processing were due to the activity of an arousal system. Kahneman did not specify the arousal system that was involved, but the state model argues that it is the pattern of electrocortical arousal that is important for information processing.

There is abundant evidence that electrocortical arousal is controlled by pathways from the reticular formation with cholinergic synapses at the cortex (Warburton 1981). This relationship is supported by the electrocortical effects of cholinergic drugs. In animals nicotine increases the release of acetylcholine at the cortex and increases cortical desynchronization (Armitage *et al.* 1969). In people, nicotine injections also produce an increase in electrocortical arousal (Kenig and Murphree 1973). Nicotine produces this effect by acting at the midbrain reticular formation (Il'yuchenok and

Ostrovskaya 1962; Kawamura and Domino 1969), and this effect can be blocked by both muscarinic and nicotinic antagonist drugs, but not by adrenergic blockade. On the other hand, scopolamine blocks acetylcholine receptor sites and decreases electrocortical arousal (Ostfeld and Aruguette 1962).

In our laboratory we have carried out an extensive series of studies of the effects of nicotine and scopolamine on human attentional processing and on electrophysiological measures. In this section the work on focussed attention and divided attention is reviewed.

Cholinolytics and focused attention

Evidence for the adverse effects of cholinolytics on attention came from early studies of atropine. Soldiers, who were given atropine, reported difficulty in concentrating and a shortened attention span (Ketchum *et al.* 1973). Atropine produced more errors on the Stroop test (Callaway and Band 1958). This test measures distraction; subjects are asked to name the colour in which a colour word is printed (e.g. the word red printed in blue). Thus, conflict exists between the print colour and the semantic information. Subjects treated with atropine also had difficulty in filtering out irrelevant parts of the design in the Gottschalk embedded figures test.

Callaway and Band (1958) hypothesized that atropine produced 'broadened attention', i.e. an increase in the influence on performance of peripheral information which is removed from the central focus of attention by space, time, or by differences of meaning (Callaway and Dembo 1958). This broadening would result in disruption of performance on tasks in which attention had to be focused.

This hypothesis was tested by examining the effects of scopolamine, another cholinolytic, on performance on a visual vigilance task in which subjects had to detect signals that were presented at the rate of two per minute in a 60-minute session (Wesnes and Warburton 1983*a*). Vigilance performance was analysed using signal detection theory and it was predicted that if attention was impaired then the index of stimulus sensitivity would be decreased. It was found that stimulus sensitivity was lowered, but the response criterion was unchanged, i.e. no change in motivation or the control of response output. Methscopolamine, which does not pass the blood-brain barrier, had no effect on either stimulus sensitivity or response bias, suggesting that only cholinergic mechanisms in the brain were involved.

A second scopolamine study examined the hypothesis using a different test of sustained attention performance (Wesnes and Warburton 1983*b*, 1984). The rapid visual information processing involved the detection of sequences of three consecutive odd or even digits from a series of digits presented visually at the rate of 100 per minute for 20 min. This test of sustained attention differs from the last one in terms of the rate of information input and the length of the session. On the basis of the effects of scopolamine in a vigilance task we expected that the drug would lower the efficiency of task performance. This prediction was confirmed; following scopolamine, correct detections were significantly lower over the 20-minute period, whereas no decrement was observed with methscopolamine.

These results were not supported by the set of experiments performed by Caine and his colleagues (Caine *et al*. 1981). They found no impairment on an auditory gap detection task after scopolamine, but did find an impairment on a short-term recall task. They also found a deficit in a more difficult task involving the discrimination of tones, a tonal order task. The reason for this discrepancy was tackled by Dunne (1985). He argued that the effect of the drug may be on the allocation of cognitive resources; the gap detection task has a very small cognitive component, while the tasks of Wesnes and Warburton (1978, 1983*a, b, c*) and tonal order task have a substantial cognitive component.

In a study of Dunne and Hartley (1985) on dichotic listening, scopolamine impaired the recall of attended words, but facilitated the recall of unattended words, while recognition was not modified. These authors suggest that the effect was due to an impairment of resources allocation to the more complex cognitive task, while the simpler one was unaffected. This possibility was tested using a divided attention task.

Cholinolytics and divided attention

Capacity theories (Kahneman 1973; Norman and Bobrow 1975) assume that the total amount of effort which can be developed at any time is limited. A person is able to allocate processing resources in many different ways, concentrating sometimes on an aspect of the sensory input, sometimes on processing of internally generated information, sometimes on the integration of information from different sources. Resource allocation can result from either the nature of the input, a 'bottom-up', data driven sequence of processing or from internal hypotheses, a 'top-down', conceptually driven sequence of processing (Rabbitt 1979). Evidence for variations in allocation of the resources of the cholinergic information processing system have come from a study of divided attention.

In this study, Dunne and Hartley (1986) investigated the detection of targets in different spatial locations. There was a probability bias with some locations having a greater probability of having a target than others. Scopolamine facilitated target detection in low probability spatial locations, while impairing detection in high probability locations.

The results of this experiment directly support the prediction that scopolamine would facilitate target detection in low probability spatial locations, while impairing detection in high probability locations. If the probability of a correct detection at a given spatial location depends upon the allocation of attentional processing resources to that location, then the present data suggests that scopolamine impairs the optimal allocation.

Parasuraman (1984) argued that sensitivity decrements in vigilance tasks occur because subjects have an impaired use of short-term memory processing resources in the control of attention. One interpretation of the present results is that scopolamine exerts its effects upon subjects' ability to use knowledge of the probability bias on the display in allocating attentional resources.

Nicotine and focused attention

The vigilance test which was used to test scopolamine was used in two nicotine studies (Wesnes *et al.* 1983). We found that nicotine prevented the decrement in stimulus sensitivity which occurred over time in the placebo condition, while having no effect on response criterion. In a study of smoking and vigilance (Wesnes and Warburton 1978), we examined the effects of the nicotine, delivered in the cigarette smoke, on the performance of a prolonged visual vigilance task. Both non-smokers and deprived smokers showed a marked vigilance decrement, whereas the smoking group maintained their initial level of attention.

In a second series of experiments (Wesnes and Warburton 1983*a*) the effects of nicotine on performance of the rapid visual information processing task was studied. Nicotine prevented both the decline in detections and the increase in reaction time which occurred over time in the placebo condition. It also improved both the speed and accuracy of performance above baseline levels, i.e. nicotine not only restored information processing performance to baseline, but improved attentional performance above baseline. In a companion study (Wesnes and Warburton 1983*b*), smoking also improved both the speed and accuracy of performance above rested baseline levels, the greatest improvement occurring with the highest nicotine delivery cigarette. Performance deteriorated over time after not smoking, as well as after smoking a nicotine-free cigarette. Thus, nicotine from cigarettes also produced absolute improvements in performance.

In a study designed to parallel that done by Callaway and Band (1958) with atropine using the Stroop Test, we tested smokers and non-smokers with nicotine tablets. The performance of both groups was improved by nicotine tablets so that they were less distracted by the semantic information and could focus their attention, and name the print colour more rapidly (Wesnes and Warburton 1978).

In order to obtain evidence for the possibility that nicotine and scopolamine were acting on a common system, we tested for pharmacological antagonism using the rapid visual information processing task and the Stroop Test. Nicotine completely counteracted the decrement in performance produced by scopolamine on both the rapid information processing task and the Stroop Test (Wesnes and Revell 1984).

Nicotine and divided attention

In one of our studies (D.M. Warburton, K. Wesnes, and M. Ansboro, unpublished) we used a divided attention task that was based on the rapid visual information task previously described. In one set of conditions, the subjects were presented with digits at a rate of 50 per minute in both the visual and auditory modality, with a different sequence for each modality. In a second set of conditions subjects were presented with numbers at the same rate in the visual modality only. It was emphasized to the subjects that they were to divide their attention equally between the two inputs and not to attend to one input for a while, then switch to the other.

Detection of sequences in the divided attention task improved significantly after smoking a cigarette. The number detected was greater in the test period compared to the baseline period. There was no improvement when subjects had not smoked a cigarette. The single modality visual task proved so easy that scores were too high to yield significant improvements. In the non-smoking condition, reaction times increased significantly in both the divided attention and single task. In both smoking conditions, reaction times became slightly shorter, but in the non-smoking conditions they were longer. It was only for the divided attention non-smoking condition that the increase from baseline to test period was significant. This provided additional evidence that nicotine was enhancing information processing capacity increased with heightened electrocortical arousal, and that subjects were able to allocate this to the two inputs.

Cholinergic drugs, electrocortical arousal, and attentional performance

In a study correlating performance with electrocortical activity, Warburton and Wesnes (1979) found that both cigarettes and nicotine tablets increased the dominant alpha frequency (11.5–13.5 Hz) and beta activity (13.5–20 Hz), and these changes were correlated with more efficient performance in the rapid visual information processing task described earlier.

In a further study, the P300 component of the event-related brain potentials produced by the target stimuli in the rapid information processing task were recorded (Edwards *et al*. 1985). This particular electrical event is related to attention and is not elicited by an input which is irrelevant to the subject i.e. an unattended stimulus (Donchin 1984). The occurrence of the P300 wave depends on the completion of stimulus evaluation processes (Kutas *et al*.

1977). McCarthy and Donchin (1981) have argued that the latency of the P300 component reflects the duration of stimulus evaluation processes and is relatively insensitive to response selection processes (p. 79). Stimulus evaluation involves the encoding, evaluating, and categorization of the input — including memory search and identification of the stimulus.

As in previous studies, we found that smoking increased both the number of correctly detected targets and the speed of detection in the rapid visual information task. In addition, we found that smoking significantly decreased the latency of the P300 to targets which were correctly detected. This finding indicates that improvements in information processing performance resulting from cholinergic stimulation are paralleled by changes in electrocortical activity which are indicative of improved stimulus evaluation.

In a study of the effect of scopolamine on information processing and the latency of the P300 component of the event-related potentials (Callaway *et al.* 1985), scopolamine was tested on a battery of tasks which combined two levels of stimulus complexity and two levels of response difficulty. Reaction times and P300 were measured. Scopolamine slowed reaction times and the P300 latency to simple stimuli more than to complex stimuli. This suggests an involvement of a cholinergic system in early stimulus preprocessing. This cannot be the only mechanism since tasks involving complex stimulus evaluation, like the Stroop Test, are also impaired by scopolamine (Callaway and Band 1958) and facilitated by nicotine (Wesnes and Warburton 1978).

Summary

In summary, these experiments provide evidence that cholinergic pathways are involved in human information processing. The cholinergic stimulation produced by nicotine increased processing efficiency while the cholinergic blockade produced by atropine and scopolamine decreased efficiency. These data are consistent with the hypothesis that cholinergic pathways in the brain determine the efficiency of attentional processing. Cortical desynchronization is controlled by a cholinergic pathway from the nucleus basalis of Meynert (Robinson 1985). This pathway is degenerated in Alzheimer's disease and other dementias (Rossor *et al.* 1984), which are characterized by a disorder of information processing. This cholinergic pathway seems to be important for the selective activation of the specific areas of the sensory cortices.

The acetylcholine system is thought to be functionally linked with the vasoactive intestinal peptide (VIP) neurones in the cerebral cortex (Emson, *et al.* 1980). VIP does not coexist with acetylcholine, but it may be co-released with acetylcholine, since cholinergic neurones terminate on the VIP neurones and cortical VIP neurones have terminals ending close to cerebral blood vessels. Regional cerebral blood-flow studies have shown that there is increased blood flow in the sensory-specific cortical regions corresponding to the modality of the information input to which attention is being directed. (Roland 1981, 1982). The VIP neurones constitute the anatomical substrate for the increased cerebral blood flow found in the cortex during information processing in co-ordination with the cholinergic enhancement of the activity of cortical sensory neurones.

MEMORY

Psychopharmacological and neuropsychological studies have strongly suggested that there are two forms of memory storage: immediate or short-term memory, and longer-term memory. Immediate memory is labile and is susceptible to change by a variety of drugs and procedures while longer-term memory is much more resistant to disruption. Longer-term information storage has been divided into procedural information (skill) and declarative (verbalizable) information (see Squire 1986 for a review). Declarative information can be subdivided into episodic (context specific) information and semantic (associated and integrated) information. We have been investigating the possibility that a cholinergic system initiates the changes which are necessary for procedural and declarative memory, but is not sufficient for the associative processes that are required for semantic memory. In contrast, a fully functioning catecholamine system (in our view a noradrenergic system) is necessary for the elaborations that characterize semantic memory.

Cholinolytics and memory processes

In the first part of this section, the effects of preaquisition doses of cholinolytics are described, which suggest that the major effects of these drugs are on the storage of information. Some of the earliest studies, e.g. Ostfeld *et al.* (1960), Ostfeld and Aruguette (1962), found that atropine and scopolamine impaired memory in terms of reading memory and card object recall, but not digit recall, a measure of short-term memory. These authors also observed impairment of the subjects' ability to 'maintain an attentive set'. However, it was unclear on which memory process the drug was acting. There were attentional deficits in these subjects suggesting impairment of information input, but equally there could have been storage or retrieval deficits. Later, Safer and Allen (1971) also noted that scopolamine only slightly altered digit 'registration', a measure of immediate memory span, but that delayed recall of digits was grossly impaired after a 20-second interval, a deficit suggestive of an impairment of storage.

Similar evidence for a storage impairment came from a study of Drachman and Leavitt (1974) who gave injections of scopolamine to young adult volunteers. Tested 1 hour later, they showed clear deficiencies on a variety of memory tasks. While immediate memory was unaffected, information storage was significantly impaired, as measured by ordered recall of digits and by free recall of word lists. Retrieval by category, which is supposed to measure both the retrieval mechanisms and encoding effects showed only slight impairment. Part of this deficit was due to the distractability of the subjects, who would often retrieve items from other categories while performing accurately.

In a study which aimed to dissect out the mode of action of scopolamine on memory, Crow (1979) tested subjects in a word list task in which it was possible to look at short-term memory and long-term memory. There was no significant reduction in the number of recently presented words which were recalled, but there was an effect on the words presented earlier. Crow proposed that scopolamine has an effect on the transition from immediate to long-term memory.

The same problem was addressed by Mewaldt and Ghoneim (1979). Scopolamine did not impair retrieval processes, but impaired the recall of information. As the recall tasks were thought to exceed the capacity of short-term storage, and retrieval was not affected by the drug, then the memory deficit results from an interference with information storage. They found that scopolamine impairs the transfer of information from the short-term to the long-term store, but does not interfere with the retrieval of information.

In a study of Drachman and Sahakian (1979), groups of normal volunteers were given scopolamine or methscopolamine. Drug effects on immediate memory, memory storage in terms of free recall, and ordered recall and retrieval from long-term storage were examined. After scopolamine, immediate memory span was unimpaired, but the ability of subjects to store new information was severely disrupted, both for serial order of digits and for free recall of words. In addition, there was some evidence in this study that retrieval from long-term memory was significantly impaired after scopolamine. Those subjects who received methscopolamine were not impaired.

Drug effects on storage and retrieval effects can be distinguished by using state dependent design in the experiments. This design was used by Petersen (1979) to study scopolamine's action on memory. The drug interfered with the storage of the information, but there was also a retrieval effect due to state dependency, the fact that the drug state is itself a cue for recall. Interestingly, these context-specific effects seemed to depend on imagery because more concrete words than abstract words were retrieved when the state was switched. This raises the intriguing possibility that scopolamine may have a differential effect on encoding processes.

A complex study has investigated the effect of scopolamine on encoding processes in terms of phonemic, semantic, and imaginal mnemonic representation (Frith *et al.* 1984). These three types of representation are believed to be implicated in the storage of information (Craik and Lockhart 1972).

The retention of concrete and abstract material, and the influence of phonemic and semantic similarity upon immediate serial recall were compared. Scopolamine impaired recall, irrespective of whether an item-recall scoring criterion or an ordered-recall scoring criterion was used, which suggests the amnesic effect was in the encoding of information.

However, the amnesic effects of scopolamine could not be definitively attributed to impairment of specific categories of encoding processes. The drug produced similar deficits on concrete and abstract words, i.e. no change in imaginal mnemonic encoding. The drug did not reduce the magnitude of the phonemic similarity effect. The phonemic similarity effect is thought to be due to the use of phonemic encoding as a means of representing serial-order information and so it must be concluded that scopolamine does not appear to disrupt this form of phonemic encoding either.

Scopolamine produced impaired performance in terms of recall on lists of unrelated words, but this impairment was reduced when the words to be remembered were phonemically or semantically related to one another. This suggests that the drug disrupted the encoding processes normally employed to represent item information in human memory, but that the subjects could compensate for this deficit by attending to the phonemic or semantic properties of the material.

These data argue for little, if any, effect on immediate memory. There is clear evidence for an information storage impairment, but the retrieval deficit could be accounted for by impaired attention. There did not seem to be a specific effect of scopolamine on different types of mnemonic encoding.

Cholinergic agonists and memory processes

A series of studies by Drachman tested the hypothesis that cholinergic systems are involved in storage, but the findings were disappointing. For example, Drachman and Leavitt (1974) obtained no significant effect with physostigmine (an acetylcholinesterase inhibitor which enhances cholinergic function), although those receiving the lower dose performed marginally better. The study of Drachman and Sahakian (1979) obtained the same result; young volunteers receiving a low dose of physostigmine performed slightly better, but those receiving a higher dose of physostigmine slightly worse than normal controls. The differences were again not statistically significant. In another study, Drachman and Sahakian (1980) gave a group of normal, aged volunteers a small dose of physostigmine. Their memory and cognitive functions were compared with those in a set of untreated aged subjects. A trend toward improvement was seen in the experimental group for each of the cognitive measures tested, particularly in memory storage, but the variance in this population was too large for statistical significance to be reached.

The interaction of physostigmine with scopolamine was studied by Mewaldt and Ghoneim (1979). Scopolamine impaired the transfer of information from the short- to the long-term store, but did not interfere with the retrieval of information. Physostigmine antagonized most features of the memory impairment produced by scopolamine. Thus, the blockade of the amnesic action of scopolamine by physostigmine supports a role of cholinergic mechanisms in the transfer of new information into long-term storage.

Studies of nicotine on learning and memory have provided further support for this hypothesis. The effect of nicotine-free cigarettes and nicotine cigarettes on learning a nonsense syllable list was studied by Andersson and Post (1974). The learning curves were identical for the two conditions prior to smoking, but after nicotine the number correct decreased and remained below the scores in the nicotine-free condition although the learning curves were parallel. After the second nicotine cigarette, the number of correct syllables increased significantly to the same level of acquisition performance as in the nicotine-free cigarette condition. Thus, relative to the previous performance, nicotine had improved learning the syllables. The data give no evidence of nicotine interfering with acquisition because the learning curves are parallel after the first nicotine cigarette and it could be argued that there was some facilitation of information storage after the second cigarette. After the first nicotine cigarette, the information stored in the non-nicotine state was less available in the drug state, i.e. state-dependent learning.

The state dependency hypothesis with nicotine has been investigated in several studies and in three of these support has been found for it (Peters and McGee 1982; Warburton *et al.* 1986). In our study (Warburton *et al.* 1986), smokers were given a nicotine cigarette or nothing immediately before serial presentation of a set of Chinese ideograms (items which could not be phonemically or semantically coded). Subjects who received nicotine prior to learning had significantly better recognition scores than the subjects who did not smoke in the first part of the experiment. This suggests that nicotine can facilitate the storage of material which has no phonemic and semantic properties. A significant interaction term indicated that changing the drug state interfered with recognition.

In a second study (Warburton *et al.* 1986), the effect of a dose of nicotine on both short- and long-term recall was examined, using a state dependent design. After a nicotine tablet or a placebo, subjects listened to a list of words. After 1 minute, a free recall test was given and a second one after 60 minutes. Before the second test they were given either nicotine or a placebo. The short-term recall data revealed a significant superiority of the nicotine group. Long-term recall was also significantly better when subjects had taken nicotine prior to learning, but not when taken prior to recall. A retrieval by category analysis was made of the recalled words to see if the drug had influenced encoding processes, but no effect was seen. The very significant interaction term again gave evidence for a state dependent effect of nicotine. Nicotine was facilitating the input of information to storage and improved storage, but had no effect on retrieval.

The fact that nicotine improved both recognition and recall is consistent with a cholinergic system being involved in information storage. Recall is a two stage process involving search and decision, while recognition involves only a decision. A change in recall only, may reflect a retrieval effect, but change in both recognition and recall suggests an effect on storage, not retrieval.

Storage effects are best examined by giving drugs after information input. Relevant to the studies that we have discussed, Squire (1969) injected rats with physostigime and found that when a treatment preceded or immediately followed Trial One on a spatial alternation task, physostigmine increased the probability of alternation. Scopolamine decreased this probability. The results suggest that a cholinergic mechanism is involved in memory storage, but is only acting for a short time after information input, i.e. in the initiation of storage, but not in the longer-term consolidation processes.

Animal studies have given some ideas about the nature of cholinergic involvement in information storage. Marczynski observed that reward of a hungry cat produced cortical synchronization and positive steady potential shift in the same region. The occurrence of these events results in an enhancement of any evoked response potentials, regardless of whether they are from 'relevant' stimuli or not (Marczynski 1969). Marczynski (1971) found that scopolamine and atropine blocked the post-reinforcement synchronization. Thus, there is a cholinergic mechanism that is activated immediately after significant events and enhances the size of sensory evoked potentials at the cortex. This would meet the specifications of a mechanism that initiates information storage (Warburton 1983).

Summary

In summary, after the input of information, activation of a cholinergic system is important for the effective storage of information and the transfer of information from immediate memory to longer-term memory. It is separate from the attentional system, but may not be independent from it.

CATECHOLAMINE SYSTEMS AND MEMORY

Most tests of the involvement of catecholamine systems in human memory have been made using the amphetamines, which increase functional catecholamine levels. Before discussing agonist studies, studies of two catecholamine antagonists are reviewed.

Catecholamine antagonists

One piece of evidence supporting the involvement of noradrenaline in human memory comes from a study of hypertensive patients, who were given the adrenergic blockers, methyldopa and propranolol (Solomon *et al.* 1983). There was highly impaired verbal memory in both hypertensive and nonhypertensive subjects taking methyldopa or propranolol as compared with controls. This study suggests the involvement of a catecholamine system in information storage.

An important study (Frith *et al.* 1985) was done with clonidine, a central noradrenergic noradrenaline release. It acts on presynaptic alpha-2-receptor sites, leading to a decrease in noradrenaline release. Clonidine did not impair short-term memory or the free recall of word lists. It did impair retrieval from remote semantic memory. There was a marked impairment of paired-associate learning, especially when novel associations had to be learned, which seemed to be partly due to interference from prior associations. Frith *et al.* (1985) conclude the results suggest that 'noradrenaline has a special role in the replacement of old by new associations' (p. 493).

Catecholamine agonists

Although amphetamines facilitate memory in animals (Hunter *et al.* 1977) the results of human experiments have been mixed. Short-term memory was tested by Talland and Quarton (1965) using a running digit span test. No differences from the no drug condition were found with methamphetamine. Similarly, Crow and Bursill (1970) found no effect of methamphetamine on short-term memory in two experiments using a related task. Drachman (1977) also studied the effects of amphetamine on free recall, but obtained no significant improvement. In contrast, Mewaldt and Ghoneim (1979) found that methamphetamine produced a small improvement of delayed recall of words, but little if any effect on immediate recall and no significant influence on retrieval of information. This pattern of findings suggests that the drug is enhancing processes that are occurring after the immediate memory stage.

Evidence of the nature of this change comes from a study of normal and hyperactive children who were compared on cognitive tasks following placebo or amphetamine (Weingartner *et al.* 1980). Both normal and hyperactive children demonstrated similar, amphetamine-related increases in the recall of semantically, and acoustically processed words. This enhancement was independent of any attentional changes. This pattern of amphetamine-induced changes in memory suggests an enhancement of associative processes, such as semantic processing and organization of the information.

In one of our own studies we used RU 24722 which releases noradrenaline at the cortex, acetylcholine in the hippocampus and, in addition, acts on protein synthesis as shown by increased activity of ornithine decarboxylase. We found that RU 24722 in young, normal volunteers produced highly significant improvements in delayed verbal recall. There was a trend for an improvement in recognition memory. In contrast, there was no evidence for any changes in attentional performance.

A *post hoc* analysis was made of the recalled words to find evidence of differences in mnemonic coding. Free recall is influenced by concreteness, imagery, and associative meaningfulness of words (Paivio *et al.* 1968) and our words could be scaled for these properties. Unfortunately, the list was not designed for this purpose and so the scale range was narrow with the scores on the three measures highly correlated. Nevertheless, the analysis revealed that the recall of words, which were lower on associative meaningfulness, was increased by RU 24722, suggesting that post-storage encoding was being improved. For the higher association items, it is possible that a 'ceiling' effect prevented further improvement.

Studies of post-acquisition doses of catecholaminergic drugs show these drugs act on processes that continue for some time after the input of information. Injections of amphetamine were given at various times before and after three trials of an appetitive visual discrimination (Krivanek and McGaugh 1969). The performance of mice, on the following day, was facilitated by the doses given after training. These data suggest that a catecholamine, either noradrenaline or dopamine, is involved in the process of information storage.

Opposite results were found using a noradrenaline synthesis blocker, diethyldithiocarbamate (McGaugh *et al.* 1975). Retention was impaired by the drug dose preceding training and by the doses after training. The involvement of dopamine can be ruled out, since dopamine is significantly increased by diethyldithiocarbamate. Persuasive support for this idea has come from a study by Stein *et al.* (1975) in which they injected diethyldithiocarbamate prior to training and the drug impaired retention. Intraventricular injections of noradrenaline but not dopamine protected against impairment.

The precise neural pathways are not known, but Crow (1968) suggested that it is the projection of the dorsal noradrenaline fibres to the neocortex which initiates the changes involved in learning. This suggestion was similar to one made by Kety (1970) who postulated that the cortical release of noradrenaline might provide mnemonic activation of cortical cells and enable information storage.

Summary

So far, memory has been considered as either short- or long-term memory storage. However, these multistore models are now considered as too simple. Craik and Lockart (1972) think of memory as a continuum of analysis from the results of sensory analyses in short-term memory to the long-term results of semantic-associative processes. The strength of the memory trace arises from perceptual processing and its persistence is a function of depth of analysis, with deeper levels of analysis associated with more elaborate, longer lasting, and stronger traces (Craik and Lockart 1972). Unless further processing occurs, the information will not leave a long lasting trace and the greater the degree of associative analysis, the better the memory. Efficient information retrieval requires that relationships be found between the newly arrived information and what a person already knows. Retrieval, therefore, will be improved by elaborate encoding (Norman 1976). It is the concluding proposal that a noradrenergic system mediates these associative elaborations.

PROBLEM SOLVING AND THINKING

Traditionally, cognitive scientists have worked on tasks that have required mental effort and the performance outcome has been the result of the balance of the states of electrocortical activity that have been maintained by the person. However, we should remember that not all cognitive activity is of this intensive sort and that it is just as important for the person to fantasize, think, plan, problem solve, and think creatively.

Problem solving occurs when a person's activity has a goal, but no learned route to the goal. Problems can either be well defined or ill defined. The former are solved by *convergent thinking* which leads to logically correct answers and the latter by *divergent thinking* which leads to innovative solutions. Problem solving calls for the manipulation of information as well as for the assembling of enough information to justify a conclusion.

'Thinking' is difficult to define, but it occurs when the person goes beyond the immediately given situation, and uses memories and previously formed concepts. It enables people to organize and reorganize information. Human thought capabilities are limited precisely because attention and memory are both intimately related to thinking. Limitation by attention occurs when concentration on one task generally causes a deterioration in the performance of other tasks. Thus, when thinking we can sometimes fail to notice the events around us and when there is a lot of environmental stimulation, we are distracted from our thoughts. In addition, short-term memory plays a critical role in thought by guiding the thought processes, but is limited by its capacity.

The processing of information depends on the brain state as defined by electrocortical activity, as we have said earlier. An investigation of the experiences of subjects when showing these EEG patterns was carried out (Brown 1974). In the experiments, people watched their cortical activity and described their thoughts and feelings. The electrocortical activity was recorded at the scalp and then filtered for the beta, alpha, and theta frequencies. There was a correlation between the introspective reports of alertness, attention, tension and anxiety and beta activity. Alpha activity was associated with relaxation and lack of concentration. As the activity slowed even more to the theta range the subjects reported 'day-dreaming', 'planning', 'problem solving', and a sense of unreality.

The hypothesis that creative thinking is done best at lower levels of cortical arousal has been examined by Martindale and Armstrong (1975). Subjects were tested on several tasks that called for imagination and their electrocortical arousal was simultaneously measured from the right hemisphere. Subjects who scored highly on these tasks had the lowest levels of arousal and the most alpha activity while the less uncreative subjects produced increased arousal. Creativity seems to be incompatible with concentration and perhaps that is why our best ideas come to us when we are not thinking about a problem but are relaxing. After cholinolytic drugs which decrease the electrocortical activity, subjects reported difficulty in concentrating and the subject appeared to be in a 'world of his own' or 'day-dreaming' (Ketchum *et al.* 1973).

In the next sections, the effects of cholinergic drugs on thinking will be discussed.

Cholinergic drugs and problem solving

In early studies of cholinolytics, impaired problem solving was noted and tests of arithmetic ability showed a decrement in performance with scopolamine (Ketchum *et al.* 1973) and atropine (Ketchum *et al.* 1973). These findings could be interpreted in terms of the Callaway and Band (1958) suggestion that cholinolytics produce 'broadened attention', an increase in the influence of peripheral information on performance. Callaway and Band had formulated this view on the basis of a study in which cholinolytics facilitated performance on the Luchins Jar Test (Luchins 1942), which consists of obtaining given volumes of water with three different measures. The same algorithm generated the correct solution throughout the test but halfway through a simpler algorithm could be used.

Groups ranging from primary school children to graduate students learned to use the three jar procedure and usually persisted in it throughout the list of problems, overlooking the simpler possibilities. However, atropine improved performance when the subjects attempted a difficult solution to the problem; in other words, subjects receiving atropine discovered the new, short method of solving the problems faster than the control group (Callaway and Band 1958). They claimed that with broadened attention subjects attended to aspects of the task which were not essential for the original test, but which helped in discovering the simpler method.

In our study of the same phenomenon using scopolamine and nicotine, we confirmed and extended these findings. We found that fewer of the subjects given scopolamine used the correct solution repeatedly in the first half of the task. This could be interpreted as an impairment of convergent thinking. In the second half of the session they used a wider variety of solutions, more divergent thinking, but there was no evidence that they switched solutions more rapidly than the control group, i.e. the relative decrease in time to solution was the same as the control subjects. Subjects who were given nicotine performed better on the first half of the test where subjects could use the same solution repeatedly (convergent thinking) and persisted longer with the old algorithm. However, the relative decrease was the same as the control subjects. Of course, it could be argued that use of the same algorithm is more efficient information processing, which would integrate these findings with those previously discussed in the section on Attention.

In a second test of problem solving we used an anagram test based on one used by Mendelsohn and Griswold (1964). They had divided subjects up into high and low creativity groups on the basis of scores on Mednick's (1962) Remote Associates Test (see later) and made both groups memorize lists of words while hearing a second word list recited continuously on a tape recorder. Subjects then had to solve a series of anagrams, the solutions to some of which had appeared on the memorized (focal) list or the tape recorded (peripheral) list. It was found that high score subjects used focal and peripheral cues to a greater extent than did other subjects. Mendelsohn and Griswold suggested that highly 'creative' subjects have a 'broadened attention', which gave greater receptivity to seemingly irrelevant stimuli and ideas, this being responsible for the greater originality of their thoughts.

If broadened attention is produced by scopolamine then we would expect analogous results to those obtained from highly creative subjects. Undergraduates were given either placebo or 1.2 mg of scopolamine. We used two sets of 24 anagrams and their words, and four sets of 12 five-letter 'filler' words. Subjects were given a *focal list* of words to memorize, consisting of eight words from the anagram list and 12 filler words. At the same time subjects were told to ignore words that they heard which were eight other anagram words and 12 more filler words: the peripheral list. Then they had to solve a list of 24 anagrams, of which eight came from the focal list, eight from the peripheral list, and eight were novel. There was also a free recall test of words from the focal list.

In the scopolamine condition, subjects memorized fewer words, solved fewer of the novel anagrams, and solved fewer of the focal anagrams. There was no difference in solution or peripheral anagrams. Within drug conditions, subjects solved more focal than novel or peripheral anagrams in the placebo condition, while they solved more peripheral than novel or focal anagrams after scopolamine. This suggests that there are differences in the information processing of the scopolamine subjects. There is clearly a memory impairment which could be due to either interference with information storage or to information input due to an attentional deficit. Some evidence for the latter is the solution of the peripheral anagrams.

Creative thinking

Ill-defined problems are more common in everyday life and need creative thinking to produce inventive solutions to problems or for the creation of novel things, like a work of art. It was hypothesized that one area where a broadened attention might be useful would be in tasks requiring 'divergent' intellectual abilities as described by Guilford (1967), where the task requirement is not the selection of the one correct answer to a problem but producing an original, unobvious answer to a more open-ended, adaptable task.

Mednick (1962) proposed that creative thinking involves the formation of associations between stimuli and responses, which are not normally associated. On the basis of this definition of creativity, Mednick designed the Remote Associates Test in which subjects are given three words and have to provide a fourth word which has some association with the other three, e.g. electric, high, wheel, ...(chair). If scopolamine induces broadening of attention, then it should improve performance. However, we found that there were significantly fewer correct solutions, where correct solutions were defined as words that could be justified by the subject rather than experimenter-defined solutions. Thus, there was no evidence from this measure of creativity that scopolamine was facilitating thinking. However, successful performance on this task demands use of focused attention and memory, and is more analogous to anagram solution which scopolamine impaired.

One standard test of creative thinking is the Torrance test (Torrance 1974) which is said to assess originality (the unusualness of responses), fluency (the total number of relevant responses), and flexibility (the number of different response categories). If scopolamine produced divergence of thinking we might expect changes in originality and flexibility. The test consists of seven activities.

Activity 1: subjects wrote any questions about activities and objects in the picture.
Activity 2: subjects listed the causes for the situation depicted in the picture.
Activity 3: subjects listed the possible consequences of the situation in the picture.
Activity 4: subjects were presented with a picture of a toy and asked to write down any suggestions for improvement.
Activity 5: subjects listed all the unusual uses for a cardboard box (in test booklet A) or a tin can (in test booklet B).

Activity 6: subjects were requested to list any unusual questions they had to ask about cardboard boxes or tin cans (whichever was appropriate from Activity 5).

Activity 7: in this section the subjects were given a bizarre situation and asked to list their ideas and guesses as to how life on earth might change.

Scopolamine produced significantly greater scores for originality and fluency.

Summary

In summary, the shift in the electrocortical arousal to slower activity results in changes in thinking. The nature of the change depends on the type of thinking. Problem solving in which the person manipulates information to arrive at a single correct answer demands focused attention and use of memory. Slowing the electrocortical activity impairs this form of thinking. In contrast to logical thinking, creative thinking can result in many answers, none of which are right or wrong. It does not require focused attention or use of memory. For this type of thinking, slower electrocortical arousal is beneficial.

HALLUCINATIONS

In the laboratory it has been found that high doses of cholinolytics induce hallucinations (Crowell and Ketchum 1967; Ketchum *et al*. 1973) which were usually visual; but auditory, tactile, olfactory, and gustatory hallucinations have been reported. The visual hallucinations were integrated and extremely realistic with familiar objects and faces (Ketchum *et al*. 1973). Some subjects smoked imaginary cigarettes (Crowell and Ketchum 1967) and drank from non-existent glasses, making appropriate drinking movements, and commenting on taste and smell (Ketchum *et al*. 1973). Subjects hear recognizable voices, and music played by single instruments and whole orchestras. Even informed subjects were convinced by the intensity of the hallucinations.

In order to investigate the phenomenon of cholinolytic hallucinations in the laboratory, we designed a sensory conditioning study in which a light and a faint 20-second tone were paired, but the onset and offset of the tone were gradual so subjects believed that it was a test of auditory acuity. They were asked to press a key when they thought that they could hear a tone and press again when they could hear it no longer. After performance had stabilized, a set of test trials were given in which the light was switched, but no tone was presented until either the subject pressed the key or 30 seconds had elapsed. A 'hallucination' was a response before the stimulus was presented and it was found that the percentage of subjects having hallucinations increased from

30 per cent with placebo to 85 per cent with scopolamine, and the mean number of hallucinations for the group increased significantly. The latency for making a response was similar, which gives no indication of a lower criterion for reporting detections.

One way of interpreting hallucinations is as internal information, irrelevant, bizarre thoughts which are normally filtered out by cholinergic mechanisms and never have the opportunity to reach consciousness. When these mechanisms are attenuated chemically, this results in cortical synchronization, thus bizarre, irrelevant thoughts which would otherwise be filtered out, reach the level of consciousness. This manifests itself as a hallucination.

CONCLUDING COMMENTS

Information processing depends on the brain state. There seem to be at least three neurochemical systems which are involved. A cholinergic system controls the electrocortical activity, and the efficiency of information throughput including problem solving depends on the state whose electrophysiological correlate is cortical desynchronization. More synchronized activity is associated with a state in which creative thinking occurs. Separate from this system is a cholinergic system which enables initial transition from immediate memory to long-term memory. A noradrenergic system seems to be involved in the associative processes which occur after this transition and results in the depth of processing which is necessary for effective retrieval.

REFERENCES

Andersson, K. and Post, B. (1974) Effects of cigarette smoking on verbal rote learning and physiological arousal. *Scandinavian Journal of Psychology* 15, 263-7.

Armitage, A.K., Hall, G.M., and Sellers, C.M. (1969). Effects of nicotine on electrocortical activity and acetylcholine release from the cat cerebral cortex. *British Journal of Pharmacology* 35, 152-60.

Brown, B.A. (1974). *New Mind — New Body*. Aldine, Chicago.

Caine, E.D., Weingartner, H., Ludlow, C.L. Cudahy, E.A., and Wehry, S. (1981). Qualitative analysis of scopolamine-induced amnesia. *Psychopharmacology* 74, 74-80.

Callaway, E. and Band, I. (1958). Some psychopharmacological effects of atropine. *Archives of Neurology and Psychiatry* 79, 91-102.

—— and Dembo, D. (1958). Narrowed attention: a psychological phenomenon that accompanies a certain physiological change. *Journal of Neurology and Psychiatry,* 79, 74-90.

——, Halliday, R., Naylor, H., and Schechter, G. (1985). Effects of oral scopolamine on human stimulus evaluation. *Psychopharmacology* 85, 133-8.

Craik, F.I.M. and Lockart, R.S. (1972). Levels of processing: a framework for memory research. *Journal of Verbal Learning and Verbal Behaviour* 11, 671-6.

Crow, T.J. (1968). Cortical synapses and reinforcement. *Nature (London)* **219**, 245-6.

—— (1979). Action of Hyoscine on verbal learning in man: evidence for a cholinergic link in the transition from primary to secondary memory? In *Brain Mechanisms in Memory and Learning: from the Single Neuron to Man* (ed. MAB. Brazier), pp. 269-75. Raven Press, New York.

Crow, T. and Bursill, A.E. (1970). An investigation into the effects of methamphetamine on short-term memory in man. In *Amphetamines and Related Compounds* (ed. E. Costa and S. Garattini), pp. 889-96. Raven Press, New York.

Crowell, E.B. and Ketchum, J.S. (1967). The treatment of scopolamine-induced delirium with physostigmine. *Clinical Pharmacology and Therapeutics* **8**, 409-14.

Donchin, E. (ed.) (1984). *Cognitive Psychophysiology: Event-Related Potentials and the Study of Cognition*. Lawrence Erlbaum, Hillsdale, New Jersey.

Drachman, D.A. (1977). Memory and cognitive function in man: Does the cholinergic system have a specific role? *Neurology* **27**, 783-90.

—— and Leavitt, J. (1974). Human memory and the cholinergic system. A relationship to ageing. *Archives of Neurology* **30**, 113-21.

—— and Sahakian, B.J. (1979). Effects of cholinergic agents on human learning and memory. In *Nutrition and the Brain* Vol. 5. (ed. A. Barbeau, J.H. Growdon, and R.J. Wurtman), pp. 351-66. Raven Press, New York.

—— and —— (1980). Memory and cognitive function in the elderly. A preliminary trial of physostigmine. *Archives of Neurology* **37**, 674-5.

Dunne, M.P. (1985). An interpretation of some conflicting evidence regarding the effects of scopolamine upon vigilance. *Psychopharmacology* **87**, 126.

—— and Hartley, L.R. (1985). The effects of scopolamine upon verbal memory: evidence for an attentional hypothesis. *Acta Psychologica* **58**, 205-17.

—— and —— (1986). Scopolamine and the control of attention in humans. *Psychopharmacology* **89**, 94-7.

Edwards, J.A., Wesnes, K., Warburton, D.M., and Gale, A. (1985). Evidence of more rapid stimulus evaluation following cigarette smoking. *Addictive Behaviours* **10**, 113-26.

Emson, P.C., Hunt, S.P., Rehfeld, J.F., Golterman, N., and Fahrenkrug, J. (1980). Cholecystokinin and vasoactive intestinal polypeptide in the mammalian CNS: distribution and possible physiological roles. In *Neural Peptides and Neuronal Communication* (ed. E. Costa and M. Trabucchi), pp. 63-74. Raven Press, New York.

Frith, C.D., Dowdy, J., Ferrier, I.N., and Crow, T.J. (1985). Selective impairment of paired associate learning after administration of a centrally-acting adrenergic agonist (clonidine). *Psychopharmacology* **87**, 490-3.

Frith, J.T.E., Richardson, M.S., Crow, T.J., and McKenna, P.J. (1984). The effects of intravenous diazepam and hyoscine upon human memory. *Quarterly Journal of Experimental Psychology* **36A**, 133-44.

Ghoneim, M.H. and Mewaldt, S.P. (1975). Effects of diazepam and scopolamine on storage, retrieval and organisational processes in memory. *Psychopharmacologia (Berlin)* **44**, 257-62.

Groll, E. (1966). Central nervous system and peripheral activation variables during vigilance performance. *Zeitschrift fur experimentelle und angewandte Psychologie* **13**, 248-64.

Guilford, J.P. (1967). *The Nature of Human Intelligence*.

Hunter, B., Zornetzer, S.F., Jarvik, M.E., and McGaugh, J.L. (1977). Modulation of learning and memory. In *Handbook of Psychopharmacology* (ed. L.L. Iversen, S.D. Iversen, and S.H. Snyder), Vol. 8, 531–77. Plenum Press, New York.

Il'yuchenok, R. Yu. and Ostrovskaya R.U. (1962). The role of mesencephalic cholinergic systems in the mechanism of nicotine activation of the electroencephalogram. *Experimental Biology and Medicine* 54, 753–7.

Kahneman, D. (1973). *Attention and Effort*. Prentice-Hall, Englewood Cliffs, N.J.

Kawamura, M. and Domino, E.F. (1969). Differential actions of m and n cholinergic agonists on the brain stem activating system. *International Journal of Neuropharmacology* 8, 105–10.

Kenig, L. and Murphree, M.B. (1973). Effects of intravenous nicotine in smokers and non-smokers. *Federation Proceedings* 32, 805.

Ketchum, J.S., Sidell, F.R., Crowell, E.B., Aghajanian, G.K., and Haines, A.H. (1973). Atropine, scopolamine and ditran: comparative pharmacology and antagonists in man. *Psychopharmacologia* 28, 121–45.

Kety, S.S. (1970). The biogenic amines in the central nervous system: their possible roles in arousal, attention and learning. In: *Neurosciences: Second Study Program* (ed. F.O. Schmitt), pp. 324–36. Rockefeller University Press, New York.

Krivanek, J. and McGaugh, J.L. (1969). Facilitatory effects of pre- and post-trial 1-amphetamine administration of discrimination learning in mice. *Agents and Actions* 1, 36–42.

Kutas, J., McCarthy, G., and Donchin, E. (1977). Augmenting mental chronometry: The P300 as a measure of stimulus evaluation time. *Science* 197, 792–5.

Lindsley, D.B. (1952). Psychological phenomenon and the electroencephalogram. *Electroencephalography and Clinical Neurophysiology* 4, 443–8.

Luchins, R. (1942). Mechanization in problem solving. *Psychological Monographs* 54.

Marczynski, T.J. (1969). Postreinforcement synchronization and the cholinergic system. *Federation Proceedings* 28, 132–4.

—— (1971). Cholinergic mechanism determines the occurrence of reward contingent positive variation (RCPV) in cat. *Brain Research* 28, 71–83.

Martindale, J. and Armstrong, L.A. (1975). What makes creative people different. *Journal of Genetic Psychology* 124, 311–20.

McCarthy, G. and Donchin, E. (1981). A metric for thought: a comparison of P300 latency and reaction time. *Science* 211, 77–80.

McGaugh, J.L., Gold, P.E., Van Buskirk, R., and Haycock, J. (1975). Modulating influences of hormones and catecholamines on memory storage. In *Progress in Brain Research Vc.* 42 (ed. W.H. Gispen, Th. B. van Wimersma Griedanus, B. Bohus, and D. de Wied), pp. 51–62. Elsevier, Amsterdam.

Mednick, S.A. (1962) The associative basis of the creative process. *Psychological Review* 69, 220–32.

Mendelsohn, G.A. and Griswold, B.B. (1964). Differential use of incidental stimuli in problem solving as a function of creativity. *Journal of Abnormal and Social Psychology* 68, 431–6.

Mewaldt, S.P. and Ghoneim, M.M. (1979). The effects and interactions of scopolamine, physostigmine and methamphetamine on human memory. *Pharmacology, Biochemistry and Behavior* 10, 205–10.

Norman, D.A. (1976). *Memory and Attention*. Wiley, New York.

—— and Bobrow, D.G. (1975). On data-limited and resource-limited processes. *Cognitive Psychology* 7, 44–64.

O'Hanlon, J.F. and Beatty, J. (1977). Concurrence of electroencephalographic and performance changes during a simulated radar watch and some implications for the arousal theory of vigilance. In *Vigilance: Theory; Operational Performance and Physiological Correlates* (ed. R.R. Mackie), pp. 189-201. (1962). Plenum Press, London.

Ostfeld, A.M. and Aruguette, A. (1962). Central nervous system effects of hyoscine in man. *Journal of Pharmacology and Experimental Therapeutics* **137**, 133-9.

——, Machne, X., and Unna, K.R. (1960). The effects of atropine on the electro-encephalogram and behaviour in man. *Journal of Pharmacology and experimental Therapeutics* **128**, 265-72.

Oswald, I. (1962). *Sleeping and Waking*. Elsevier, Amsterdam.

Paivio, A., Yuille, J.C., and Madigan, S.A. (1968). Concreteness, imagery and meaningfulness values for 925 nouns. *Journal of Experimental Psychology Monograph Supplement*. **76**, No.1 , Part 2.

Parasuraman, R. (1984). Sustained attention in detection and discrimination. In *Varieties of attention*. (ed. R. Parasuraman and D.R. Davies), pp. 243-71. Academic Press, London.

Peters, R. and McGee, R. (1982). Cigarette smoking and State-Dependent Memory. *Psychopharmacology* **76**, 232-5.

Petersen, R.C. (1979). Scopolamine state-dependent memory processes in man. *Psychopharmacology* **64**, 309-14.

Rabbitt, P. (1979). Current paradigms and models in human information processing. In *Human Stress and Cognition: An Information Processing Approach* (ed. V. Hamilton and D.M. Warburton), pp. 115-40. John Wiley, London.

Robinson, S.E. (1985). Cholinergic Pathways in the Brain. In *Central Cholinergic Mechanisms and Adaptive Dysfunctions* (ed. M.M. Singh, D.M. Warburton, and H. Lai), pp. 37-62. Plenum Press, New York.

Roland, P.E. (1981). Somatotopical tuning of postcentral gyrus during focal attention in man. A regional cerebral blood flow study. *Journal of Neurophysiology* **46**, 744-54.

—— (1982). Cortical regulation of selective attention in man; A regional cerebral blood flow study. *Journal of Neurophysiology* **48**, 1059-77.

Rossor, M.N., Iversen, L.L., Reynolds, G.P., Mountjoy, C.Q., and Roth, M. (1984). Neurochemical characteristics of early and late onset types of Alzheimer's disease. *British Medical Journal* **288**, 961-4.

Safer, D.J. and Allen, R.P. (1971). The central effects of scopolamine in man. *Biological Psychiatry* **3**, 347-55.

Sitaram, N., Weingartner, H., and Gillin, J.C. (1978). Human serial learning: Enhancement with arecholine and choline and impairment with scopolamine. *Science* **201**, 274-6.

Solomon, F., Hotchkiss, E., Saravay, S.M., Bayer, C., Ramsay, P., and Blum, R.S. (1983). Impairment of memory function by antihypertensive medication. *Archives of General Psychiatry* **40**, 1109-12.

Squire, L.R. (1969). Effects of pretrial and posttrial administration of cholinergic and anticholinergic drugs on spontaneous alternation. *Journal of Comparative and Physiological Psychology* **69**, 69-75.

—— (1986). Mechanisms of memory. *Science* **232**, 1612-19.

Stein, L., Belluzzi, J.D., and Wise, C.D. (1975). Memory enhancement by central administration of norepinephrine. *Brain Research* **84**, 329-35.

Talland, G.A. and Quarton, G.C. (1965). The effects of methamphetamine and pentobarbital on the running memory span. *Psychopharmacologia* **7**, 379–82.

Torrance, E.P. (1974). *Torrance Tests of Creative Thinking*. Scholastic Testing Service, Bensenville, Ill.

Warburton, D.M. (1979). Neurochemical basis of consciousness. In *Chemical Influences on Behaviour* (ed. K. Brown and S.J. Cooper), pp. 421–62. Academic, London.

—— (1981). Neurochemical bases of behaviour. *British Medical Bulletin* **37**, 121–5.

—— (1983). Towards a neurochemical theory of learning and memory. In *Physiological Correlates of Human Behaviour*, (ed. R. Gale and J. Edwards), Vol. 1 pp. 143–58. Academic Press, London.

—— and Wesnes, K. (1979). The role of electrocortical arousal in the smoking habit. In *Electrophysiological Effects of Nicotine* (ed. A. Remond and C. Izard) pp. 183–200. Elsevier, Amsterdam.

—— and —— (1984). Drugs and human information processing. In *Psychology Survey No. 5*, (ed. J. Nicholson and H. Beloff), pp. 129–55. British Psychological Society, Leicester.

——, ——, Shergold, K., and James, M. (1980). Facilitation of learning and state dependency with nicotine. *Psychopharmacology* **89**, 55–9.

Weingartner, H., Rapoport, J.L., Buchsbaum M.S., Bunney, W.E., Jr, Ebert, M.H., Mikkelsen, E.J., and Caine, E.D. (1980). Cognitive processes in normal and hyperactive children and their response to amphetamine treatment. *Journal of Abnormal Psychology* **89**, 25–37.

Wesnes, K. and Revell, A. (1984). The separate and combined effects of scopolamine and nicotine on human information processing. *Psychopharmacology* **84**, 5–11.

—— and Warburton, D.M. (1978). The effect of cigarette smoking and nicotine tablets upon human attention. In *Smoking Behaviour: Physiological and Psychological Influences* (ed. R.E. Thornton), pp. 131–47. Churchill-Livingstone, London.

—— and —— (1983a). Smoking, nicotine and human performance. *Pharmacology and Therapeutics* **21**, 189–208.

—— and —— (1983b). Effects of scopolamine on stimulus sensitivity and response bias in a visual vigilance task. *Neuropsychobiology* **9**, 154–7.

—— and —— (1983c). The effects of smoking on rapid visual information processing performance. *Neuropsychobiology* **9**, 223–9.

—— and —— (1984). Effects of scopolamine and nicotine on human rapid information processing performance. *Psychopharmacology* **82**, 147–50.

——, —— and Matz, B. (1983). Effects of nicotine on stimulus sensitivity and response bias in a visual vigilance task. *Neuropsychobiology* **9**, 41–4.

9

Psychopharmacological studies of arousal and attention

T.W. ROBBINS AND B.J. EVERITT

There has been considerable interest generated in the nature and functions of the central monoaminergic and cholinergic transmitter systems following the discovery of specific post-mortem neurochemical pathologies suggesting their likely involvement in conditions such as Parkinson's and Alzheimer's diseases (Edwardson *et al.* 1985), Korsakoff's syndrome (Joyce, this volume), depression (van Praag 1982) and schizophrenia (Owen *et al.* 1985).

The functional implications of such discoveries are profound because it appears that there is a need to re-appraise even classical neuropsychological syndromes with the realization that there may be both cortical and subcortical contributions to their neuropathology. Thus, for example, Alzheimer's disease involves not only the disruption of cortical function through the development of plaques, neurofibrillary tangles and progressive neuronal loss, but also through considerable depletion in cortical transmitter systems of subcortical origin, including for example, the cholinergic projection from the nucleus basalis of Meynert, the noradrenergic projection from the pontine locus coeruleus and the serotoninergic projection from the mesencephalic raphé nuclei. A similar broad pattern may also be present in Korsakoff's syndrome, where specific diencephalic damage may be accompanied by degeneration of cholinergic cells in the basal forebrain and of noradrenergic cells within the locus coeruleus. Such complex patterns of neuropathological change contrast with those seen following focal neocortical damage and are often paralleled by the breadth of the functional impairments observed. Several important questions can thus be framed. For example, are the changes in subcortical transmitter systems causal in the cortical pathology and the cognitive deficits, or are they merely secondary to the cortical changes, occurring, for example as a result of retrograde degeneration (cf. Edwardson *et al.* 1985)? And in the latter case, do the ancillary subcortical changes nevertheless contribute to the functional deficits? Such questions are difficult to answer from the correlational perspective of post-mortem studies and can probably best be tackled by investigations of the functional sequelae of experimental manipulations of specific

135

neurotransmitter systems. Although it is both possible and informative to pursue this aim through psychopharmacological studies of the effects of systemically administered drugs in normal human subjects (e.g. Frith *et al.* 1985; Clark *et al.* 1986; Sahakian 1987; Warburton, this volume) this approach has obvious limitations. The main alternative is to study the effects of selective central manipulations in animals, but the challenge then is to relate any functional changes observed to those seen in a clinical setting. This chapter will illustrate the advances we have made in addressing such, inevitably complex, issues.

NEUROBIOLOGICAL CLUES TO FUNCTION

Some idea of the roles of transmitter-defined sub-cortical systems can be gleaned from a consideration of the steadily accumulating information on their organization. As can be seen from Fig. 9.1, the location of the neuronal cell groups and their projection to terminal regions has been well-studied and this has been discussed in detail by several authorities (Lindvall and Bjorklund 1983: Mesulam *et al.* 1983). However, several points can be raised which have direct functional implications. First, each of the systems depicted in Fig. 9.1 arises in brain regions associated with what was classically termed the 'ascending reticular activating system', and thereby linked with non-specific processes such as arousal. This is particularly obvious in the case of the noradrenergic neurons arising in the locus coeruleus and the serotoninergic neurons of the raphé nuclei. Secondly, they are diverse in their terminal projection sites, the coeruleal noradrenergic projection innervating, for example, the spinal cord, cerebellum, hypothalamus, thalamus, hippocampus, and the entire neocortical mantle. Thirdly, although there is a degree of point-to-point topographical organization in each of the systems, a given cell (e.g. in the case of noradrenergic coeruleal cells) may, through its repeatedly branching axon, ramify to distinct terminal locations, so that its activity can potentially effect a simultaneous modulation of the different forms of processing that may be occurring within these regions. The diffuse nature of this projection obviously suggests that it has a rather 'general' role. Fourthly, there is obvious topographical organization in the sense that each of these systems has a number of distinct cell groups of origin. Thus, in the case of dopaminergic neurons, there are projections from the ventral tegmental area to limbic frontal cortex and the nucleus accumbens, whereas the cells of the more laterally situated substantia nigra predominantly innervate the caudate-putamen. For the cholinergic projections, the sources of innervation of the neocotex and amygdala from the nucleus basalis of Meynert (nbM) are distinct from those of the hippocampus, which receives a cholinergic input from more rostral cell groups in the medial septum. The ascending serotoninergic projections arise from distinct cell groups in the dorsal and median

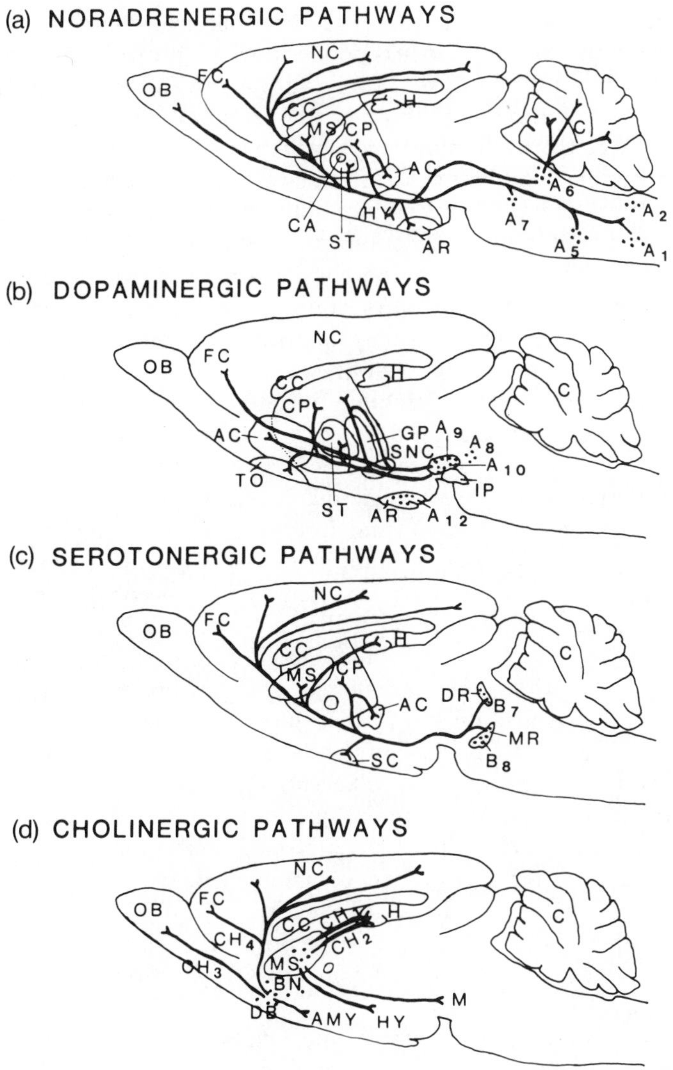

(a) NORADRENERGIC PATHWAYS

(b) DOPAMINERGIC PATHWAYS

(c) SEROTONERGIC PATHWAYS

(d) CHOLINERGIC PATHWAYS

FIG. 9.1. Sagittal section through rat brain to show the comparative anatomy of the ascending monoaminergic (including cholinergic) projections. Abbreviations: NC, neocortex; OB, olfactory bulb; FC, frontal cortex; CC, corpus callosum; MS, medial septum; H, hippocampus; CP, caudate-putamen; CA, anterior commissure; AC, nucleus accumbens; ST, stria terminalis; GP, globus pallidus; SNC, substantia nigra, pars compacta; AR, arcuate nucleus; C, cerebellum; IP, interpeduncular nucleus; SC, suprachiasmatic nucleus; DR, dorsal raphe nucleus; MR, median raphe nucleus; AMY, amygdala; HY, hypothalamus; DB, diagonal band; BN, basal nucleus of Meynert; M, midbrain. A1–A12; catecholamine cell body groups. B7–B8; indole-amine cell groups. CH1–CH4; cholinergic cell groups according to the nomenclature of Mesulam *et al.* (1983). (Reproduced from Robbins (1986) with permission from Martinus Nijhoff.)

raphé nuclei. Finally, the ascending noradrenergic projections can be divided into those arising from the locus coeruleus, which project more dorsally in the brainstem, enter the medial forebrain bundle and innervate primarily telencephalic structures, and those arising from cell groups in the medulla oblongata, which project more ventrally in the brainstem and, via the medial forebrain bundle, innervate subcortical, largely diencephalic structures. This complexity makes it unlikely that monolithic constructs of arousal are going to be useful in defining the functions of these systems. Fifthly, it is evident from Fig. 9.1 that, although there is considerable overlap in the distributions of these different subcortical projection systems, there are also different domains of influence. Thus, the only significant dopaminergic innervation of the neocortex is to prefrontal regions, whereas the otherwise profuse nora-drenergic innervation of the forebrain is relatively sparse within the caudate-putamen. This suggests that the different systems participate in very different behavioural processes and raises the possibility that their functions will be largely reflected in those of the terminal regions to which they project. Finally, and what is not evident from the gross projections shown in Fig. 9.1, there are important differences in the ways in which the different ascending projections connect to the intrinsic circuitry of a particular region, even where there is considerable overlap in their gross innervation of that area. For example, in the visual cortex, noradrenergic neurones predominantly ter-minate in layer 6, while the serotoninergic inputs are preferentially to layer 4, where they are in a position to influence the specific sensory afferents arising in the lateral geniculate nucleus in the thalamus (Morrison and Magistretti 1983). This type of information will eventually prove vital in distinguishing among the functions of the different systems when they all converge onto a common area, as in the case of the prefrontal cortex. Despite the present emphasis on the differences in function among these systems, this should not be taken to imply that they necessarily have specific roles in information pro-cessing. Rather, the diffuse nature of the projections suggests that it may be more profitable to consider the possibility that they subserve a range of diffe-rent 'non-specific' influences.

Electrophysiological and neurochemical measures can be used to indicate the types of stimuli, environmental conditions or endogenous states which modulate the activity of the various systems. Such evidence again generally confirms that these systems do not carry specific information, such as the spatio-temporal patterns carried by the sensory (or motor) systems and also that they are unlikely to be directly implicated in processes of memory trace consolidation or retrieval. For example, noradrenergic cells in the locus coe-ruleus do not respond to visual stimuli of specific orientations, like complex cells in the visual cortex, but do respond to intensive properties, such as brightness (Watabe *et al.* 1982). Such cells may also respond polymo-dally, especially to novel stimuli (Aston-Jones and Bloom 1981). Recent comparison of the electrophysiological properties of the major monoaminergic systems showed that they were differentially dependent upon sleep-waking state (Jacobs 1984). Thus, noradrenergic cells in the locus coeruleus show monotonic increases in spontaneous firing rate over stages of REM and slow-wave sleep, and waking, an ordering which also corresponds to the degree of the animal's behavioural responsiveness. Serotoninergic cells of the raphé nuclei show a similar state-dependency with respect to the sleep-wake cycle, but midbrain dopaminegic cells do not, although they are sensitive to stimuli which act as cues to action (Jacobs 1984).

In the case of the more-recently characterized cells of the nucleus basalis, there is perhaps less information, although it has long been known that the medial septal cholinergic innervation of the hippocampus probably plays a role in the generation of the theta rhythm (Krnjevic and Ropert 1981). The ascending cholinergic and noradrenergic hippocampal projections also influence the phenomenon of long-term potentiation (Halliwell and Adams 1982; Hopkins and Johnston 1984).

A reasonable conclusion from these electrophysiological investigations is that the systems under survey become active under various, somewhat ill-defined, states of arousal. This impression is strengthened by a brief consideration of the neurochemical evidence. For example, Robbins and Everitt (1982) review evidence showing how the turnover of the coeruleal-cortical noradrenaline (NA) system is enhanced by a variety of environmental stressors, not necessarily aversive in nature. Bannon and Roth (1983) review the evidence that stress increases central dopamine (DA) turnover, particularly in the mesocorticolimbic regions. Certain of these areas, such as the nucleus accumbens, also exhibit elevated DA turnover in conditions such as in the presence of food following deprivation (Heffner *et al.* 1980), suggesting it might reflect the behavioural excitability that can occur in the presence of rewarding stimuli (incentive motivation).

CONCEPTS OF AROUSAL

Even today, the construct of arousal is somewhat nebulous both in experimental psychology as well as in neuroscience, largely because of its many different connotations. Yet it seems the key to understanding how the monoaminergic and cholinergic systems affect behavioural function. One way of resolving this difficulty would be to examine the supportive data and relative utility of some of the different concepts of arousal, while considering the possibility that what has been termed arousal is, in fact, a collection of different, dissociable processes. Three distinct concepts of arousal are: (i) as a source of drive or energy for behaviour (e.g. Hull 1949); (ii) as a measure of neural activity in the reticular formation which maintains efficient neuronal function in the neocortex (Hebb 1955); and (iii) as a relatively undifferentiated source of visceral afferent information serving, among other functions, as the substrate of emotional experience (e.g. Mandler 1975). Some of these connotations have been preserved in more modern contexts. For example, the role of peripheral autonomic stimuli as a prominent mechanism of memory enhancement by peptide hormones has recently been advanced (Le Moal *et al.* 1981).

The main type of result lending credibility to arousal theories has been the bitonic, inverted U-shaped function (see e.g. Hebb 1955; Hockey 1979; Eysenck 1982) relating efficiency of performance to the effect of some treatment which hypothetically produces linear increases in activity of the reticular formation. Thus, there is an optimum level of arousal for maximally efficient performance, but levels either lesser or greater than this optimum lead to performance decrement. It is this type of relationship, for example, which underlies Hebb's (1955) theory. It receives rather general support from a number of pharmacological dose-response behavioural determinations, which, of course, have analogous bitonic functions. When the details of this relationship between hypothesized arousal and performance are examined further, however, it runs into a number of difficulties.

One of the first complications was the realization that the concept of optimal levels of arousal was task dependent, so that very difficult tasks seemingly required much lower levels of arousal for their efficient performance than easier ones (see Eysenck 1982). This suggested that a given level of arousal could both improve and impair performance. Analogous effects of motivation had been observed by Yerkes and Dodson as early as 1908, when studying discrimination by mice swimming under different levels of motivation, and so this type of effect has come to be known as the Yerkes–Dodson law. The use of different tasks also begs the question of the nature of the processing affected by changes in arousal level. This could involve processes as diverse as response preparation, selective attention, short-term memory or long-term retention. Evidence exists to show that short-term memory is impaired by levels of arousal that promote long term retention (see Eysenck 1982 for a review), but it is difficult to see how this can be explained by the relative 'difficulty' of the two modes of recall, or indeed in what units this can be measured.

A further problem for arousal theory, raised especially by Hockey (1979), relates to the measures of 'efficiency' on the Y axis of the classical inverted U-shaped function. Even for the same task, different indices of performance might behave in different ways. Thus, high levels of arousal favour the speed of responding, but not accuracy of immediate retention, in a letter transformation task (see Hockey 1979). This likely reflects the trade-off in functioning between different component processes of any complex cognitive task and the fact that they may be differentially affected by arousal level as described above.

Another complication came when the effects of different stressors thought to affect arousal in predictable ways were compared, either alone or in combination. The five-choice serial reaction time task, which became standard for this type of investigation is described by Leonard (1959). Human subjects were required continuously to report in which of five locations a light occurred and performance was monitored in terms of the accuracy of detecting the targets, missed targets, and reaction time. The effects of white noise and sleep deprivation did behave more-or-less in accordance with arousal theory. Thus, performance was impaired by both of these stressors alone, at least late in the test session, but their effects in combination tended to cancel out so that performance remained close to control levels (Wilkinson 1963; see Fig. 9.2). This form of interaction would be predicted by an arousal account which assumed that sleep deprivation places the subject on the ascending limb of the inverted U-shaped curve because of its propensity to reduce arousal, whereas noise has the opposite effect. Greater problems were provided by the effects of another manipulation of arousal, knowledge of results, which can act like a form of incentive. This, too, counteracted the deleterious effects of sleep deprivation, but by itself, improved performance (Wilkinson 1961; see Fig. 9.2). However, this could be reconciled with arousal theory by the further observation that noise and incentive together led to further declines in performance, if one assumed that the two stressors together were producing supra-optimal effects on performance in accordance with the inverted U-shaped function. Another striking discrepancy concerned the interaction between incentives and alcohol (Wilkinson and

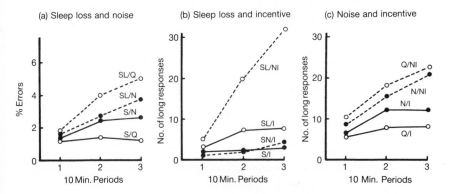

FIG. 9.2. Evidence for interactions among stressors on the five-choice task of Leonard (1959). Abbreviations used in the figure are: S, normal sleep; SL, sleep loss; Q, quiet; N, noise; I, incentive; NI, no incentive. Number of long responses = frequency of responses with very long latencies. [Reproduced from Hockey (1979) with permission from John Wiley (Copyright 1979). The data are redrawn from articles by Wilkinson (1961, 1963). Copyright 1961 and 1963 by the American Psychological Association. Reprinted by permission.]

Colquhoun 1968). The prediction that incentives should attenuate the deleterious effects of alcohol on the grounds that these two manipulations
should, respectively, increase and decrease arousal was confounded by the
fact that incentives exaggerated the performance decrement produced by
alcohol. In attempting to account for these and other results, Broadbent
(1971) proposed that there might be two arousal processes, in which sub- or
supra-activity of a 'Lower' mechanism is compensated by activity in an
'Upper' mechanism, and in which different stressors preferentially affect
processing in the Upper or Lower mechanism. The observation that performance decrement was most obvious towards the end of a test session (see
Fig. 9.2) could now be explained by invoking the compensatory processes set
into play by the Upper mechanism which, however, could only operate over
fairly restricted periods to maintain performance in the face of stress. Figure
9.3 provides a diagrammatic representation of this model in the context of
animal research to be described.

Trying to make sense of the effects of different drugs, as well as of more
conventional stressors, on human performance, Broadbent (1971) speculated
shrewdly about the possibility that the Upper and Lower mechanisms were

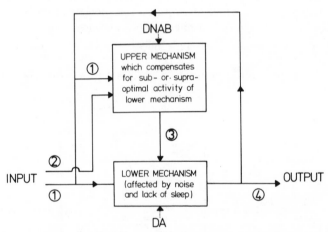

① non-specific arousal, drive (intensive aspects)
② informational (discriminative aspects)
③ response selection or choice
④ speed and likelihood of response

FIG. 9.3.   Diagram to show Broadbent's (1971) dual arousal model interpreted in
terms of the possible contributions to the 'Upper' and 'Lower' arousal mechanisms by
the cortical NA and the central DA systems, respectively. The cortical ACh projections may also contribute to the upper mechanism. [Reproduced from Robbins (1984)
with permission from Cambridge University Press.]

modulated by different neurotransmitter systems. Of course, pharmacologists should hardly despair of the fact that the effects of different forms of stress seem to add and antagonize in different ways—much as the effects of increasing dose and of agonist-antagonist interactions in psychopharmacological studies. A direct demonstration of the potential of this pharmacological specificity has been provided by Drachman's studies of the relative effects of physostigmine and amphetamine on the deleterious cognitive effects of scopolamine. Physotigmine antagonized the impairment produced by scopolamine, while amphetamine exaggerated it; as perhaps expected, the effects of physostigmine and amphetamine cannot necessarily be encapsulated by appealing to their effects operating along a single continuum of arousal (see Drachman and Sahakian 1979).

CONCEPTS OF ATTENTION AND MEMORY

Just as the construct of arousal should probably be fractionated into several components, so too should that of attention, as the writings of Posner (1978), and Kahneman and Treisman (1984) have made clear. One major distinction is between divided and selective attention. The aim of studies of divided attention is to assess the capacity to perform more than one task simultaneously, and the constraints on this ability. Tasks involving selective attention, on the other hand, typically measure the resistance to some form of distraction. However, this resistance might be manifested either at the level of ensuring adequate perceptual processing of the important sensory signals, via a filtering mechanism (Broadbent 1958) or to ensure adequate selection and execution of the appropriate actions from among the many alternatives (e.g. Posner 1978; Norman and Shallice 1980). One way of facilitating either stimulus or response selection is to bias or prime various options, for example, by instructions, or by previous experience. Thus, in the 'selective set' paradigm, the subject is instructed to respond to a particular set of stimuli (see Kahneman and Treisman 1984), whereas attention can be summoned by priming to particular locations in Posner's 'covert orienting' paradigm (see Posner 1978), and animals can be biased against learning about particular stimuli if these have been rendered irrelevant by failing to predict reinforcement (Mackintosh 1983). Kahneman and Treisman (1984) characterize the different interpretations of selective attention afforded by the filtering and selective set paradigms, respectively, as preventing or reducing perceptual processing of unattended stimuli, and as selecting and speeding responses to expected targets.

A further distinction is possible, between controlled and automatic forms of processing (see Eysenck 1982). Posner and Snyder (1975) explicitly distinguish between automatic processes and conscious attention of limited capacity. Automatic processes have been characterized by several criteria,

including their involuntary and possibly unconscious nature, lack of susceptibility to interference from activities involving attention, and lack of their
ability to interfere with such activities (see Kahneman and Treisman 1984).
Automatic processes can be monitored by priming phenomena, when, for
example, the prior presentation of a letter facilitates its own recognition. In
contrast, conscious attention is biased by expectancies, and retards decision-
making when an unexpected or novel stimulus occurs. However, the exact
relationship of automatic processing to attention (and indeed, to consciousness) is controversial, as made clear by Kahneman and Treisman (1984). The
dynamic relationship between these two forms of processing is however well
exemplified in the theory of Treisman and Gelade (1980) which argues that
features of a stimulus (such as the colour blue) are registered automatically
and in parallel, whereas objects which consist of conjunctions of features
(perhaps including semantic and affective aspects) are identified by a serial
search process that can be called focussed attention.

Like Broadbent's earlier account (Broadbent 1971), an important part of
Posner's (1978) theory is to recognize the contribution of arousal-like processes to attention. Posner refers to alertness, detection, and orientation as
three basic factors in attention. Of these, alertness can be divided into phasic
and tonic types, and seems most clearly equated to fluctuations in arousal,
being defined 'as the overall activation level of the central processing system.
The higher the activation level, the more readily mechanisms could be
brought to bear on any new stimulus event' (Posner 1978, p. 204). Phasic
alertness is identified with the processes occurring in the preparation of a response to a specific signal, which can be inhibitory as well as excitatory in
nature, as expressed, for example, in the inhibition of spinal reflexes associated with the muscle that is eventually to respond. An electrophysiological
correlate of this preparation may be evoked potentials such as contingent
negative variation (see Picton *et al.* 1978, for a review). Tonic alertness
changes more slowly and is suggested by Posner to be subject, for example, to
circadian fluctuations in electrocortical arousal and hormonal rhythms.
Detection refers to the entry of a signal into the central processing mechanism, and orientating signifies a particular response that may be made to a
signal, generally to point the peripheral or central processing mechanisms in
the direction of the target. According to Posner, orientating, which can
involve covert components as well as obvious behavioural ones, facilitates
signal detection and is particularly beneficial in the visual modality. Orientating, whether overt or covert, thus enables attention to be selective in nature,
ignoring portions of stimulus input, while processing the rest.

While the integration of arousal and attentional theory attempted by
Posner is valuable, it should be clear from our preceding discussion that
'arousal' may take more than one form, and also affect different types of
processing, if only on the neuroanatomical grounds argued above. Thus, it

seems possible that 'arousal' may affect, for example, not only stimulus detection, but also response choice, and also that electrocortical and hormonal changes may be reflections of different types of arousal process.

After a stimulus has been detected, of course, prior experience of it will determine whether it will be recognized and what aspects of it will be stored in memory. Evidence from experimental psychology and neuropsychology has given impetus to the possibility that there are qualitatively different forms of memory. The old 'modal model' of Atkinson and Shiffrin (1971) of dual short- and long-term memory processes has now given way to theories of discrete short-term memory processes [e.g. in the 'working memory' model of Baddeley (1983)] and different forms of representation in long-term memory [e.g. of declarative and procedural, or alternatively, of episodic and semantic distinctions, (see for example, Squire 1982)].

From this brief survey of work in human experimental psychology it will be evident that monolithic terms such as 'attention' and 'memory' are in danger of becoming as empty as old concepts of arousal. Hence, there is a growing need for neuroscientists to describe the effects of various manipulations of the brain or drugs on cognitive processes with much greater care and precision than has been the case hitherto.

NEUROBIOLOGICAL IMPLICATIONS

Our current conceptions of processes such as memory and attention make it clear that these are complex processes, subserved and co-ordinated over diverse regions of the forebrain. It is not the task of this chapter to attempt to detail this complexity, but rather to indicate how these processes are modulated by changes in activity of ascending noradrenergic, dopaminergic, and cholinergic systems. It would seem to be futile now to refer to such changes in terms of a single 'arousal' process. Each of the ascending systems described has different afferent connections and different patterns of termination. Thus, despite the non-specific nature of its activity, each system is likely to be modulated by different influences and to contribute to disparate forms of processing in the cortex, or elsewhere. In the regions receiving overlapping projections from the various systems, it is also likely, though presently unclear, that transmitter-specific synaptic events affect local information-processing in different, and perhaps complementary, ways.

Although NA has often been considered as an inhibitory transmitter from electrophysiological investigation of its actions following iontophoresis into terminal regions, work by Segal and Bloom (1976) has shown its action may subserve an 'enabling' or biasing function, such that locus coeruleal activity will tend to accentuate the effects of other inputs mediated by the actions of other local transmitters (Bloom 1979). Even when its influence is primarily inhibitory, it appears more drastically to affect spontaneous firing than evoked responses to signals such as light flashes, touch or vocalizations, when studied in the sensory cortices of various species. Such findings have led to the notion that activity in the coeruleo-cortical NA neurons alters the 'signal-to-noise' (S/N) ratio for inputs onto a target (e.g. Foote *et al.* 1975).

Although its precise mechanism of action is almost certainly different (for example, including neuronal excitatory effects, Krnjevic and Phillis 1963; Stone 1972; Foote *et al.* 1975), cortical acetylcholine (ACh) may well have a similar action, which Marczynski (1978) has argued could affect the generation of surface negative slow potentials, including components that could be involved in attentional processing. Some evidence has recently been provided which suggests that dopamine in the striatum also increases S/N ratios, via an action that appears analogous to that of NA (Rolls *et al.* 1984).

In speculating about the consequences of enhancing S/N ratios throughout the forebrain, it is evident that the subsequent improvement in information transmission could impinge on processes of attention, memory, learning, or action preparation and selection, to name but a few. However, it is evident from the discussion of the evidence from human psychology described above that such enhancement need not always prove beneficial and that improvements in some functions may be offset by impairments of others. Thus, although the elucidation of the synaptic action of the ascending monoaminergic and cholinergic transmitter pathways continues to be an important goal of research, it is one that should proceed in parallel with an experimental analysis of the behavioural effects of selective manipulations of these systems. The challenge is to link these effects to the normal functioning of these neural systems on one hand, and to their derangement in the clinical disorders described previously, on the other.

PSYCHOPHARMACOLOGICAL CONTRIBUTIONS TO AROUSAL THEORY

The approach we have used will be illustrated by considering the function of the ascending DA projections as revealed by both animal and human studies, and special attention will also be given to the possible functions of the cortical NA and ACh projections. In general, the central DA systems appear to be involved in processes which determine response vigour, whereas the NA and ACh projections have subtle roles in modulating processes of attention and response selection according to certain conditions.

Dopamine and activation

In rats, profound bilateral depletion of striatal DA, using the neurotoxin 6-hydroxydopamine, produces a life-threatening aphagia with attendant problems of akinesia and catalepsy which resemble some of the symptoms of

Parkinson's disease in humans (see Marshall and Teitelbaum 1977). Stressful circumstances, such as tail-pinching or cold water can temporarily reverse these symptoms, as they apparently also can do in man (Schwab 1972). The behavioural deficits underlying the deficits produced by striatal DA depletion are nevertheless difficult to analyse, and a less drastic approach involving unilateral injection of the neurotoxin, has been employed (e.g. Ungerstedt 1971). One of the sequelae of such treatment is an apparent polymodal 'sensory-motor neglect' in which rats exhibit deficits in orienting towards contralateral sensory events. This has been termed an 'inattentional' syndrome (c.f. Marshall *et al.* 1980), but the question remains whether this is truly a deficit of sensory detection as defined above by Posner (1978), or whether it is the behavioural expression of the detection that is impaired, by, for example, a contralateral hemi-akinesia. Carli *et al.* (1985) recently attempted to answer this question by measuring the accuracy, side bias, and reaction times of rats with unilateral striatal DA depletion in detecting a brief visual event presented to either side of the head when held in a central location. One group of rats was trained to move their heads to the location of the visual target; the other group was trained instead to respond opposite to the side of the presentation. This design helped to distinguish the contribution of stimulus versus response factors in the 'neglect' syndrome.

The results were most compatible with the hypothesis that unilateral DA depletion produces an ipsilateral response bias, irrespectve of the side of presentation of the visual target. Reaction time measures showed that the rats were slower to withdraw their heads from a central position to make responses contralateral to the side of the lesion, but not to execute the rest of the required lateral head movement (see Fig. 9.4). The results can be interpreted as indicating that unilateral striatal DA depletion leads to a lateralized deficit in the ability to initiate, but not execute contralateral actions. This interpretation was supported by the results of a second experiment in which rats were required to detect peripheral events by a non-lateralized response, there being no effects on accuracy of choice following unilateral DA depletion under these conditions (Carli *et al.* 1985).

We see our results as being broadly consistent with recent reaction time studies of Parkinsonian patients. For example, Bloxham *et al.* (1984) reported that Parkinsonian patients failed to show the usual improvement in reaction time produced by providing an informative signal prior to the cue for action, although choice reaction times were no different from normal. These results could be interpreted as showing deficits in response preparation and initiation, rather than response selection. Rafal *et al.* (1984) similarly did not show specific impairments in attentional switching by Parkinsonian patients, either on or off L-dopa medication, in Posner's 'covert orienting' paradigm, although overall speed of responding was clearly hastened by the drug.

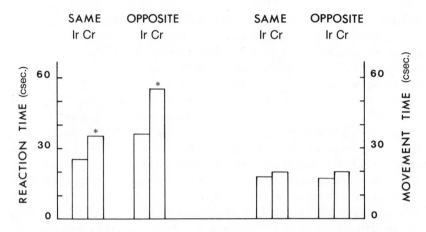

FIG. 9.4.   Effects of unilateral striatal DA depletion on measures of reaction time and movement time in groups of rats trained to respond on the same side of a brief visual target presented to either side of the head or on the opposite side. Left: time taken to initiate either ipsilateral (Ir) or contralateral (Cr) responses when required to the visual stimulus for the two groups SAME and OPPOSITE. The SAME group was trained to respond on the same side as the stimulus, the OPPOSITE group to the opposite side. Thus, the common effect of unilateral striatal DA depletion in the two groups was to retard the initiation of responding to contralateral space, regardless of the side on which the visual stimulus was presented. By contrast, movement time (the time taken to complete the response into the side hole) was unaffected (see Carli *et al.* 1985).

Dopamine depletion from mesolimbic regions, such as the nucleus accumbens, also reduces the speed and probability of responding in a 5-choice reaction task [analogous to Leonard's (1959) task for humans] designed for rats in which they have to report the location of brief visual events. Although resulting in increased errors of omission, DA depletion in the n. accumbens does not increase errors of commission, visual discrimination being unaffected (Robbins *et al.* 1982). By contrast, the indirect DA agonist, *d*-amphetamine, increases the incidence of premature responses (reducing omissions), but has no effect on discrimination, when injected either systemically, or directly into the n. accumbens (Cole and Robbins 1987*a*, see Fig. 9.5).

It is by now quite well-established that most of the behavioural effects of amphetamine depend upon DA release in various forebrain regions. Two of the most prominent effects of the drug on unconditioned behaviour in rats, locomotor hyperactivity, and behavioural stereotypy, have been shown to depend upon DA release in the n. accumbens and caudate-putamen, respectively (Kelly *et al.* 1975), which suggests that the postulated process of response activation itself has different components, or at least, is controlled by different areas of the forebrain. As the dose of amphetamine increases, there is an increasing rate of responding which leads to a competition among the various forms of behaviour stimulated (Lyon and Robbins 1975). It is typically the short elements of behaviour, such as repetitive sniffing and fragmented oral movements which continue to be initiated and impair the sequencing of behaviour. This is illustrated quite dramatically by the perseverative, but counterproductive, activities of human amphetamine addicts (Kramer *et al.* 1967). The so-called rate-dependent effects of stimulant drugs on operant behaviour can also be explained in these terms, high baseline rates being especially susceptible to interruption by competing activities and low baseline rates being sufficiently low for the stimulant effect to be observed. This dose- and baseline-dependent succession of inverted U-shaped dose-effect functions is strongly reminiscent of the Yerkes–Dodson principle described above. Apparent deficits in response selection and sequencing produced by amphetamine can be seen as secondary to excessive DA release in mesolimbic and striatal sites leading to supra-optimal levels of activation for efficient behavioural output.

It is salient to compare the effects of amphetamine in animals with some of the effects of arousal produced by various stressors, such as white noise, in man. As mentioned above, white noise tends to increase response speed, while impairing short term recall (specifically by disrupting item sequencing, see Hockey 1979), but it generally improves long-term retention. Comparable effects of systemic amphetamine on short-term memory (Kesner *et al.* 1981) and of intra-caudate amphetamine on long term consolidation (Carr and White 1984) have been reported for the rat. Thus, clues about the neurochemical mechanisms underlying effects of forms of 'arousal' on specific forms of processing in man may be obtained by study of the behavioural effects of drugs in rats.

The dorsal noradrenergic bundle and cortical arousal

Almost complete depletion of cortical NA following 6-hydroxydopamine (6-OHDA) infusions into the midbrain producing lesions of the dorsal noradrenergic ascending bundle (DNAB) has very different effects from lesions produced by damage to central DA neurons. Rats with such lesions have no obvious ingestive, or indeed any other motor or motivational disabilities, and there is little sign of any obvious sensory deficit (Mason and Iversen 1979). The latter conclusion is highlighted by the lack of effect of DNAB lesions in rats on their performance in the five-choice visual reaction time task described above, even when the stimuli are dimmed to produce graded decrements in performance (Carli *et al.* 1983). However, DNAB-lesioned rats can exhibit deficits on this task in certain circumstances which may provide some insights into the normal functions of this system.

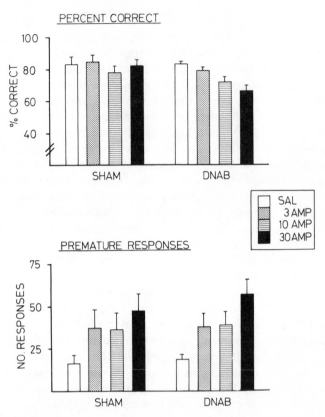

FIG. 9.5.   Effects of *d*-amphetamine injected into the nucleus accumbens (3–30 μg) on percent correct and premature responses in the five choice visual localization task for DNAB and sham operated rats (Cole and Robbins (1987*a*), reproduced with permission from Springer Verlag).

Figure 9.5 shows the effects of infusions of amphetamine into the nucleus accumbens on the performance of rats with DNAB lesions as assessed by two parameters, accuracy (percentage correct) and number of premature responses (i.e. inappropriate responses made prior to the onset of the visual discriminanda). The basic result for the sham-operated rats has already been described; amphetamine increases premature responses, but does not affect choice accuracy. However, in rats with DNAB lesions, the similar increase in premature responses is paralleled by a dose-dependent decline in choice accuracy (Cole and Robbins 1987*a*). The enhancement in premature responses is primarily dopaminergic, being blocked by the DA receptor antagonist, alpha-flupenthixol (Cole and Robbins 1987*a*) and by DA depletion from the n. accumbens (B.J. Cole and T.W. Robbins, unpublished results). One interpretation of such data might be that activity in the DNAB normally preserves response accuracy under conditions of dopaminergic activation.

An earlier experiment using the same paradigm had provided analogous results (Carli *et al*. 1983). This time, white noise was interpolated at various points in the inter-trial interval, prior to, or simultaneous with, the presentation of the visual targets. As Figure 9.6 shows, the white noise produced some effects that were similar to those of amphetamine. When presented just prior to the visual stimuli, it produced similar increases in premature responses, together with some quickening of reaction time in the two groups. (Similar effects are seen when noise is used as an accessory stimulus in human visual reaction time studies, see Posner 1978). Despite this disruption, normal rats maintained high levels of discrimination, but DNAB-lesioned rats once more showed impairments in choice accuracy. The rats with DNAB lesions seem less able to cope with such disruptions, and this leads to the adoption of riskier response criteria and more rapid responding. These results were originally interpreted as being consistent with the type of theorizing advanced by Broadbent (1971). Thus, under conditions engaging the 'Lower Arousal' mechanism, which was hypothesized to involve dopaminergic activation and lead to changes in response vigour, the 'Upper Arousal' mechanism (modulated by the locus coeruleus, among other influences) was assumed to compensate for any deleterious effects on performance.

Further experiments were designed to identify the nature of the interaction

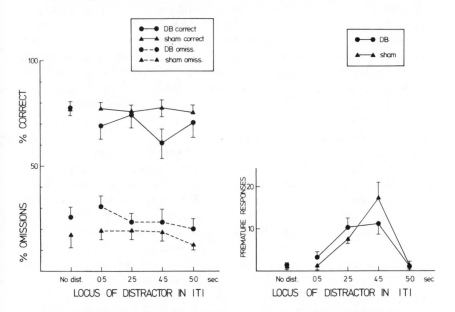

FIG. 9.6.   Effects of distracting bursts of white noise on accuracy, errors of omission and premature responses in sham-operated or DNAB lesioned rats [redrawn from Carli *et al*. (1983), with permission from Elsevier Science Publishers].

of activating white noise with mechanisms of sensory attention or response selection. Thus, bursts of white noise were interpolated just before visual targets, which were also varied in brightness. The white noise again made the DNAB-lesioned rats less accurate than controls, but this effect was not exaggerated by dimming the visual discriminanda (B. J. Cole and T. W. Robbins, unpublished results). This result suggests that the effects on performance of degrading the sensory input and interpolating white noise are only additive, and thus appear to depend upon separate mechanisms, only one of which is affected by DNAB lesions. The impairment of response accuracy by white noise in this group does not seem to depend upon a degradation of the visual sensory input, but operates at a later stage of processing, perhaps on response preparation and selection.

The hypothesis of an effect at the level of response preparation is further strengthened by observations that DNAB lesioned rats exhibit impairments in choice accuracy when the visual discriminanda are presented at unpredictable intervals, rather than at a fixed intertrial interval (ITI) of 5 secs (Carli *et al.* 1983). Further experiments (B. J. Cole and T. W. Robbins, unpublished observations) have confirmed these results, and show that they occur both when the ITIs are much shorter than usual and when they are longer. Again, under special circumstances, the accuracy of the DNAB-lesioned rats can be exposed when there is no sign of impairment in the control condition. This separation cannot be attributed simply to the use of a more difficult test because the manipulation of dimming the stimuli, which produces similar degrees of decrement in control performance, fails to produce a differential impairment in the DNAB-lesioned group. This pattern of results described for the five-choice task, suggests that rats with cortical NA depletion are deficient in implementing control processes which appear to be recruited in certain conditions, notably when performance is affected by endogenous (i.e amphetamine) or exogenous (i.e. white noise) stressors, or in conditions of uncertainty produced by variable ITIs.

Can this hypothesis apply to the effects of central NA manipulations on processes such as memory and learning? On the face of it, the evidence implicating NA in these processes is, at best, controversial. Early claims of effects on learning and memory (Anlezark *et al.* 1973; Crow and Wendlandt 1976) have not been consistently borne out (e.g. Mason and Iversen 1979; Pisa and Fibiger 1983*a*). However, the hypothesis proposed above provides us with a clear prediction; stressful conditions which hasten performance will lead to less accurate performance in rats with DNAB lesions. Recently, B. J. Cole and T. W. Robbins (unpublished observations) made a direct test of this hypothesis, using the Morris place navigation task. This is a swimming test in which the rat has to locate a submerged platform fixed in a constant position with respect to spatial cues in the testing room. The rat is trained to find the platform from a variety of different starting points until it learns the position of the platform, as evidenced by a progressively shorter latency to reach it and a shorter distance swum. The task has already been used to show a profound deficit in rats with hippocampal lesions, which can be interpreted as a deficit in spatial memory (Morris *et al.* 1982), but no effect of DNAB lesions (Hagan *et al.* 1983). However, in these experiments, the swim-maze used water or other liquid kept at a warm temperature. Cole and Robbins repeated the experiment on DNAB-lesioned rats, but in addition to a warm water condition also used a cold water maze as a direct manipulation of stress. As Hagan *et al.* (1983) had earlier reported, there was no difference in swim speed or accuracy in the warm water between the two groups. The cold water made both sham-operated and DNAB-lesioned rats swim more quickly than in the warm water, but to similar extents. However, the DNAB-lesioned group took longer and swam further before finding the hidden platform in cold water (Fig. 9.7). These results were counterpointed by a completely different pattern shown in rats with bilateral caudate DA depletion. The caudate group swam more slowly and further, to find the hidden platform in warm water compared with a sham-operated group. However, in cold water the rats with caudate lesions swam as quickly as normal rats, although their

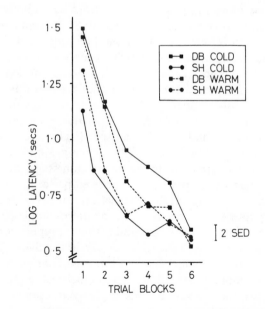

FIG. 9.7. Effects of lesions of the DNAB on place navigation learning on the latency to find the submerged platform in either warm or cold water over blocks of daily sessions. Progressively reducing latencies indicate learning. Similar results were also found for distance swam (Cole and Robbins, unpublished observations).

learning deficit was unaffected. This double dissociation, of results for DNAB and caudate DA lesions on swim speed and accuracy parallels some of the results described earlier which suggested that (i) rats with DNAB lesions have no deficits of response activation or vigour, but under conditions of more vigorous responding (resulting from stress) exhibit deficits in discrimination, whereas (ii) rats with caudate DA lesions show deficits in response vigour (or activation), which can be ameliorated by stress.

The learning deficits shown by the DNAB-lesioned group in the swim maze have been paralleled by acquisition deficits shown by rats in other situations. Thus, Cole and Robbins (1987*b*) have recently shown that DNAB-lesioned rats are impaired in the acquisition of conditioned suppression to an aversive CS and that this result cannot be due to changes in performance factors such as motivating anxiety because there is no effect of the lesions on conditioned suppression that has been established pre-operatively by prior training. Also of considerable interest is the earlier demonstration by Everitt *et al.* (1983) of impairments in acquisition by DNAB-lesioned rats in learning a left-right conditional visual or auditory discrimination, based on temporal frequency. A replication of this finding is shown in Fig. 9.7. In terms of both errors and sessions to criterion, the DNAB-lesioned group took longer to reach successively more stringent criteria of learning, although, in cumulative terms this only became significant when an 85 per cent correct criterion was used. This result is interesting for several reasons. First, it is one of the few demonstrations of a learning deficit in an appetitive paradigm with DNAB lesions, but it is not clear exactly what aspect of the task is responsible for this. It could, for example, depend on the fact that it is a conditional, rather than simple discrimination. Such discriminations may depend far more readily upon processes of response selection than simple Pavlovian conditioning situations, and may depend specifically upon frontal cortical regions (e.g. Petrides 1985). Secondly, it is once again of interest that the DNAB deficit is only apparent in the acquisition stage, because it is not observed when the rats are pretrained to the 85 per cent correct criterion. These results tentatively suggest that the locus coeruleus is specially engaged during learning rather than performance of both complex appetitive and aversive tasks, and may reflect the possibility that the type of process affected by the DNAB lesion is particularly important in the dynamic, novel, and relatively unpredictable circumstances of learning. If one assumes that learning situations are more stressful than ones in which well-established routines are available, then the learning deficits for DNAB-lesioned rats described here are compatible with the evidence described earlier.

The hypothesis of DNAB function described above is relatively novel and fits with much of the neurobiological data. Amaral and Sinnamon (1977) advanced a similar idea when they suggested that the coeruleal NA projection preserved normal physiological and behavioural function during stress. The main differences from an earlier formulation, of an involvement in selective attention, is that we propose no special role in attention *per se*, as the influence of this system may impinge upon other processes, and also that we specify the types of circumstance when deficits will be observed following DNAB lesions. Thus, we are not surprised that reports of impairments in paradigms thought to measure selective attention in animals, such as latent inhibition and non-reversal-shift (Mason and Lin 1980) have not been reliably replicated (Tsaltas *et al.* 1984: Pisa and Fibiger 1983*b*) as there is no reason to postulate that such functions will necessarily be affected. The important point to grasp is that we do not see the locus coeruleus as mediating cortical arousal, rather its noradrenergic projections become active under conditions of cortical arousal and this action helps to preserve discriminative selectivity of responding. In simple terms, it may be part of a mechanism which helps us to maintain concentration and efficient cognitive function under stress and which may have its experimental counterpart in the data reviewed by Easterbrook (1959) (see also Eysenck 1982) suggesting that high levels of arousal normally narrow the focus of attention.

We have discussed the complex issues related to the interpretation of these lesion effects elsewhere (Robbins *et al.* 1986). It has proven difficult to relate the data to results of experiments on human subjects because of the limited amount of data available from the rather narrow range of options open to human psychopharmacology in exploring the functions of the central noradrenergic system. Frith *et al.* (1985) and Clark *et al.* (1986) have recently reported results suggesting that the alpha-2 agonist, clonidine, impairs learning in man. Mair and McEntee (1983) have found similar results for normal human subjects on measures of both memory and attention, although Korsakoff patients show a surprising boost in cognitive function in response to the same doses of the drug. These latter results have recently been paralleled by the finding of improved performance in a delayed alternation procedure in aged monkeys (Arnsten and Goldman-Rakic 1985; Goldman-Rakic, this volume) following low doses of clonidine. These authors explained their results in terms of a post-synaptic alpha-2 effect superimposed upon deficient presynaptic NA activity, and this would be compatible with the position we have outlined. However, both Frith *et al.* (1985) and Clark *et al.* (1986) also see their results as compatible with a positive NA involvement in cognitive function by emphasizing the presynaptic, sedative effects of clonidine. Clearly, we need to study how such drugs affect central NA function in more detail, concentrating on their facilitatory effects, in order to extend our hypothesis to encompass these data.

The cortical cholinergic systems and cortical arousal

There seems to be little doubt that the central cholinergic systems are involved in cognitive function, but the difficulty has been in deciding what the nature of this involvement might be. The availability of rather specific muscarinic and nicotinic agonists (such as arecoline and nicotine, respectively) and antagonists (such as the anti-muscarinics scopolamine and atropine) as well as acetylcholinesterase inhibitors (e.g. physostigmine) has provided a battery of tools for exploring the functions of the central cholinergic systems in man as well as other animals, given that their peripheral effects can be controlled.

The problem has been that the effects of say, scopolamine, have been experimentally inconsistent and theoretically controversial, some authors preferring to emphasize its amnesic effects (e.g. Kopelman 1985) while others have pointed out its effects on basic attentional processes (Sahakian 1987). The facilitatory effects of cholinergic drugs upon cognitive processes, although real and clearly not to be ignored, are also fragile, at least in the case of muscarinic drugs and physostigmine (Sahakian 1987), though perhaps more robust in certain aspects of human information processing (Warburton, this volume). Whereas in principle, greater specificity and size of effects might be expected from studies which attempt to pinpoint their neural location, progress has been slow because of the lack of availability of a specific cholinergic neurotoxin that can be used in behavioural studies in animals. This section, therefore, cannot aspire to the formulation of detailed hypotheses about central cholinergic function, but offers some directions towards this end, based on a necessarily selective view of the literature.

It is now clear that scopolamine does not appear specifically to antagonize inhibited responses (as had been earlier suggested by Carlton 1963) or to produce time-dependent declines in short-term memory. For example, a recent, elegant experiment by Spencer, *et al.* (1985) has shown that the drug produces a dose-dependent decline in retention on a continuous non-matching to sample schedule which is independent of temporal delay between the sample and matching stimulus. Its capacity to disrupt the correction procedure suggested that scopolamine may not only affect stimulus discrimination *per se* [as suggested by Warburton and Brown's (1972) signal detection studies in rats], but also the retrieval of response rules from reference memory. This conclusion is interesting from the perspective of the well-known short-term memory impairments shown by patients with Alzheimer's disease (see for example, the chapter by Morris *et al.* this volume) and the cholinergic hypothesis of this disorder (Sahakian 1987). Lesions of the cholinergic cells of the nucleus basalis in the ventral pallidum/substantia innominata in experimental animals using the excitotoxin ibotenic acid similarly do not appear to produce delay-dependent declines in memory in either simple matching-to-sample or non-matching-to-sample procedures (Dunnett 1985; R.E.

Etherington and T.W. Robbins, unpublished results). Again, basic deficits in discrimination are observed which are consistent with the results of several other studies in the rat (e.g. Murray and Fibiger 1986) and in primates such as the marmoset (Ridley *et al.* 1986). However, these conclusions must be tempered by doubts about the cholinergic specificity of ibotenic acid lesions of the ventral pallidum/substantia innominata, although both of the latter studies provided pharmacological evidence to support the disruption of cholinergic mechanisms.

An important issue is the comparison of the relative contributions of the cortical cholinergic and noradrenergic systems to efficient cortical processing. The effects we have described of (presumed) cortical cholinergic manipulations are still consistent with the idea that cholinergic activity in the cortex enhances S/N ratios (cf. Drachman and Sahakian 1979) leading to accurate discrimination between different responses and retrieval of response rules appropriate to a particular situation. However, it is unclear how such

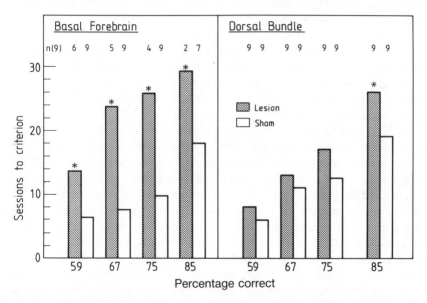

FIG. 9.8. Comparative effects of ibotenic acid lesions of the basal forebrain, aimed to destroy cells of cholinergic cells of the nucleus basalis of Meynert, and 6-OHDA lesions of the DNAB on learning of a conditional appetitive discrimination, at several different criteria of acquisition. Fifty-nine per cent is just better than chance. Eighty-five per cent correct generally represents asymptotic performance. The numbers at the top refer to number of animals progressing to that stage of acquisition within the 30 test sessions. The asterisks represent significant differences, compared with the sham operated controls. Errors to criterion provides a similar set of results. [See Everitt *et al.* (1983) for further experimental details. Unpublished observations of Robbins, Salamone and Everitt.]

effects would differ from those produced by DNAB lesions, except in the case of the cholinergic system where we have no reason, as yet, to believe that the effects are stress-dependent. One comparison is shown in Fig. 9.8. Ibotenic acid lesions of the basal forebrain profoundly impaired the acquisition of the conditional discrimination task described earlier, the effects clearly being much greater than in the case of DNAB lesions (Robbins, Salamone and Everitt unpublished results). These effects have been replicated with controls for the neuroanatomical specificity of the nbM lesions, (Everitt *et al*. 1987), but it has also been shown by the same authors that there are severe deficits on performance of the discrimination if the animal is pre-trained to criterion prior to surgery, unlike the effects of cortical NA depletion. Thus, cortical ACh may be involved in memory retrieval to a much greater extent than cortical NA.

Deutsch (1983) also suggested that the ACh system was involved in memory consolidation or retrieval, but his hypothesis had some novel aspects which are relevant to the notion that ACh affects memory by modulating cortical activity via some non-specific process, perhaps akin to arousal. What interested Deutsch particularly was the fact that the effect of cholinergic manipulations seemed to depend upon the age (and hence the strength) of the memory trace. Thus, paradoxically, physostigmine would improve the discrimination of rats in a Y-maze choice task if injected after a lengthy training-delay interval, but would impair performance if injected, at the same dose, after much shorter intervals.

These results have been essentially replicated and extended in a fine series of studies by Stanes *et al*. (1976), the results of which are shown in Fig. 9.9. These results are important in defining the very narrow range of conditions under which cholinergic drugs facilitate performance and may be relevant to understanding the marginal effects seen in man. The results are also interesting in the context of a Yerkes–Dodson interpretation (see above), as applied to memory trace retrieval. Thus, 'strong' traces are already likely to be retrieved with close to optimum efficiency and so are capable of little further improvement. Weak traces, by contrast could readily exhibit improvement if their S/N ratios were improved. Why, then, should strong traces suffer impaired retrieval under conditions in which weak ones are strengthened? One possible explanation would be that the strengthening of weak traces has the deleterious effect of incidentally interfering with the retrieval of strong ones. This 'competition' between traces is analogous to that postulated to occur between responses under DA activation. It would indeed be of interest if similar principles could be applied to the effects of different forms of 'arousal' mediated by different transmitter systems on different types of processing in forebrain regions.

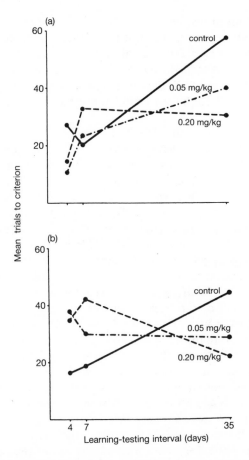

FIG. 9.9.   Effects of two doses of physostigmine on maze retention by (a) slow and (b) fast learners at different learning-testing intervals. The drug was administered 30 minutes prior to retention testing and the numbers of trials taken to re-attain criterion performance were measured. Note that the direction of effect of the drug (i.e. facilitation or impairment) depends on the age (and hence probably strength), of the memory trace. [Reproduced from Stanes *et al.* (1976) with permission from Springer Verlag.]

CONCLUSIONS

We have pursued some parallels between the effects of stressors, either alone or in combination, with the effects of drugs on human performance, mainly on different aspects of attention. The parallel between stressors and drugs may stretch to more than a simple analogy, as it seems that they may have overlapping neurochemical effects. In a sense, stressors may be akin to rather non-specific drugs. The effects of certain stressors, such as white noise, are more complex than drugs, because they have exteroceptive, informational properties, in addition to a pattern of non-specific interoceptive effects. A stressor may also affect a range of different neurotransmitter systems, as compared to pharmacologically more specific drugs. Because of this greater specificity and the ease of administering graded doses, drugs may be more useful research tools for investigating the effects of altered arousal on attention.

In attempting to explain the effects of both drugs and stressors, whether alone, or in combination, we have found energetic constructs such as arousal useful, but only in theoretical schemes which allow more than one type of arousal process. This fractionation of the arousal construct is consistent with neurobiological evidence that the old ascending reticular activating system has many components, including the ascending noradrenergic, dopaminergic, serotoninergic and cholinergic systems. From the evidence we have reviewed, the action of these systems is likely to be non-specific, in the sense of not providing the highly patterned spatio-temporal information required by forebrain computational systems. However, activity in these systems appears necessary for the efficiency of such computations. We have provided evidence that certain of the transmitter systems appear to mediate rather different aspects of arousal. For example, we identified a role for forebrain DA-ergic projections to the striatum in response activation, which facilitates the speed and vigour of responding, whereas we postulated that the coeruleo-cortical NA system was engaged under conditions of elevated arousal or activation to preserve attentional selectivity (perhaps operating at the level of response preparation and selection) and learning capacity.

It is important to realize that these transmitter-defined, subcortical systems do not play a passive enabling role in the support of complex forebrain processing. Thus, for example, the functions of the coeruleo-cortical NA system may be best observed, and manipulations of its activity most easily observed, when it is brought into play, as we have postulated may be the case during 'stress'. An important question arises as to whether the DA-ergic, NA-ergic, and cholinergic systems actually have similar functions, but different effects because of their different pattern of forebrain terminations. Thus, the role of DA in response activation might then simply reflect the importance of the rich DA-ergic innervation of the basal ganglia. The answer to this key question may depend upon an analysis of different manipulations of these systems where they anatomically overlap, as in the case of the prefrontal cortex (see Morrison and Magistretti 1983). However, it is clearly important to be able to specify whether the consequences of NA or ACh depletion, for example, in the neocortex, differ, perhaps as a function of environmental circumstances, such as the presence of stressors.

Figure 9.10 illustrates in schematic and highly speculative form, the type of contribution that each of the major systems we have discussed could poten-

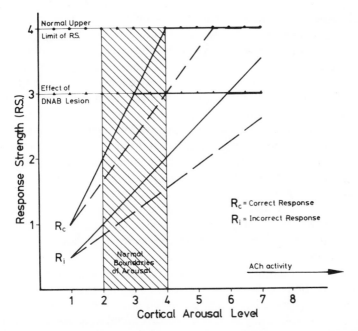

FIG. 9.10. Speculative scheme for integrating some of the results described following manipulations of the cortical noradrenergic and cholinergic systems. Cortical arousal (abscissa) is measured in arbitrary units. Response strength (ordinate) is also measured in arbitrary units and refers to the hypothetical values or weightings attached to particular responses, or memory traces, which are compared and used by a computational mechanism for response selection (see text for further description).

tially contribute to mechanisms of selective attention, each by exerting a somewhat non-specific, but separate influence, upon response choice. It was stimulated by the theory of Broen and Storms (1961), concerning the effects of arousal of response selection, in the specific context of schizophrenia. Evenden (1983) has developed this model for explaining drug effects, and we have further modified the scheme here. The diagram also helps to summarize some of the results we have described above. On the ordinate is represented the tendency to respond, or response strength (R.S.) which we assume has a normal upper limit. This hypothetical value represents the weighting or bias allocated to a response, perhaps on the basis of its association with reward, which is used by a response selection mechanism to compute which response should be initiated and executed. Thus, it is more relevant to choice accuracy than response vigour.

Shown in the diagram are two sets of correct and incorrect responses, which vary in strength as a function of some hypothetical level of 'cortical arousal', along the abscissa, and which also vary in slope, so that the solid lines, representing one set of correct and incorrect responses, both increase at an equivalent, steeper rate than the dashed lines, which represent another set. The different slopes (1/2 for the solid lines and 1/3 for the dashed lines) model motivational influences. These response tendencies could represent, for example, response choice in a discrimination task, or memory trace retrieval, and their performance is assumed to depend upon some function of their numerical difference in response strength. By imposing a hypothetical ceiling on response strength, it is evident that, as cortical arousal increases from a low level, discrimination will be first improved, and then impaired.

Here we restrict ourselves to a consideration of the separate contributions to response choice by noradrenergic and cholinergic activity which can be distinguished within the diagram. Thus, for example, assume that cortical arousal varies monotonically with cholinergic activity. Boundaries of arousal within which an individual normally operates are cross-hatched. If the subject is at the upper boundary of normal arousal (at the arbitrary value of 4 on the abscissa) and discriminating between responses represented by the solid lines, it is evident that further increases in cholinergic activity (for example, following physostigmine) will impair discrimination. However, if the subject is at the lower boundary (2) then the same dose will improve discrimination. For weaker response strengths (e.g. dashed lines) the facilitatory effects of higher doses of physostigmine will be correspondingly greater, (as can be observed from actual data shown in Fig. 9.9).

Some of the effects of DNAB lesions can be represented by allowing NA activity to alter the limits on the hypothetical ceiling of response strength, which will thus affect choice accuracy, but only under certain conditions. Thus, if this ceiling is lowered following a DNAB lesion, it is evident that: (i) discrimination at relatively low levels of cortical arousal will be left unaffected; (ii) discrimination at relatively higher levels will not show much benefit of increasing arousal level, and in fact, will begin to show impairments instead. The fact that DNAB lesions affect acquisition of behaviour more deleteriously than previously established performance could perhaps be incorporated by assuming that training elevates the hypothetical response strength ceiling, although another way of explaining the effect would postulate that arousal is higher during acquisition, rather than performance.

The effects of a motivational influence in Fig. 9.10 can be modelled by the gradient of the slopes, as mentioned above. Thus, the gain-amplifying influence on response strength of enhanced motivation (produced, for example, by using the cold water rather than the warm water maze) will result in the response strength ceiling being reached more quickly, and hence in the case of DNAB lesions, impairment rather than improvement in discrimination.

An important point here concerns the long-suggested role for DA in incentive-motivational processes (see Robbins and Everitt 1982) and its effect on both choice accuracy (modelled in Fig. 9.10 by differences in response strength) and response speed or vigour. We argued earlier that DA mediates some of the activational influence that results in changes in response vigour, but a detailed consideration of the possibly complementary roles of dorsal and ventral striatal DA systems is not represented in Fig. 9.10, and is beyond our present scope.

We emphasize that this type of modelling is often salutary for clarifying notions about the possible role of the various transmitter systems rather than necessarily bearing any relation to reality. One shortcoming of the scheme is that it does not comfortably incorporate our distinction between speed or vigour of responding, and choice accuracy. From this point of view, therefore, it is evident that this model may apply more readily to the 'Upper' type of mechanism shown in Fig. 9.3, as changes in response vigour were held there to result from altered activity of a 'Lower', possibly, DA-modulated mechanism. However, Fig. 9.10 does depict how NA-ergic and cholinergic influences could exert interactive, but different influences upon discrimination (including processes as diverse as action selection and memory trace retrieval).

It is tempting to take the speculation into neuropsychological domains. The 'so-called' 'Upper' mechanism has clear affinities to the types of psychological process called 'controlled' (as opposed to automatic) and 'effortful' (see Kahneman and Treisman 1984, and above). In what way could this type of processing be distributed in the brain? One possibility is that it depends on regions such as the pre-frontal cortex (c.f. Norman and Shallice 1980). It is then of considerable interest that this area not only has important cortico-cortical connections (see Goldman-Rakic, this volume), but is apparently one of the few neocortical regions to be able to influence activity of NA cells in the locus coeruleus, 5-HT cells in the raphe, and cholinergic cells of the nbM (Arnsten and Goldman-Rakic 1984; Mesulam and Mufson 1984). Thus, the prefrontal cortex can regulate the separate influences of each of these subcortical inputs to the entire neocortical mantle.

Despite our ability to define some dissociable effects of specific manipulations of the central neurotransmitter systems in animal experiments, a critical question would be: how comparable are the results of human and animal studies? A few positive answers have emerged from our review, but we now take the opportunity to mention some more. The tendency for amphetamine and scopolamine in combination to produce a potentiation of cognitive disturbance in human subjects (Drachman and Sahakian 1979) has a parallel in a study by Glick and Jarvik (1969) of delayed matching to sample in monkeys. Thus, the two drugs had a synergistic, disruptive effect on response accuracy, but mutually antagonistic effects on response rate.

Considering the animal analogue of the Leonard five-choice task, there have been no directly comparable studies of human subjects with altered central NA-ergic activity. However, it is of considerable interest, that in a task, impairments were linked, not with altered DA-ergic activity, but correlated instead with alterations in NA turnover, as indexed by changes in the NA metabolite MHPG in the CSF (Stern *et al*. 1985). The work of Clark *et al*. (1986), comparing the effects of methylphenidate and clonidine is similarly germane. They found that clonidine impaired tasks involving both divided and focused attention, whereas methylphenidate had little effect, except to produce changes in elation and response rate. Clark *et al*. (1986) hypothesized that the subjects are exerting effort to prevent deleterious effects of the drug. In our five-choice task (which, in its basic form, is probably measuring divided attention), intra-accumbens amphetamine similarly enhanced response output, but had no effect on discrimination in normal rats. However, this capacity to 'cope' with the enhanced rate was impaired in rats with DNAB lesions, which exhibited some disruption of discriminative control.

The work and ideas described also have some clinical implications. To take two examples. First, it is unclear whether pharmacological attempts to treat disorders such as Alzheimer's disease in its later stages will be successful. The relatively non-specific facilitatory action of drugs affecting the cortical monoamine or cholinergic projections may be ineffective, if operating on absent, or severely disrupted neuronal activity resulting from cell loss or disruption of cortical circuitry. If the disease is not far progressed, then it is possible that some benefits on cognitive function will result, but these will be constrained by the types of consideration emerging from experimental work on the interactions of drugs with arousal, and summarized in Fig. 9.9. Secondly, in terms of rehabilitation, the exacerbation of cognitive deterioration seen in many elderly, dementing subjects following the stress of institutionalization (see Jorm 1986) may result from the loss of a coping capacity reflected by degeneration of central NA-ergic neurons. Further work is required to establish under what circumstances deficits in other transmitter systems may be maximally expressed.

ACKNOWLEDGEMENTS

The published and unpublished research reported here was supported by an MRC Project Grant, and by a grant from the Parkinson's Disease Society. We thank Dr J.L. Evenden for discussion, B.J. Cole for permission to refer to her unpublished findings, H. Marston for figure preparation, and J. Hulyer for photographic assistance.

REFERENCES

Amaral, D.G. and Sinnamon, H.M. (1977). The locus coeruleus: neurobiology of a central noradrenergic nucleus. *Progress in Neurobiology* **9**, 147–96.

Anlezark, G.M., Crow T.J., and Greenway, A.P. (1973). Impaired learning and decreased cortical norepinephrine after bilateral locus coeruleus lesions. *Science* **181**, 682–4.

Arnsten, A.F.T. and Goldman-Rakic, P.S. (1984). Selective prefrontal cortical projections to the region of the locus coeruleus and raphé nuclei in the rhesus monkey. *Brain Research* **306**, 9–18.

—— and —— (1985). α-2-adrenergic mechanisms in prefrontal cortex associated with cognitive decline in aged non-human primates. *Science* **230**, 1273–6.

Aston-Jones, G. and Bloom, F.E. (1981). Norepinephrine-containing neurons in behaving rats exhibit pronounced responses to non-noxious environmental stimuli. *Journal of Neuroscience* **1**, 887–90.

Atkinson, R.C. and Shiffrin, R.M. (1971). The control of short term memory. *Scientific American* **225**, 82–90.

Baddeley, A.D. (1983). Working memory. *Philosophical Transactions of the Royal Society.* **B302**, 311–24.

Bannon, M.J. and Roth, R.H. (1983). Pharmacology of mesocortical dopamine. *Pharmacological Reviews* **35**, 53–67.

Bloom, F.E. (1979). Is there a neurotransmitter code in the brain? In *Neurotransmitters. Advances in Pharmacology and Therapeutics, Vol. 2* (ed. P. Simon), pp. 205–13. Pergamon Press, Oxford and New York.

Bloxham, C.A., Mindel, T.A., and Frith, C.D. (1984). Initiation and execution of predictable and unpredictable movements in Parkinson's disease. *Brain* **107**, 371–84.

Broadbent, D.E. (1958). *Perception and Communication.* Pergamon Press, London.

—— (1971). *Decision and Stress.* Academic Press, London.

Broen, W.E. and Storms, L.H. (1961). A reaction potential ceiling and response decrements in complex situations. *Psychological Review* **68**, 405–15.

Carli, M., Robbins, T.W., Evenden, J., and Everitt, B.J. (1983). Effects of lesions to ascending noradrenergic neurones on performance of a 5-choice serial reaction task in rats: implications for theories of dorsal noradrenergic function based on selective attention and arousal. *Behavioural Brain Research* **9**, 361–80.

——, Evenden, J., and Robbins, T.W. (1985). Depletion of unilateral striatal dopamine impairs initiation of contralateral actions and not sensory attention. *Nature* **313**, 679–82.

Carlton, P. (1963). Cholinergic mechanisms in the control of behaviour by the brain. *Psychological Review* **70**, 19–39.

Carr, G.D. and White, N.M. (1984). The relationship between stereotypy and memory improvements produced by amphetamine. *Psychopharmacology* **82**, 203–9.

Clark, C.R., Geffen, G.M., and Geffen, L.B. (1986). Role of monoamine pathways in attention and effort: effects of clonidine and methylphenidate in normal adult humans. *Psychopharmacology* **90**, 3–9.

Cole, B.J. and Robbins, T.W. (1987*a*). Effects of *d*-amphetamine in a five-choice serial reaction task in DNAB-lesioned and sham control rats: new evidence for dopamine-noradrenaline interactions. *Psychopharmacology* **91**, 458–66.

—— and —— (1987*b*). Dissociable effects of cortical and hypothalamic noradrenaline on the acquisition, performance and extinction of aversive conditioning. *Behavioral Neuroscience* (in press).

Crow, T.J. and Wendlandt, S. (1976). Impaired acquisition of a passive avoidance response after lesions of the locus coeruleus. *Nature (London)* **259**, 42–4.

Deutsch, J.A. (1983). The cholinergic synapse and the site of memory. In *The Physiological Basis of Memory*, 2nd edn (ed. J.A. Deutsch), pp. 367–86. Academic Press, London.

Drachman, D.R. and Sahakian, B.J. (1979). The effects of cholinergic agents on human learning and memory. In *Nutrition and the Brain*, Vol. 5 (ed. A. Barbeau, J.H. Growdon, and R. Wurtman), pp. 351–66. Raven Press, New York.

Dunnett, S.B. (1985). Comparative effects of cholinergic drugs and lesions of the nucleus basalis or fimbria fornix on delayed matching in rats. *Psychopharmacology* **87**, 357–63.

Easterbrook, J.A. (1959). The effect of emotion on cue utilization and the organization of behaviour. *Psychological Review* **66**, 183–201.

Edwardson, J.A., Bloxham, C.A., Candy, J.M., Oakley, A.E., Perry, R.H., and Perry, E.K. (1985). Alzheimer's disease and Parkinson's disease: pathological and biochemical changes associated with dementia. In *Psychopharmacology: Recent Advances and Future Prospects* (ed. S.D. Iversen), pp. 131–45. Oxford University Press, Oxford.

Evenden, J.L. (1983). A behavioural and pharmacological analysis of response selection. Unpublished Ph.D. thesis, University of Cambridge.

Everitt, B.J., Robbins, T.W., Gaskin, M., and Fray, P.J. (1983). The effects of lesions to ascending noradrenergic neurones on discrimination learning and performance in the rat. *Neuroscience* **10**, 397–410.

——, Robbins, T.W., Evenden, J.L., Marston, H.M., Jones, G.H., and Sirkiä, T. (1987). The effects of excitotoxic lesions of the substantia innominata, ventral and dorsal globus pallidus on the acquisition and retention of a conditioned visual discrimination: implications for cholinergic hypotheses of learning and memory. *Neuroscience* (in press).

Eysenck, M.W. (1982). *Attention and Arousal*. Springer-Verlag, Berlin.

Foote, S., Friedman, R., and Oliver, A.P. (1975). Effects of putative neurotransmitters on neuronal activity in monkey cerebral cortex. *Brain Research* **86**, 229–42.

Frith, C., Dowdy, J., Ferrier, N., and Crow, T.J. (1985). Selective impairment of paired associate learning after administration of a centrally acting adrenergic agonist (clonidine). *Psychopharmacology* **87**, 490–3.

Glick, S.D. and Jarvik, M. (1969). Amphetamine, scopolamine and chlorpromazine interactions on delayed matching performance in monkeys. *Psychopharmacologia (Berlin)* **16**, 147–55.

Hagan, J., Alpert, J.E., Iversen, S.D., and Morris, R.G.M. (1983). The effects of central catecholamine depletion on spatial learning in rats. *Behavioural Brain Research* **9**, 83–104.

Halliwell, J.V. and Adams, P.R. (1982). Voltage clamp analysis of muscarinic excitation in hippocampal neurons. *Brain Research* **250**, 71–92.

Hebb, D.O. (1955). Drives and CNS (conceptual nervous system). *Psychological Review* **62**, 243–54.

Hopkins, W.F. and Johnston, D. (1984). Frequency-dependent noradrenaline modulation of long-term potentiation in the hippocampus. *Science* **226**, 350–2.

Heffner, T., Hartman, J.A., and Seiden, L.S. (1980). Feeding increases dopamine metabolism in the rat brain. *Science* **208**, 1168–70.

Hockey, R. (1979). Stress and the cognitive components of skilled performance. In *Human Stress and Cognition* (ed. V. Hamilton and D. Warburton), pp. 141–78. Wiley, Chichester.

Hull, C. (1949) Stimulus intensity dynamism (V) and stimulus generalization. *Psychological Review* **56**, 67–76.

Jacobs, B.L. (1984). Single unit activity of brain monoaminergic neurons in freely moving animals: a brief review. In *Modulation of sensorimotor activity during alterations in behavioral states*. (ed. R. Bandler), pp. 99–120. A.R. Liss, New York.

Jorm, A. (1986). Controlled and automatic processing in senile dementia: a review. *Psychological Medicine* **16**, 77–88.

Kahneman, D. and Treisman, A. (1984). Changing views of attention and automaticity. In *Varieties of Attention* (ed. R. Parasuraman and D.R. Davies), pp. 29–61. Academic Press, Orlando.

Kelly, P.H., Seviour, P., and Iversen, S.D. (1975). Amphetamine and apomorphine responses in the rat following 6-OHDA lesions of the nucleus accumbens septi and corpus striatum. *Brain Research* **94**, 507–22.

Kesner, R.P., Bierley, R.A., and Pebbles, P. (1981). Short term memory: the role of *d*-amphetamine. *Pharmacology, Biochemistry and Behavior* **15**, 673–6.

Kopelman, M.D. (1985). Multiple memory deficits in Alzheimer-type dementia: implications for pharmacotherapy. *Psychological Medicine* **15**, 527–41.

Kramer, J.C., Fischman, V.S., and Littlefield, D.C. (1967). Amphetamine abuse. *Journal of the American Medical Association* **201**, 305–9.

Krnjevic, K. and Phillis, J.W. (1963). Acetyl choline sensitive cells in the cerebral cortex. *Journal of Physiology* **166**, 296–327.

—— and Ropert, N. (1981). Septo-hippocampal pathway modulates hippocampal activity by a cholinergic mechanism. *Canadian Journal of Physiology and Pharmacology* **59**, 911–4.

Le Moal, M., Koob, G.F., Koda, L.Y., Bloom, F.E., Manning, M., Sawyer, W.H., and Rivier, J. (1981). Vasopressor receptor antagonist prevents behavioural effects of vasopressin. *Nature (London)* **291**, 491–3.

Leonard, J.A. (1959). 5-choice serial reaction apparatus. Applied Psychology Research Unit Report No. 326/59, Medical Research Council, Cambridge.

Lindvall, O. and Bjorklund, A. (1983). Dopamine- and norepinephrine-containing neuron systems: their anatomy in the rat brain. In *Chemical Neuroanatomy* (ed. P.C. Emson), pp. 229–55. Raven Press, New York.

Lyon, M. and Robbins, T.W. (1975). The action of central nervous system stimulant drugs: a general theory concerning amphetamine effects. In *Current Developments in Psychopharmacology* Vol. 2 (ed. W. Essman and L. Valzelli), pp. 79–163. Spectrum, New York.

Mackintosh, N.J. (1983). *Conditioning and Associative Learning*. The Clarendon Press, Oxford.

Mair, R.G. and McEntee, W.J. (1983). Korsakoff's pychosis: noradrenergic systems and cognitive improvement. *Behavioural Brain Research* **9**, 1–32.

Mandler, G. (1975). *Mind and Emotion*. Wiley, New York.

Marczynski, T.J. (1978). Neurochemical mechanisms in the genesis of slow potentials: a review and some clinical implications. In *Multi-disciplinary Perspectives in Event-related Brain Potential Research* (ed. D.A. Otto), pp. 25–35. US Government Printing Office, Washington.

Marshall, J.F., Berrios, N.G., and Sawyer, S. (1980) Neostriatal dopamine and sensory inattention. *Journal of Comparative and Physiological Psychology* **94**, 833–46.

Marshall, J. and Teitelbaum, P. (1977). New considerations in the neuropsychology of motivated behavior. In *Handbook of Psychopharmacology*, Vol. 7 (ed. L.L.

Iversen, S.D.Iversen and S.H. Synder), pp. 201-29. Plenum Press, New York.

Mason, S.T. and Iversen, S.D. (1979). Theories of dorsal bundle extinction effect. *Brain Research Reviews* **1**, 107-37.

Mason, S. and Lin, D. (1980). Dorsal noradrenergic bundle and selective attention. *Journal of Comparative and Physiological Psychology* **94**, 819-32.

Mesulam, M-M and Mufson, E.J. (1984). Neural inputs into the nucleus basalis of the substantia innominata (Ch4) in the rhesus monkey. *Brain* **107**, 253-74.

——, ——, Levey, A.I., and Wainer, B.H. (1983). Cholinergic innervation of cortex by the basal forebrain: cytochemistry and cortical connections of the septal area, diagonal band nuclei, nucleus basalis (substantia innominata) and hypothalamus in the rhesus monkey. *Journal of Comparative Neurology* **214**, 170-97.

Morris, R.G.M., Garrud, P., Rawlins, J.N.P., and O'Keefe, J. (1982). Place navigation impaired in rats with hippocampal lesions. *Nature (London)* **297**, 681-3.

Morrison, J.H. and Magistretti, P.J. (1983). Monoamines and peptides in cerebral cortex. *Trends in Neuroscience* **6**, 146-51.

Murray, C.L. and Fibiger, H.C. (1986). Pilocarpine and physostigmine attenuate spatial memory impairments produced by lesions of the nucleus basalis magnocellularis. *Behavioral Neuroscience* **100**, 23-32.

Norman, D.A. and Shallice, T. (1980). Attention to Action: Willed and automatic control of behavior. *Center for Information Processing*, Technical Report No. 99, University of California, San Diego.

Owen, F., Crawley, J., Cross, A.J., Crow, T.J., Oldland, S.R., Poulter, M., Veall, N., and Zanelli, G.D. (1985). Dopamine D2 receptors and schizophrenia. In *Psychopharmacology; Recent Advances and Future Prospects* (ed. by S.D. Iversen), pp. 216-27. Oxford University Press, Oxford.

Petrides, M. (1985). Deficits on conditional associative-learning tasks after frontal- and temporal-lobe lesions in man. *Neuropsychologia* **23**, 601-14.

Picton, T.W., Campbell, K.B., Baribeau-Braun, J., and Proulx, G.B. (1978). The neurophysiology of attention: a tutorial review. In *Attention and Performance VII* (ed. J. Requin), pp. 429-67. Erlbaum, Hillsdale.

Pisa, M., and Fibiger, H.C. (1983*a*). Evidence against a role of the rat's dorsal noradrenergic bundle in attention and place memory. *Brain Research* **272**, 319-29.

—— and —— (1983*b*). Intact selective attention in rats with lesions of the dorsal noradrenergic bundle. *Behavioral Neuroscience* **97**, 519-29.

Posner, M. (1978). *Chronometric Explorations of Mind*. Erlbaum, Hillsdale.

—— and Snyder C.R.R. (1975). Attention and cognitive control. In *Information Processing and Cognition: the Loyola Symposium* (ed. R.L. Solo). Erlbaum, Hillsdale.

Rafal, R.D., Posner, M.I., Walker, J.A., and Friedrich, F.J. (1984). Cognition and the basal ganglia. *Brain* **107**, 1083-94.

Ridley, R.M., Murray, T.K., Johnson, J.A., and Baker, H.F. (1986). Learning impairment following lesion of the basal nucleus of Meynert in the marmoset: modification by cholinergic drugs. *Brain Research* **376**, 108-16.

Robbins, T.W. (1984). Cortical noradrenaline, attention and arousal. *Psychological Medicine* **14**, 13-21.

—— (1986). Psychopharmacological and neurobiological aspects of the energetics of human information processing. In *Energetic Aspects of Human Information*

Processing (ed. R. Hockey, A. Gaillard, and M. Coles). Martinus Nijhoff, Dordrecht, The Netherlands.

—— and Everitt, B.J. (1982). Functional studies of the central catecholamines. *International Review of Neurobiology* **23**, 303–65.

——, ——, Fray, P.J., Gaskin, M., Carli, M., and de la Riva, C. (1982). The roles of the central catecholamines in attention and learning. In *Behavioural Models and the Analysis of Drug Action* (ed. M.Y. Spiegelstein and A. Levy), pp. 109–14. Elsevier, Amsterdam.

——, ——, Cole, B.J., Archer, T. and Mohammed, A. (1985). Functional hypotheses of the coeruleo-cortical noradrenergic projection: a review of recent experimentation and theory. *Physiological Psychology* **13**, 127–50.

Rolls, E.T., Thorpe, S.J., Boytim, M., Szabo, I., and Perrett, D.I. (1984). Responses of striatal neurons in the behaving monkey. 3. Effects of iontophoretically applied dopamine on normal responsiveness. *Neuroscience* **12**, 1201–12.

Sahakian, B.J. (1987). Cholinergic drugs and their effects on human cognitive function. In *Handbook of Psychopharmacology*, vol. 20 (ed. L. Iversen, S.D. Iversen, and S.H. Synder). Plenum Press, New York (in press).

Schwab, R.S. (1972). Akinesia paradoxica. *Electroencephalography and Clinical Neurophysiology* (Suppl.) **31**, 87–92.

Segal, M. and Bloom, F.E. (1976). The action of norepinephrine in the rat hippocampus IV. The effects of locus coeruleus stimulation on evoked hippocampal activity. *Brain Research* **107**, 513–25.

Spencer, D.G., Pontecorvo, M.J., and Heise, G.A. (1985). Central cholinergic involvement in working memory: effects of scopolamine on continuous non-matching and discrimination performance in the rat. *Behavioural Neuroscience* **99**, 1049–65.

Squire, L.R. (1982). The neuropsychology of human memory. *Annual Review of Neuroscience* **5**, 241–73.

Stanes, M.D., Brown, C.P., and Singer, G. (1976). Effect of physostigmine on Y-maze retention in the rat. *Psychopharmacologia* **46**, 269–76.

Stern, Y., Mayeux, R., and Côté, L. (1985). Reaction time and vigilance in Parkinson's disease: possible role of noreprinephrine metabolism. *Archives of Neurology* **41**, 1086–9.

Stone, T.W. (1972). Cholinergic mechanisms in the rat somatosensory cortex. *Journal of Physiology* **225**, 485–99.

Treisman, A. and Gelade, G.A. (1980). A feature integration theory of attention. *Cognitive Psychology* **12**, 97–136.

Tsaltas, E., Preston, G.C., Rawlins, J.N.P., Winocur, G., and Gray, J.A. (1984). Dorsal bundle lesions do not affect latent inhibition. *Psychopharmacology* **84**, 549–55.

Ungerstedt, U. (1971). Striatal dopamine release after amphetamine or nerve degeneration revealed by rotational behaviour. *Acta Physiologica Scandinavica Supplementum* **367**, 49–68.

van Praag, H.M. (1982). Neurotransmitters and CNS disease. *Lancet* ii, 1259–64.

Warburton, D.M. and Brown, K. (1972). The facilitation of discrimination performance by physostigmine sulphate. *Psychopharmacologia* (*Berlin*) **27**, 275–84.

Watabe, K., Nakai, K., and Kasamatsu, T. (1982). Visual afferents to norepinephrine-containing neurones in cat locus coeruleus. *Experimental Brain Research* **48**, 66–80.

Wilkinson, R.T. (1961). Interaction of lack of sleep with knowledge of results, repeated testing and individual differences. *Journal of Experimental Psychology* **62**, 263–71.

—— (1963). Interaction of noise with knowledge of results and sleep deprivation. *Journal of Experimental Psychology* **66**, 332–7.

—— and Colquhoun, W.P. (1968). Interaction of alcohol with incentive and with sleep deprivation. *Journal of Experimental Psychology* **76**, 623–9.

Yerkes R.M. and Dodson, J.D. (1908). The relation of strength of stimulus to the rapidity of habit formation. *Journal of Comparative and Neurological Psychology* **18**, 459–82.

10

The neuropsychology of emotion and personality

JEFFREY A. GRAY

Most conditions in psychiatry can be regarded either from the perspective of the so-called 'medical model' or as a point in a multi-dimensional personality space (Eysenck 1960). In the former case, one looks for a specific aetiology (e.g. an infectious agent or a brain lesion) that causes a specific dysfunction in behaviour and, inferentially, in the brain systems that control behaviour. In the latter, one supposes that individuals who manifest the condition in question are simply those who occupy an extreme position along one or more smoothly varying distributions of behavioural propensities known as personality 'traits' or 'dimensions'. However, such personality dimensions must themselves reflect individual differences in the functioning of the brain systems that control behaviour (Gray 1968). Thus, whether one adopts the medical model or the framework of personality dimensions, one must in the end ask what kind of brain activity (or lack of activity) gives rise to the observed behavioural phenomena. For our present purposes, then, we may ignore the long-standing controversy between proponents of the medical and personality models (usually psychiatrists and psychologists respectively). Whoever is right (and they may both be, since the same condition might arise in some cases because of, say, perinatal hypoxia and in others because of the accumulation of polygenes), we need to understand the brain systems that control the kind of behaviour involved, disturbance in which gives rise to psychiatric disorder.

In many cases (e.g. anxiety, depression, mania), the kind of behaviour concerned is clearly 'emotional'. What we require, therefore, is a neuro-psychology of emotion, that is, a description of the brain systems that control human emotional behaviour. In attempting to construct such a description, we largely lack the type of data that has buttressed similar attempts in other areas of psychology, such as cognitive or language function, that is, data relating behavioural abnormalities to abnormalities in the human brain (see, e.g. Butters, this volume); though, with the advent of non-invasive techniques of *in vivo* brain monitoring such as positron-emission tomography

171

('PET scanning'; Stahl *et al.* 1986), such data are now becoming available (see below). For the moment, therefore, the neuropsychology of human emotion rests largely upon data gathered in experiments on animals. This situation is not without its advantages. Above all, if one chooses to study emotion in animals, one is forced to define 'emotional behaviour'; whereas, with human subjects, one can too often get by with an intuitive understanding of concepts such as 'fear', 'anger', etc.

How, then, should we define 'emotional behaviour'? In general terms, this consists of behaviour elicited by primary reinforcing events (i.e. reward, punishment, frustrative non-reward, relieving non-punishment), and stimuli (secondary or conditioned reinforcers) associated with primary reinforcing events (Mowrer 1960; Amsel 1962; Gray 1972, 1975). The separate systems that control emotional behaviour can then be construed as each responding to a subset of the full set of types of reinforcing events, both primary and secondary (Gray 1972). These are the systems, then, that require analysis in terms both of their behavioural manifestations and the neuronal machinery of which they are constituted.

It is possible, I believe, to delineate and separate three such neuro-psychological systems that are important in the control of emotional behaviour (Gray 1972, 1975, 1987). What follows is a brief description of each in both behavioural and neurological terms as they have emerged from experiments with animals. Where they are known, the neurotransmitters involved are mentioned. Each system consists of a multiplicity of pathways, and since few of the relevant transmitters have as yet been identified, the leap from system to single transmitter should be undertaken with extreme caution, if at all. At the end of the chapter I shall offer some brief speculations about possible interactions between the three systems and about their relations to dimensions of normal personality.

THE APPROACH SYSTEM

To a first approximation, the approach system mediates behaviour elicited by rewards (where 'rewards' are defined as events that increase the probability of future emission of responses upon which they are made contingent). However, this statement needs three qualifications.

We must first distinguish between reinforcement and incentive motivation. Rewards provide reinforcement for the acquisition and maintenance of specific motor responses. Recent evidence suggests that, for at least some forms of response, this reinforcing effect is mediated at brainstem (Huston and Borbely 1973) and/or cerebellar (Polenchar *et al.* 1985) levels, the exact neural location varying according to the particular response the animal has learned. However, the systems involved in emotional behaviour ought plausibly to be concerned with some more general motivational function than the acquisition of specific motor responses. The most plausible candidate for such a function is that of incentive motivation. There is a wealth of evidence (Gray 1975) that, as well as reinforcing specific responses, rewards provide a general source of motivation for the performance of such responses, a source whose intensity is proportional to the size, quality, frequency, etc., of the reward and to the subject's proximity to it. It is possible that the mechanisms that mediate reinforcement and incentive motivation are closely linked; but the 'approach system' as defined here is concerned only with the latter.

The second qualification is that the term 'reward' must be understood to include 'relieving non-punishment'. Let us first define punishment as any event which, if made contingent upon a response, decreases the probability of future emission of that response. 'Relieving non-punishment' is then the non-occurrence of punishment at a time when the subject (owing to previous experience and current environmental stimuli) has some reason to expect punishment (Mowrer 1960). This event, when made contingent upon responding, gives rise to active avoidance (i.e. the acquisition of new responses), as distinct from the passive avoidance (i.e. the inhibition of existing responses) that is observed when punishment is made contingent upon behaviour. There is a substantial (though by no means overwhelming) body of evidence that active avoidance learning is mediated and motivated by the same mechanisms that mediate learning to obtain rewards (Gray 1975, 1987). Thus, the approach system is parsimoniously considered to deal with both kinds of learning: in both cases, the animal learns to approach or produce by some means the desired event (reward or non-punishment)—hence the term 'approach system'. (An alternative term, now in common use, is the 'behavioural activation system'; Fowles 1980.)

The third qualification concerns the distinction between unconditioned (primary) and conditioned (secondary) rewarding stimuli. The stimuli which elicit approach behaviour are *conditioned* stimuli which have come to be associated (probably by the process of Pavlovian conditioning) with primary reinforcers, be these rewards or non-punishment. In the latter case of non-punishment, the secondary reinforcers are often called 'safety signals' (Mowrer 1960; Gray 1987). The primary reinforcers themselves elicit a variety of fixed action patterns (e.g. eating, drinking, copulation) depending upon their particular quality. Such fixed action patterns are too specific to be of central concern in the analysis of emotional behaviour; and like the reinforcement of specific motor responses (see above), they depend upon mechanisms located at brainstem and hypothalamic levels, rather than higher brain regions. Note, however, that the approach system (as defined here) mediates the effects of conditioned rewarding stimuli once these have been established, not the conditioning process itself (though it may no doubt influence this). The basic mechanisms that give rise to Pavlovian conditioning are apparently present in neural circuits at widely differing levels both of the neuraxis and of phylogeny (Alkon and Farley 1984; Thompson *et al*. 1984; Yeo *et al*. 1985).

With these qualifications in mind, then, we may define the approach system as one that responds to conditioned rewarding (including safety) stimuli by facilitating behaviour that maximizes further exposure to such stimuli (Gray 1975).

It is more difficult to make such crisp statements when one turns to the brain. When Olds and Milner (1954) first reported that rats will electrically self-stimulate the brain, it appeared that it would be only a short time before one had constructed a detailed map of the neural machinery that mediates the behavioural effects of reward. Thirty years later is is becoming ever clearer that the interpretation of such experiments is fraught with difficulty. About the only firm statement that can at present be made is that the neural machinery that corresponds to the approach system as defined above is likely to include the dopaminergic projections (1) from nucleus A 10 in the ventral tegmentum to the ventral striatum (nucleus accumbens), and (2) from the substantia nigra to the dorsal striatum (see the discussion by Willner 1985). The former projection appears to relate more closely to incentive motivation as such; the latter, to the selection of the relevant motor programme which is to benefit from this form of motivation. Consistent with this view (and with the importance of this type of analysis for understanding behavioural disorders), there is evidence that the projection from A 10 to the nucleus accumbens plays a central role in mediating the hedonic effects of opiate and other drugs of abuse (Stewart *et al*. 1984).

THE FIGHT/FLIGHT SYSTEM

We distinguished above between conditioned and unconditioned reward. Similarly, we must distinguish between conditioned and unconditioned punishment. The fight/flight system mediates the behavioural effects only of unconditioned punishing events; it is the third of our systems (the behavioural inhibition system; see below) which mediates the effects of conditioned punishing stimuli. Moreover, just as we have taken 'reward' to include relieving non-punishment, so we must understand 'punishment' to include frustrative non-reward, i.e. the omission of reward at a time when, given previous experience and current environmental stimuli, the subject has reason to expect reward (Amsel 1962).

The behaviour that is elicited by unconditioned punishment and non-reward includes components of both aggression and escape. The type of aggression observed is that qualified as 'defensive' (Adams 1979); and the type of escape is unconditioned (e.g. running, jumping) rather than the skilled escape behaviour observed in an animal trained, say, to press a bar to terminate electric shock. Studies employing electrical stimulation and/or lesions of the brain indicate that the mechanisms that produce unconditioned escape and defensive aggression are so closely intermingled that it is often impossible to distinguish between them. For this reason, at least as a provisional tactic, it is parsimonious to postulate a single system concerned with the motivation and organization of both types of behaviour, rather than two separate systems. The detailed organization of the particular motor patterns that go to make up escape or aggressive behaviour (and different types of escape or aggressive behaviour) is likely to depend upon mechanisms in the brainstem and cerebellum. Thus, to lump together unconditioned escape and defensive aggression as functions of a single system is in some respects a similar strategy to that of lumping together different types of rewards (each eliciting its own fixed action pattern) into a single approach system; in both cases the line drawn between the presumed overarching emotional system and the more detailed motor patterns corresponds to a line between higher and lower brain centres.

The chief brain structures which instantiate the fight/flight system are the amygdala, the ventromedial hypothalamus and the central grey of the midbrain (Adams 1979; Gray 1987). These appear to be organized hierarchically: neurons in the ventromedial hypothalamus inhibit the final common pathway in the central grey, and neurons in the amygdala activate this pathway by disinhibition at the level of the hypothalamus (De Molina and Hunsperger 1962). Particularly at the level of the central grey, there is a close relationship between the fight/flight system and mechanisms that mediate the central perception of pain (Melzack and Wall 1983); these mechanisms may indeed form an integral part of the fight/flight system (Graeff, in press).

The central grey component of the fight/flight system is under three kinds of inhibition, mediated by γ-aminobutyrate (GABA), serotonin, and endorphins, respectively. Graeff (in press) has suggested that part of the clinical action of anxiolytic drugs (including benzodiazepines, barbiturates, and alchohol; Gray 1977) is mediated by facilitation of the GABAergic inhibition of central grey neurons; and others (e.g. Iversen 1983) have proposed a role for the amygdala in the action of these drugs. If these proposals are correct, it may follow that anxiolytic drug action depends upon a reduction in the activity of the fight/flight system, and that anxiety is mediated by activity in this system. However, these inferences are opposed by an important double dissociation between the behavioural effects of anti-anxiety drugs, on the one hand, and opiates on the other: the former, but not the latter, block the effects of conditioned aversive stimuli, suggesting an action on the behavioural inhibition system (see below); the latter but not the former, block the effects of unconditioned aversive stimuli, suggesting an action on the fight/flight system. The effects of opiates on fight/flight behaviour may be mediated by opiate receptors in the central grey or amygdala.

THE BEHAVIOURAL INHIBITION SYSTEM

As already indicated, the behavioural inhibition system (Gray 1976, 1982) organizes responses to conditioned aversive stimuli, whether these have been formed in association with punishment or non-reward. The reactions that occur in response to such stimuli have three aspects. First, all ongoing behaviour is brought to a halt, though this may only be for an instant; it is this effect which motivates the term 'behavioural inhibition system'. Secondly, there is increased attention to the environment and especially novel features of the environment. Thirdly, there is an increase in the level of arousal, so that the next action (which may be identical to the one interrupted by behavioural inhibition) is undertaken with increased vigour or speed. These same types of reaction are also seen in response to novel stimuli; these are accordingly also deemed to act by way of the behavioural inhibition system (though not exclusively so; Gray *et al.* 1982). All three of these 'outputs' of the behavioural inhibition system (behavioural inhibition, increased attention, increased arousal), regardless of the 'input' (novelty, stimuli associated with punishment, stimuli associated with non-reward) to which they are a response, are blocked by the anti-anxiety drugs (Gray 1977). For this reason, I have proposed that activity in the behavioural inhibition system constitutes the human state of anxiety, as well as corresponding states in animals. This proposal has recently been supported by a number of lines of evidence from patients suffering from anxiety (see below; and Gray, in press).

The major structures making up the behavioural inhibition system consist of the septal area and hippocampal formation; the cholinergic projection from the medial septal area to the hippocampus and the hippocampal projection to the lateral septal area connecting these regions; the entorhinal cortex (in the temporal lobe), which provides the main specific afferents to the septohippocampal system; the subicular cortex, which is the main route by which the septohippocampal system transmits information to other brain regions; the Papez circuit, which consists of a chain of connections from the subicular cortex to the mammillary bodies in the hypothalamus, from there to the anteroventral thalamus, thence to the cingulate cortex, and finally back to the subiculum; the prefrontal cortex, which is connected to both the cingulate and the entorhinal cortex; ascending noradrenergic fibres (from the locus coeruleus) and ascending serotonergic fibres (from the median raphe nucleus), which provide a diffuse input to the septohippocampal system; and ascending dopaminergic fibres (from A 10) to the prefrontal and cingulate cortices (Fig. 10.1). In addition, other noradrenergic projections (e.g. to the hypothalamus) from the locus coeruleus (Redmond 1979), and perhaps also from other brainstem nuclei, participate in the mediation of the 'increased arousal' output of the behavioural inhibition system; and other serotonergic

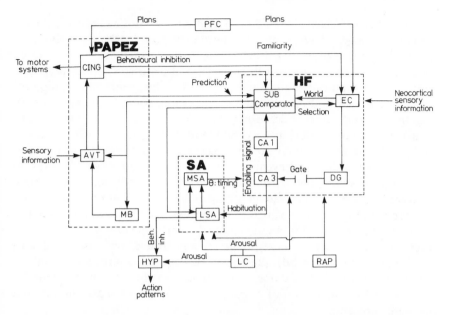

FIG. 10.1. A summary of the neurology of the behavioural inhibition system, as proposed by Gray (1982). The three major building blocks are shown in heavy print and outlined by dashed boxes: HF, the hippocampal formation, made up of the entorhinal cortex, EC, the dentate gyrus, DG, areas CA 3 and CA 1 of the hippocampus proper; SUB, the subicular area; SA, the septal area, containing the medial (MSA) and lateral septal areas (LSA); and the Papez circuit, which receives projections from and returns them to the subicular area via the mammillary bodies (MB), anteroventral thalamus (AVT), and cingulate cortex (CING). Other structures shown are the hypothalamus (HYP), the locus coeruleus (LC), the raphe nuclei (RAP), and the prefrontal cortex (PFC). Arrows show direction of projection; the projection from SUB to MSA lacks anatomical confirmation. Words in lower case show postulated functions; beh. inh., behavioural inhibition. For further explanation, see Gray (1982).

projections (e.g. to the central grey, the nucleus accumbens, and the substantia nigra) from both the median and the dorsal raphe help mediate the 'behavioural inhibition' output proper (Soubrié 1986). The major action of the anxiolytic drugs upon this ramified set of brain structures is probably due to enhanced GABAergic inhibition (Haefely *et al.* 1985) of activity in the ascending noradrenergic and serotonergic fibres (Lidbrink *et al.* 1973; Stein *et al.* 1973; Gray 1977; Redmond 1979) which innervate the septohippocampal system; this has the effect of reducing the capacity of this system to handle information entering from the entorhinal cortex (Segal 1977*a, b*).

In speculating about the information-processing ('cognitive') functions discharged by this neural system, I have proposed that it plays a general monitoring role in checking that behaviour goes 'according to plan'. To this end the system generates predictions (instant by instant, at a rate of about ten times a second) as to the next likely event. In doing so, it draws upon stored information (perhaps located in the temporal neocortex) concerning past regularities (stimulus-stimulus or response-stimulus, generated by the processes of classical and instrumental conditioning, respectively; Gray 1975), together with information describing the current state of the animal's 'world' (entering via the entorhinal cortex) and the animal's current motor programme (derived from the prefrontal cortex). The system then compares the predicted with actual events in the next instant of time. If things go according to plan (that is, actual events match prediction), the system continues to operate simply in 'checking mode', behaviour remaining under the control of other, unspecified brain mechanisms. If, however, there is a mismatch (novelty, that is, an unpredicted event or the failure of a predicted event to occur) or if the next predicted event is aversive (associated with punishment or non-reward), the system switches to 'control mode', taking control over behaviour and operating the outputs of the behavioural inhibition system (i.e. inhibition of ongoing behaviour, increased attention, and increased arousal).

These functions can be mapped onto the chief clinical phenomena of human anxiety (Gray 1982, in press) by the suppositions that phobic behaviour largely represents the behavioural inhibition output of the behavioural inhibition system when in control mode; and that obsessional symptoms reflect excessive checking when the system is in checking mode. Certain clinical phenomena, however, require the further postulate that, in Man, the ascending monoaminergic control over the septohippocampal system is supplemented by a descending control via the projection from the prefrontal to the entorhinal cortex, allowing neocortical language systems to generate anxiety to verbally coded material. The autonomic symptoms of anxiety are probably largely mediated by noradrenergic fibres descending from the locus coeruleus into the spinal cord (Redmond 1979) (but see the discussion of interactions between the fight/flight and behavioural inhibition systems below).

This is a highly condensed form of a complex theory of the neuropsychology of anxiety (for further details, the reader is referred to Gray 1982, 1987; Gray *et al.* 1982). It will be apparent that, although it deals with an emotion, it is a highly cognitive theory. There is no contradiction in this, since emotion and cognition are closely intertwined (Gray 1984). A further indication of this close connection comes from an unexpected quarter: the pathology underlying Alzheimer's disease. This is the commonest form of senile and pre-senile dementia, and it is characterized in particular by disturbances in memory and other cognitive functions. From the present point of view, therefore, it is at first disconcerting to discover that the neuropathology of Alzheimer's disease (Rossor, this volume) includes many of the structures (e.g. the hippocampal formation, the subicular and entorhinal cortices, the septohippocampal cholinergic pathway, the ascending noradrenergic and serotonergic projections to the forebrain) that have been allocated roles in the neuropsychology of anxiety. Nonetheless, recent data strengthen my confidence in the correctness of these allocations. Thus, the first PET scan of patients with panic disorders has shown them to differ from controls only in that region of the brain that contains the major input to the hippocampus (the entorhinal cortex) and its major output station (the subicular cortex) (Reiman *et al.* 1984); while, in a pharmacological study, Charney *et al.* (1984) have implicated in this same group of patients the central noradrenergic fibres whose importance in anxiety has been stressed both by the present writer (Gray 1982) and by Redmond (1979). These considerations therefore pose an interesting question for further research: what form does the relationship between emotion and cognition take, and how do disturbances in the neural systems that mediate this relationship give rise, on the one hand, to pathological anxiety and, on the other, to the cognitive disturbances of Alzheimer's disease?

INTERACTIONS BETWEEN SYSTEMS

As in the case of the individual systems outlined above, we may consider interactions between the three systems from both a behavioural and a neurological point of view.

Behaviourally, the very specification of the behavioural inhibition system requires that it should be able to inhibit activity in each of the other two systems. There is indeed a wealth of evidence that stimuli of the appropriate kind (secondary punishing, secondary frustrative, and novel stimuli) are able to inhibit approach behaviour (whether to stimuli associated with reward or to stimuli associated with non-punishment; Gray 1975, 1987). There is similar, though less abundant, evidence that the behavioural inhibition system can inhibit reactions to unconditioned punishment. In the latter case, the inhibition appears to include the sensation of pain itself. At any rate, a rat's reaction to a painful stimulus is reduced if it is simultaneously presented with a conditioned punishing stimulus, an effect that is reversed by the administration of the opiate antagonist, naloxone, implying that it is mediated by opiate receptors (Bolles and Fanselow 1980; Fanselow and Baackes 1982). In addition to these effects of the behavioural inhibition system upon the activities of the other two systems, certain phenomena observed in conflict situations (Gray and Smith 1969; Gray 1987) strongly suggest that the approach system can inhibit the behavioural inhibition system.

Neuronally, likely sites of interaction between the behavioural inhibition

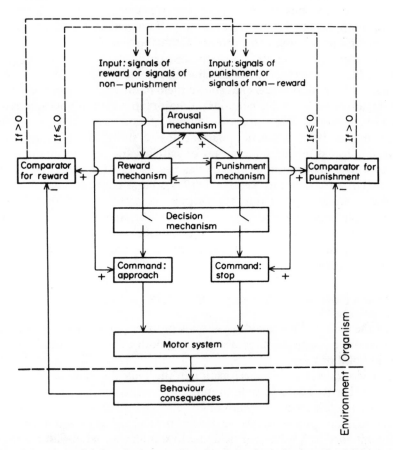

FIG. 10.2.   A model for the interactions between the behavioural inhibition and approach systems, as applied to the analysis of approach-avoidance conflict by Gray and Smith (1969). Stimuli associated with reward or the omission of anticipated punishment activate the reward mechanism to operate approach to, or production of, such stimuli; stimuli associated with punishment or the omission of anticipated reward activate the punishment mechanism to inhibit all ongoing behaviour. These mechanisms act via behaviour commands to 'approach' (on the reward side) or to 'stop' (on the punishment side). Conflict between the reward and punishment mechanisms is resolved in the decision mechanism according to whichever input to this mechanism is stronger; the reciprocally inhibitory links between the reward and punishment mechanisms ensure a stable outcome to such conflicts. The consequences, rewarding or punishing, of the behaviour that occurs are then fed back to comparator mechanisms. These compare actual to expected reward (left-hand side), or actual to expected punishment (right-hand side). Dashed lines from the comparator mechanisms indicate inputs to the reward and punishment mechanisms that occur on trial $n + 1$ as a result of classical conditioning of exteroceptive, interoceptive or pro-

and approach systems lie in the ventral striatum and the lateral septal area. As we have seen, the dopaminergic input from A 10 in the ventral tegmental area to the ventral striatum is the best candidate for the neural instantiation of incentive motivation. It is intriguing, therefore, that the ventral striatum is also a major target area for fibres directed from the subicular cortex (Kelley and Domesick 1982). This, then, is an ideal site at which the behavioural inhibition system could inhibit striatal output destined to facilitate motor behaviour directed towards reward or non-punishment. The subiculo-striatal projection is perhaps co-ordinated with the serotonergic projection from the dorsal raphe to the substantia nigra (capable of inhibiting the dopaminergic selection of motor programmes in the dorsal striatum), implicated by Soubrié (1986) in punishment-induced suppression of rewarded instrumental behaviour. The reverse interaction perhaps takes place in the lateral septal area. This region is the recipient of dopaminergic afferents from A 10. Furthermore, a recent report (Galey *et al.* 1985) has demonstrated that these fibres exercise inhibitory control over the major cholinergic projection from the medial septal area to the hippocampus. Thus, in an approach-avoidance conflict of the kind analysed by Miller (1959), the outcome could perhaps be determined by (1) the balance of forces between the dopaminergic and subicular projections to the ventral striatum, (2) the strength of the serotonergic inhibition over nigral neurons, and (3) the balance of forces between the dopaminergic and hippocampal projections to the lateral septal area (though of course, many other possibilities exist, either in place of or in addition to these).

It is a measure of the rapid advances that have been made during the last two decades in the analysis of brain function that it is possible to entertain these possibilities at all. Figure 10.2 shows a model of the interactions between the behavioural inhibition and approach systems as deduced from purely behavioural and pharmacological data by Gray and Smith in 1969. At that time, it would have been impossible even to speculate intelligibly about the neural systems involved in such interactions. Yet today one can frame testable hypotheses (as I hope to have done in the preceding paragraph) concerning the brain loci where these interactions take place, and even concerning the transmitter substances that mediate them.

We next turn to ways in which the behavioural inhibition system might exert inhibitory control over the fight/flight system. There would seem to be

---

prioceptive conditioned stimuli to the consequences of behaviour on trial *n*. It will be seen that a smaller reward than expected conditions an input to the punishment mechanism, and a smaller punishment than expected conditions an input to the reward mechanism. The arousal mechanism increases the intensity of whatever behaviour follows conflict resolution. For further explanation, see Gray and Smith (1969).

at least the following three possibilities.

First, there is a descending pathway from the lateral septal area to the ventromedial hypothalamus (Albert and Chew 1980). Rather than exerting simple inhibitory control over the fight/flight system, it is possible that this route switches the aggression of social dominance into submissive behaviour (Adams 1979).

Secondly, as noted above, neurons in the central grey are under inhibitory serotonergic control. Graeff (in press) suggested: (1) that this route plays an important role in controlling the strong autonomic discharge that accompanies fight/flight behaviour; and (2) that it is a phasic failure in such control which gives rise to panic attacks in the syndromes of anxiety state and agoraphobia. Such attacks do not typically respond to medication with anxiolytic drugs (Klein 1981); they do, however, respond to anti-depressants that block serotonin uptake. Graeff proposed that the latter effect is due to enhancement of serotonergic inhibition over the central grey neurons responsible for autonomic outflow. If this analysis is correct, it implies that panic attacks characterize patients who chronically experience intense activity in both the fight/flight and behavioural inhibition systems, and that the attacks actually take place when the serotonergic inhibition exerted by the latter system upon the former temporarily breaks down. Considerations of this kind may go some way to resolve the dispute between those who base their analysis of human anxiety upon the functions, as observed in animal experiments, of the hypothalamus (Panksepp 1982) or amygdala (Ursin 1982; Iversen 1983) and those who prefer as their starting point the septohippocampal system (Gray 1982; Gorenstein and Newman 1980).

The third route by which the behavioural inhibition system can inhibit the fight/flight system includes an opiate step: as noted above, inhibition of pain by conditioned aversive stimuli can be blocked by opiate antagonists (Fanselow and Baackes 1982). The relevant opiate receptors may be those located in the central grey; at this level of the fight/flight system, and also in the spinal cord, there appears to be close coordination between endorphinergic and serotonergic neurons in the inhibition of pain (Melzack and Wall 1983).

A final form of interaction between the behavioural inhibition system and the other two systems is excitatory rather than inhibitory. This type of interaction is implicit in the description given above of the arousal output of the behavioural inhibition system. This output seems to prepare the animal for especially vigorous action when action next becomes possible. This function of the behavioural inhibition system is apparently mediated largely by noradrenergic fibres, especially those originating in the locus coeruleus (Redmond 1979; Gray 1982). These fibres innervate target organs that are widely distributed throughout the brain, and they appear in general to facilitate the reaction of their targets to afferents arriving from other sources

(McNaughton and Mason 1980). An action of this kind could prime approach (see Fig. 10.2), fight/flight or appetitive consumatory behaviour, depending upon the functions discharged by the innervated target organ. In accordance with this analysis, anxiogenic stimuli increase noradrenergic activity throughout the brain, an effect that is blocked by administration of anti-anxiety drugs (Lidbrink *et al.* 1973; Redmond 1979) while, behaviourally, there is evidence that, under certain conditions, anxiety can enhance responding of many different kinds (Gray 1987).

PERSONALITY

As noted above, the approach adopted here assumes that the major personality dimensions reflect consistent patterns of individual differences in the functioning of separate emotion systems in the brain (Gray 1968). Ideally, therefore, in the description of personality dimensions one would focus on behaviour likely to reflect the activities of the three neuropsychological systems outlined above. However, a vast volume of research has gone into the construction of existing personality descriptions, and these cannot therefore be lightly ignored. In any case, these descriptions and especially the one developed by Eysenck and Eysenck (1969, 1976), provide convenient and useful frameworks within which to situate different diagnostic groups or conditions. These descriptions of personality have relied heavily upon the methods of multivariate statistics and especially factor analysis. This technique can establish how many independent dimensions of variation there are in a given data set, but it cannot fix the location of these dimensions in any non-arbitrary way. The system developed by the Eysencks has three independent dimensions (extraversion, neuroticism, and psychoticism). Other analyses of a similar kind generally end up with factors that closely resemble at least the first two of the Eysenckian dimensions; and it is rare for many more than three major independent dimensions to emerge in the data. The recurrence of the number 'three' may be no more than coincidence; nonetheless, it suggests the possibility that the three Eysenckian dimensions reflect (but probably indirectly) the activities of the three neuropsychological emotion systems delineated above (Gray 1972). If so, we might light upon dimensions that map more directly upon these brain systems by the simple expedient of rotating the Eysenckian factors into new (but still mutually orthogonal) positions within the same overall three-dimensional space.

Within the more tractable two-dimensional space that is bounded by the axes of introversion-extraversion and neuroticism, I have proposed a rotation of this kind before (Gray 1970), giving rise to one dimension (trait anxiety) that directly reflects individual differences in the reactivity of the behavioural inhibition system; and to a second dimension (impulsivity) which similarly reflects individual differences in the reactivity of the

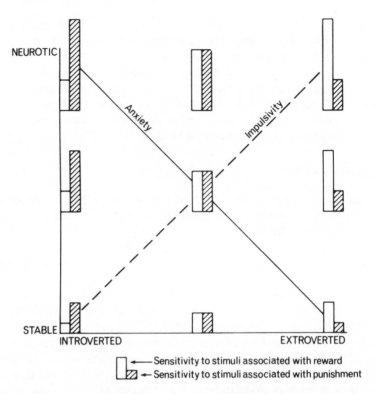

NEUROTIC

STABLE

INTROVERTED   EXTROVERTED

— Sensitivity to stimuli associated with reward
— Sensitivity to stimuli associated with punishment

FIG. 10.3. Rotation of Eysenck and Eysenck's (1969) dimensions of neuroticism and introversion-extraversion proposed by Gray (1970). The dimension of trait anxiety represents the steepest rate of growth in susceptibility to stimuli associated with punishment or with nonreward (i.e. inputs to the behavioural inhibition system); the dimension of impulsivity represents the steepest rate of growth in susceptibility to stimuli associated with reward or with non-punishment (i.e. inputs to the approach system). Introversion-extraversion now becomes a derived dimension, reflecting the balance of susceptibilities to inputs to the behavioural inhibition and approach systems, respectively; and neuroticism similarly reflects the sum of these two types of susceptibility.

approach system. The resulting diagram (Fig. 10.3) shows that one can then treat introversion-extraversion as a derived dimension, reflecting the balance between the reactivities of the behavioural inhibition and approach systems. Given our discussion of interactions between these two systems, this notion of 'balance' may have a concrete neurological instantiation, in the form of the various inputs to the ventral striatum, the substantia nigra and the lateral septal area outlined above. Figure 10.3 also suggests that one may similarly treat neuroticism as a derived dimension, reflecting the summation of the reactivities of the behavioural inhibition and approach systems.

Neurologically, this might correspond in part to a summation between the ascending dopaminergic activity responsible for incentive motivation and the ascending noradrenergic activity responsible for the arousal output of the behavioural inhibition system (see above).

A diagram similar in principle to that of Fig. 10.3 can be constructed using the behavioural inhibition system and, in place of the approach system, the fight/flight system. Such a diagram is shown in Fig. 10.4. There are no good names at present for a dimension reflecting individual differences in the reactivity of the fight/flight system (though the Eysenckian dimension, psychoticism, may have some of the appropriate characteristics; Gray 1972); individuals high on such a dimension ought to be particularly prone to excessive unconditioned escape behaviour (terror?) and/or defensive aggression

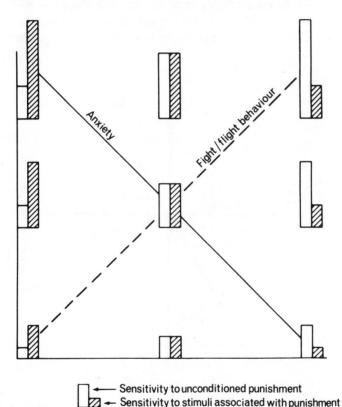

Sensitivity to unconditioned punishment
Sensitivity to stimuli associated with punishment

FIG. 10.4. Possible relations between (1) sensitivity to stimuli associated with punishment and other adequate inputs to the behavioural inhibition system (underlying a dimension of trait anxiety, as in Fig. 10.3), and (2) sensitivity to unconditioned punishment and non-reward, activating fight/flight behaviour. For further explanation, see text.

(rage?). Nor are there good names for a dimension reflecting the balance between the reactivities of the behavioural inhibition and fight/flight systems; though, given the behavioural interactions between these two systems noted above, there is some reason to consider the balance between their reactivities as possessing a claim to neurological reality. As in the case of the behavioural inhibition and approach systems (Fig. 10.3), we may also consider the possibility that summation between activity in the fight/flight and behavioural inhibition systems contributes to some personality dimension or other. Indeed, if Graeff's (in press) analysis of panic attacks is correct (see above), then the individuals who are susceptible to such attacks may be precisely those who are high on such a dimension of summed reactivities.

One can obviously use the approach and fight/flight systems to construct yet another two-dimensional diagram; but at some stage one needs to put these diagrams together to form a more complete three-dimensional hypothesis that will describe the axes of neuropsychological causation in per-

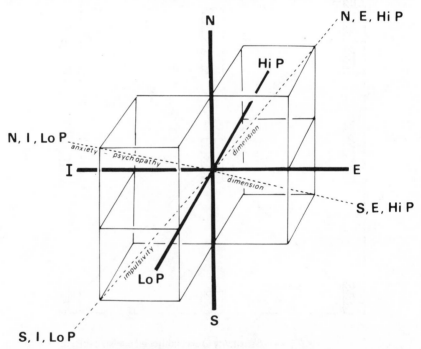

FIG. 10.5. The locations of dimensions of anxiety-psychopathy (reflecting the reactivity of the behavioural inhibition system) and impulsivity (reflecting the reactivity of the approach system) as proposed by Gray (1983), within the overall Eysenckian three-dimensional personality space. High trait anxiety is conceived as occurring at the high neuroticism (N), high introversion (I), and low psychoticism (P) pole; low anxiety, at the stable (S), extravert (E), high P pole. High impulsivity is conceived as occurring at the N, E, high P pole.

sonality space. However, the range of possibilities in constructing such a hypothesis is dauntingly large and one needs some kind of anchor point from which to start.

The best anchor presently available is probably that provided by the dimension of trait anxiety, both because we have a reasonably good understanding of the neural basis of this emotion (Gray 1982), and because the personality characteristics of the individuals susceptible to it have been relatively well described (Eysenck and Eysenck 1969). Furthermore, a good case can be made for the supposition that individuals low in trait anxiety are those termed 'primary psychopaths' (Hare and Schalling 1978). These individuals are at the opposite pole to those who suffer from pathological anxiety on every one of the three Eysenckian dimensions (that is, they are extraverted, low on neuroticism, and high on psychoticism). In addition, they differ from anxious individuals on a key behavioural index (they are especially poor at passive avoidance) and on measures of psychophysiological reactivity (Fowles 1980). Taking these various observations into account, therefore, one can anchor within the overall Eysenckian three-dimensional space at least one dimension that plausibly reflects individual differences in the reactivity of one of our neuropsychological systems (namely, the behavioural inhibition system). More speculatively, one can guess at a possible location of a dimension (impulsivity), orthogonal to the first, that reflects individual differences in the reactivity of the approach system. The resulting diagram is pictured in Fig. 10.5 (Gray 1983). Clearly, this figure (even supposing that it is correct in its major assumptions) leaves us a long way to go before we can locate within the Eysenckian space all three of the axes of neuropsychological causation that presumably exist there, but at least it is a start.

REFERENCES

Adams, D.B. (1979). Brain mechanisms for offence, defense, and submission. *Behavioral and Brain Sciences* **2**, 201–41.

Albert, D.J. and Chew, G.L. (1980). The septal forebrain and the inhibitory modulation of attack and defence in the rat: a review. *Behavioral and Neural Biology* **30**, 357–88.

Alkon, D. and Farley, J. (ed.) (1984). *Primary Neural Substrates of Learning and Behavioural Change*. Cambridge University Press, Cambridge.

Amsel, A. (1962). Frustrative nonreward in partial reinforcement and discrimination learning: some recent history and a theoretical extension. *Psychological Review* **69**, 306–28.

Bolles, R.C. and Fanselow, M.S. (1980). A perceptual defensive-recuperative model of fear and pain. *Behavioral and Brain Sciences* **3**, 291–323.

Charney, D.S., Heininger, G.R., and Breier, A. (1984). Noradrenergic function in panic anxiety. *Archives of General Psychiatry* **41**, 751–63.

de Molina, F.A. and Hunsperger, R.W. (1962). Organisation of the subcortical system governing defence and flight reactions in the cat. *Journal of Physiology* 160, 200–13.

Eysenck, H.J. (1960). Classification and the problem of diagnosis. In *Handbook of Abnormal Psychology* (ed. H.J. Eysenck), pp. 1–31. Pitman, London.

—— and Eysenck, S.B.G. (1969). *Personality Structure and Measurement*. Routledge and Kegan Paul, London.

Eysenck, S.B.G. and Eysenck, H.J. (1976). *Psychoticism as a Dimension of Personality*. Hodder and Stoughton, London.

Fanselow, M.S. and Baackes, M.P. (1982). Conditioned fear-induced opiate analgesia on the formalin test: evidence for two aversive motivational systems. *Learning and Motivation* 13, 200–21.

Fowles, D. (1980). The three arousal model: implications of Gray's two-factor learning theory for heart rate, electrodermal activity and psychopathy. *Psychophysiology* 17, 87–104.

Galey, D., Durkin, T., Sitakis, G., Kempf, E., and Jaffard, R. (1985). Facilitation of spontaneous and learned spatial behaviours following 6-hydroxydopamine lesions of the lateral septum: a cholinergic hypothesis. *Brain Research* 340, 171–4.

Gorenstein, E.E. and Newman, J.P. (1980). Disinhibitory psychopathology: a new perspective and a model for research. *Psychological Review* 87, 301–15.

Graeff, F.G. (in press). The anti-aversive action of drugs. In *Advances in Behavioral Pharmacology*, Vol. 5 (ed. T. Thompson, P.B. Dews, and J. Barrett). Erlbaum, Hillsdale, N.J.

Gray, J.A. (1968). The Lister Lecture, 1967: The physiological basis of personality. *Advancement of Science* 24, 293–305.

—— (1970). The psychophysiological basis of introversion-extraversion. *Behaviour Research and Therapy* 8, 249–66.

—— (1972). Causal theories of personality and how to test them. In *Multivariate Analysis and Psychological Theory* (ed. J.R. Royce), pp. 409–63. Academic Press, London.

—— (1975). *Elements of a Two-Process Theory of Learning*. Academic Press, London.

—— (1976). The behavioural inhibition system: a possible substrate for anxiety. In *Theoretical and Experimental Bases of Behaviour Modification* (ed. M.P. Feldman and A.M. Broadhurst), pp. 3–41. Wiley, London.

—— (1977). Drug effects on fear and frustration: possible limbic site of action of minor tranquillizers. In *Handbook of Psychopharmacology* Vol. 8 (ed. L.L. Iversen, S.D. Iversen, and S.H. Snyder), pp. 433–529. Plenum, New York.

—— (1982). *The Neuropsychology of Anxiety: An Enquiry into the Functions of the Septo-Hippocampal System*. Oxford University Press, Oxford.

—— (1983). Where should we search for biologically based dimensions of personality? *Zeitschrift für Differentielle und Diagnostische Psychologie* 4, 165–76.

—— (1984). The hippocampus as an interface between cognition and emotion. In *Animal Cognition* (ed. H.L. Roitblat, T.G. Bever, and H.S. Terrace), pp. 607–26. Erlbaum, Hillsdale, N.J.

—— (1987). *The Psychology of Fear and Stress*, 2nd edn. Cambridge University Press, Cambridge.

—— (in press). The neuropsychological basis of anxiety. In *Handbook of Anxiety Disorders* (ed. C.G. Last and M. Hersen). Pergamon, New York.

—— *et al.* (1982). Precis and multiple peer review of 'The Neuropsychology of

Anxiety: An Enquiry into the Functions of the Septo-hippocampal System'. *Behavioral and Brain Sciences* **5**, 469–525.

—— and Smith, P.T. (1969). An arousal-decision model for partial reinforcement and discrimination learning. In *Animal Discrimination Learning* (ed. R. Gilbert and N.S. Sutherland), pp. 243–72. Academic Press, London.

Haefely, W., Kyburz, E., Gerecke, M., and Möhler, H. (1985). Recent advances in the molecular biology of benzodiazepine receptors and in the structure-activity relationships of their agonists and antagonists. In *Advances in Drug Research* (ed. B. Testa), pp. 165–322. Academic Press, London.

Hare, R.D. and Schalling, D. (ed.) (1978). *Psychopathic Behaviour: Approaches to Research*. Wiley, Chichester.

Huston, J.P. and Borbely, A.A. (1973). Operant conditioning in forebrain ablated rats by use of rewarding hypothalamic stimulation. *Brain Research* **50**, 467–72.

Iversen, S.D. (1983). Where in the brain do the benzodiazepines act? In *Benzodiazepines Divided* (ed. M.R. Trimble), pp. 167–83. Wiley, Chichester.

Kelley, A.E. and Domesick, V.B. (1982). The distribution of the projection from the hippocampal formation to the nucleus accumbens in the rat: an anterograde- and retrograde-horseradish peroxidase study. *Neuroscience* **7**, 2321–35.

Klein, D.F. (1981). Anxiety reconceptualized. In *Anxiety: New Research and Changing Concepts* (ed. D.F. Klein and J. Rabkin), pp. 235–63. Raven Press, New York.

Lidbrink, P., Corrodi, H., Fuxe, K., and Olson, L. (1973). The effects of benzodiazepines, meprobamate, and barbiturates on central monoamine neurons. In *The Benzodiazepines* (ed. S. Garattini, E. Mussini, and L.O. Randall), pp. 203–23. Raven Press, New York.

McNaughton, N. and Mason, S.T. (1980). The neuropsychology and neuropharmacology of the dorsal ascending noradrenergic bundle—a review. *Progress in Neurobiology* **14**, 157–219.

Melzack, R. and Wall, P.D. (1983). *The Challenge of Pain*. Basic Books, New York.

Miller, N.E. (1959). Liberalization of basic S-R concepts: extensions to conflict behavior, motivation and social learning. In *Psychology: A Study of a Science*, Study 1, Vol. 2 (ed. S. Koch), pp. 196–292. McGraw-Hill, New York.

Milner, B. (1987). Neuropsychological evaluation of cortical lesions. This volume.

Mowrer, O.H. (1960). *Learning and Behavior*. Wiley, New York.

Olds, J. and Milner, P. (1954). Positive reinforcement produced by electrical stimulation of septal area and other regions. *Journal of Comparative and Physiological Psychology* **47**, 419–27.

Panksepp, J. (1982). Towards a general psychobiological theory of emotions. *Behavioral and Brain Sciences* **5**, 407–67.

Polenchar, B.E., Patterson, M.M., Lavond, D.G., and Thompson, R.F. (1985). Cerebellar lesions abolish an avoidance response in rabbits. *Behavioral and Neural Biology* **44**, 221–7.

Redmond, D.E., Jr (1979). New and old evidence for the involvement of a brain norepinephrine system in anxiety. In *Phenomenology and Treatment of Anxiety* (ed. W.G. Fann, I. Karacan, A.D. Pokorny, and R.L. Williams), pp. 153–203. Spectrum, New York.

Reiman, E.M., Raichle, M.E., Butler, F.K., Hersovitch, P., and Robins, E. (1984). A focal brain abnormality in panic disorder, a severe form of anxiety. *Nature* **310**, 683–5.

Segal, M. (1977a). The effects of brainstem priming stimulation on interhemispheric hippocampal responses in the awake rat. *Experimental Brain Research* **28**, 529–41.

Segal, M. (1977b). Excitability changes in rat hippocampus during conditioning. *Experimental Brain Research* **55**, 67–73.

Soubrie, P. (1986). Reconciling the role of central serotonin neurons in human and animal behavior. *Behavioral and Brain Sciences* **9**, 319–63.

Stahl, S.M., Leenders, K.L., and Bowery, N.G. (1986). Imaging nerotransmitters and their receptors in living human brain by positron emission tomography. *Trends in Neuroscience* **9**, 241–5.

Stein, L., Wise, C.D., and Berger, B.D. (1973). Anti-anxiety action of benzodiazepines: decrease in activity of serotonin neurons in the punishment system. In *The Benzodiazepines* (ed. S. Garattini, E. Mussini, and L.O. Randall), pp. 299–326. Raven Press, New York.

Stewart, J., de Wit, H., and Eikelboom, R. (1984). Role of unconditioned and conditioned drug effects in the self-administration of opiates and stimulants. *Psychological Review* **91**, 251–68.

Thompson, R.F. *et al.* (1984). Neuronal substrates of associative learning in the mammalian brain. In *Primary Neural Substrates of Learning and Behavioral Change* (ed. D. Alkon and J. Farley), pp. 71–7. Cambridge University Press, Cambridge.

Ursin, H. (1982). Substrates of anxiety: but if the starting point is wrong? *Behavioural and Brain Sciences* **5**, 503–4.

Willner, P. (1985). *Depression: a Psychobiological Synthesis*. Wiley, New York.

Yeo, C.H., Hardiman, M.T., and Glickstein, M. (1985). Classical conditioning of the nictitating membrane response of the rabbit. *Experimental Brain Research* **66**, 87–98.

11

Neuropsychological approaches to cognitive disorders—a discussion

PAUL BROKS AND G.C. PRESTON

Three forms of inquiry are made possible by the intersection of psychology and the neurosciences. One, essentially biological in character, takes psychological indicators as a guide to biological function. Another, essentially psychological, brings biological considerations to bear on psychological theorizing. The third is concerned with identifying relationships between the component parts of biological and psychological systems. In practice these distinctions are often blurred, but in considering approaches to cognitive neurochemistry, it will serve us to keep them in mind.

Human neuropsychology provides illustrative examples of all three of the above modes of investigation. In the clinical sphere, neuropsychological assessment is routinely undertaken to help determine the presence or absence, the probable location and the extent of neurological damage. More than this, however, it is becoming apparent that neuropsychology will play an increasingly important role not only in the clinical evaluation of neurological disorders (through the development of more sophisticated testing procedures—see Morris *et al.* this volume), but also, perhaps, in delineating subcategories of disease. For example, neuropsychological studies of Alzheimer's disease (e.g. Filley *et al.* 1986) might augment neurochemical evidence (e.g. Rossor *et al.* 1984) suggesting that early and late-onset forms of the disease may constitute fairly distinct neuropathological conditions.

This first form of inquiry might be characterized as psychology at the service of neuropathology since psychological measures may be selected for their practical efficacy in discriminating neurological disorders rather than for any inherent relevance to an understanding of psychological function *per se*.

As an illustration of the converse, consider Shallice's persuasive arguments in support of the view that neuropsychological data can have a significant bearing on the formulation of psychological theory (Shallice 1979). In essence, Shallice maintains that evidence on the psychological consequences of different forms of neurological disorder acts to constrain theories of normal cognitive function. Since a similar proposition also constitutes one of the central tenets of the flourishing school of cognitive neuropsychology (see Editorial, *Cognitive Neuropsychology*, 1984, 1(1), 1–8, it is worth elaborating here. The case presented by Shallice rests squarely on evidence for the neuropsychological 'fractionation' of memory systems. The theoretical distinction between short- and long-term verbal memory, posited by Waugh and Norman (1965), and Atkinson and Shiffrin (1968) among others, became widely accepted during the 1960s. According to this view, incoming information is held in a limited-capacity short-term store whence it either rapidly decays (after a period of a few seconds) or alternatively undergoes further processing to gain entry into a longer-term store where, potentially, it may remain indefinitely. There were alternative proposals; for example, that short- and long-term storage reflected different forms of activation of the same store (e.g. Norman 1968), but the dual system notion of memory was so widely adopted that it came to be referred to as 'the modal model'. In the 1970s neuropsychological evidence was presented which bore directly on the two-store theory. It had been known for some time that amnesic patients perform extremely poorly on tasks which demand relatively long-term storage and retrieval, but that, in contrast, they can perform well on tasks which require only short-term memory (e.g. Milner 1966; Baddeley and Warrington 1970). Patients were now identified who, in the absence of any general intellectual impairment, showed extremely impoverished ability on short-term memory tasks, but who could perform at normal levels on tasks of long-term memory (e.g. Shallice and Warrington 1970; Warrington *et al.* 1971; Saffran and Marin 1975).

This demonstration of a neuropsychological double dissociation between short- and long-term verbal memory systems provides compelling evidence for the existence of at least two distinct memory stores. However, while clearly substantiating the two-store model in this respect, it also forces theorists to consider an important modification. Since it is evidently possible to store and retrieve information using long-term memory despite severe disruption of the short-term system, then it becomes necessary to think of the two processes operating in parallel rather than in series. The message conveyed by Shallice is that cognitive theorists will ignore such evidence at their peril. Indeed, it is partly in consequence of neuropsychological observations such as these that the 'modal model' has come to be replaced by more sophisticated notions of short-term and long-term memory (Baddeley 1982). We can characterize this approach as neuropathology at the service of psychology because, for the cognitive psychologist, the details of the neuropathological state giving rise to such double dissociations of function are largely irrelevant. From a purely psychological standpoint, all that matters is the extent to which models of normal cognition are able to accommodate observed patterns of cognitive disorder.

The third approach to the study of brain-behaviour relationships concerns the activity in which most experimental neuropsychologists would imagine themselves to be engaged. That is, establishing which systems of the brain subserve which psychological functions. Of course, both 'brain system' and 'psychological function' are rather elastic concepts. Recognizing this, Luria (1966) distinguished between 'functions' (which might be characterized, say, by the generation of motor impulses from the giant Betz cells in the motor cortex) and 'complex functional systems' (say, the integrated involvement of distinct cortical regions in coordinating an action in response to a verbal command). Furthermore, the ways in which psychological functions might conceivably be mapped onto physical systems of the brain are multifarious. For different cognitive functions the relationship might be characterized as one-to-one (i.e. one cognitive function to one neurophysiological state), one-to-many, or many-to-one.

In the face of such complexity a valid strategy (among others perhaps) is to attempt to analyse compound functions, such as 'memory' or 'attention', into their constituent functional elements. As an example of such an approach in the realm of attention, Posner and his colleagues (e.g. Posner *et al*. 1984) have explored the mechanisms underlying shifts of attention across the visual field. Employing procedures for the analysis of 'covert orientation', they find evidence to suggest that certain deficits of visual attention associated with parietal injury result from an impairment of the ability to disengage from an attended location rather than, as might have been the case, difficulties in the process of moving the focus of attention through visual space. Since no such impairment is evident in frontal, midbrain or temporal patients, Posner *et al*. (1984) suggest that this component of visual attention is subserved by the parietal lobe.

The distinctions between the three forms of inquiry outlined above primarily in relation to lesion studies also apply to the neuropsychology of neurodegenerative disorders where neural damage may be defined not only by the area of the brain affected, but also, as for Parkinson's disease, in relation to the neurotransmitter systems involved (see Morris *et al*., this volume). This further extends to studies involving the manipulation of neurotransmitters using drugs in healthy human volunteers or laboratory animals where the accent may be on psychology, on neurochemistry/pharmacology or on the relationship between them. For example, in work on state-dependent learning (e.g. Overton 1984) the nature of the drug used is close to being irrelevant; any drug which produces state-dependency, preferably with few other effects, can be used to investigate the psychological principles underlying the phenomenon.

By contrast, in drug development the psychology is almost invariably secondary to the pharmacology; for example, drug discrimination procedures in animals have been used to good effect for classifying drugs and examining interactions between them even though the psychological basis of such procedures is often obscure. Likewise, in human psychopharmacology, relatively *ad hoc* measures may be used to investigate drug effects; for example, critical flicker fusion is commonly used as an index of sedation even though its relationship to mechanisms underlying sedation is not fully understood.

The third approach, which attempts to bring together psychology and pharmacology, can be illustrated by Gray's early work on partial reinforcement effects in animals (see Gray 1977). He was able to show that anxiolytic drugs affect extinction. This finding was incorporated into models of anti-anxiety drug action. He was also able to show at the same time that superficially similar partial-reinforcement effects were dissociable pharmacologically; these findings added weight to the view that partial reinforcement effects are multiply-determined, and that only a subset of the processes involved are sensitive to anxiolytics. The approach is also illustrated by Robbins's elegant work on ascending monoaminergic pathways in which he uses a combination of drug studies, lesion studies, and psychological innovation to determine some of the functions of particular pathways in the brain (see Robbins and Everitt, this volume). Pharmacological manipulation can, then, be used to examine brain-behaviour relationships in an analogous way to anatomical lesion studies. There is, however, the intriguing possibility that, since neurotransmitter systems cut across anatomical boundaries, pharmacological manipulation may produce quite different effects from damage defined anatomically.

As an example, spatial and verbal long-term memory are anatomically dissociable; studies of focal lesions have established that verbal memory is more commonly affected by damage to the left hemisphere whereas nonverbal memory is generally more sensitive to right hemisphere damage (Milner 1971). Preliminary data from our own studies have established that certain measures of verbal and spatial long-term memory are statistically independent in undrugged subjects, yet both appear to be sensitive to the action of the cholinergic antagonist scopolamine at certain doses. (Neither verbal nor spatial short-term memory are affected.) The implication is that the two forms of long-term memory, despite being dissociable anatomically, nevertheless have common features to the extent that they are affected analogously by a single drug. The site of action (neurological and psychological) remains an open question (Broks *et al.*, in preparation). This notion, that pharmacological manipulation may afford information at a different level to that afforded by focal lesions, is a satisfying one, but it is only in animal studies that pharmacological findings can readily be followed up with the necessary anatomical investigations.

Cognitive neurochemistry may also have certain advantages over traditional neuropsychology. For example, the ability to examine the same subject with and without a drug offers a tremendous increase in power; in neuropsychological studies information about the subject before the damage occurred is frequently difficult to come by. Also the ability to manipulate dosage offers, at least potentially, a solution to some theoretical problems encountered in neuropsychology. For example, it is accepted in neuropsychology that little weight can be put on simple differences between patient groups in a single test. Better are double dissociations between groups on two tasks. However, as Weiskrantz (1968) has pointed out, even double dissociation may be trivial or uninterpretable if performance on the tasks in question is not monotonically related to whatever psychological variables the brain damage is thought to affect. By manipulating dose it is possible for the pharmacologist to investigate this assumption of monotonicity.

In this chapter we have attempted to draw parallels between current studies of neuropsychology and studies of cognitive neurochemistry. We have tried to show that both examinations of pharmacology and of neurology in relation to cognition may give unique information about the influence of neural systems over behaviour, and have suggested that the interaction between neuropsychology and pharmacology provides a powerful strategy for interpreting cognition. That cognitive neurochemistry is necessarily a multi-disciplinary field is well illustrated by the contributions to the present volume.

The work described by Warburton reflects the traditional concern of human experimental psychopharmacology. That is, to interpret the psychological effects of pharmacological manipulation in terms of available models of normal functioning. The very possibility of a 'cognitive neurochemistry' rests on the assumption that different neurotransmitter systems have identifiable forms of influence over information processing. Warburton adopts a strong version of this assumption as a working hypothesis (i.e. that different transmitter systems can be identified with different types of processing). His survey of the differential cognitive effects of cholinergic and noradrenergic manipulation provides interesting evidence in its favour. There seems little doubt that human experimental psychopharmacology will become increasingly influential. With respect to clinical neuropathology, for example, pharmacological models of pathological states can be expected to play a significant role in the development of strategies for the evaluation of putative therapeutic agents. As noted previously, theoretical psychology might also be expected to benefit from analyses of drug-induced cognitive change.

The work of Butters and his associates has its roots firmly in clinical neuropsychology in that its central concern is the description of psychological features associated with different types of neurological disorder. However, in recognizing the limited analytical power of standard testing procedures, and choosing to develop procedures derived more directly from the theoretical constructs of cognitive psychology, their approach reflects an important trend among clinical neuropsychologists. It is acknowledged that advances in our understanding of cognitive dysfunction will depend on the specification of *process* in the description of impaired abilities. In this way it becomes possible to specify the pathologies of cognition in relatively fine detail. As Butters *et al.* suggest, the benefits that clinical neuropsychology enjoys in its relationship with cognitive psychology are reciprocated in so far as neuropathological dissociations of cognitive function serve to validate the original theoretical constructs. As an example from their own work they are able to cite the double dissociations of function observed for procedural and declarative memory in comparing patients with Huntington's disease and Korsakoff's syndrome.

Funnell's approach to the neuropsychological assessment of dementia is quite explicitly derived from information processing analyses of cognitive deficit. Clinical rating scales and psychometric instruments typically used in the detection and differentiation of dementing states yield assessments of a rather 'global' nature. More fine-grained analyses of cognitive function are required in order to identify relevant sub-categories of pathology and to evaluate putative therapeutic interventions. The methodology advocated by Funnell, with its concern for detailed description of cognitive processes and its capacity to accommodate individual differences in profiles of observed dysfunction, is of proven worth in elucidating the effects of focal brain lesions. It can be expected to prove equally valuable in neuropsychological investigations of chronic, neurodegenerative disorders.

It is generally acknowledged that micro-computerized testing procedures will feature large in clinical and experimental neuropsychology. Morris *et al.* present the case for computer-aided assessment persuasively, and their own work on Alzheimer-type dementia and Parkinson's disease provides a convincing illustration of the potential of such methods. In choosing paradigms successfully employed in animal studies they illustrate an important option for neuropsychological research. Certainly, the value of comparative studies will be optimized to the extent that the phenomena observed in animal models of neurological disease (such as those discussed by Robbins and Everitt, and by Rupniak and Iversen, this volume) are directly referrable to those observed in clinical settings.

By taking arousal and attention as their central topic, Robbins and Everitt are forced to address a broad span of issues crucial to the development of cognitive neurochemistry. It is clear from their review that a knowledge of the neuroanatomical configuration of transmitter systems is fundamental to an understanding of the role they play in modulating neuropsychological function. However, such knowledge will be of value only to the extent that appropriate conceptual analyses in the psychological domain can be achieved. Although attempts to identify different transmitter systems with specific cognitive processes have met with some success (see Warburton, this volume) there are, as Robbins and Everitt point out, limitations to such an approach. The relationship between neurochemistry and cognition may be better conceived in terms of a range of 'non-specific' influences subserved by different transmitter systems. It is here that concepts such as arousal and attention begin to assume a position of prime importance.

Rupniak and Iversen evaluate the status of primate models of dementia. Attempts to induce in non-human primates a pathological condition directly akin to Alzheimer's disease have largely met with failure. This has been the case whether metallic neurotoxins, such as aluminium, have been considered or whether the viral hypothesis of Alzheimer's disease has been tested by inoculating animals with tissue from demented patients. However, much stands to be gained from the systematic study of aged monkeys and from the pharmacological manipulation of cognition in both aged and young animals. As in man, certain neuropathological changes occurring in Alzheimer's disease are also evident to a lesser degree in normal aged primates. Concomitant with such changes there also appears to be a characteristic decline in cognitive function. As Rupniak and Iversen note, there are already encouraging signs that a range of pharmacological interventions may facilitate cognitive function in the aged monkey. Pharmacological models of dementia, using drugs such as scopolamine to induce transient cognitive deficit, are potentially of great value in primate studies not least because they allow for fairly direct comparisons with data from similar studies in human psychopharmacology.

Much of the inspiration for the work on animal models of amnesia discussed by Rawlins and by Gaffan comes from Milner's classic studies of the patient H.M. who suffered profound amnesia following bilateral temporal lobe surgery (Milner 1966). Although H.M.'s amnesia has been widely attributed to hippocampal damage, a number of other structures were implicated. The challenge for comparative neuropsychology, then, has been to establish the critical neuropathological determinants of memory deficit. Such endeavours have met with mixed success. Although experimental lesions of the hippocampus and related structures are associated with an identifiable pattern of deficits, including disruption to certain forms of memory, it cannot be claimed that they bear a clear, qualitative resemblance to the amnesic syndrome.

What remains to be determined is the extent to which the observed patterns of deficit are *analogous* to those found in amnesic patients. Gaffan, referring to work with non-human primates, argues that the conceptual distinction between 'personal memory' (i.e relating to an animal's own interactions with its environment) and 'non-personal memory' may be crucial to the development of a valid animal model. This distinction can be considered somewhat analogous to that between 'declarative' and 'procedural' memory derived from studies of patients (see Butters *et al.*, this volume). Rawlins, drawing largely on studies of the rat, refers to the effects of hippocampal damage in terms of the ability to associate new information with the context in which it was received.

In his analysis of the functions of the septo-hippocampal system, Gray offers a neurobiological perspective on a fundamental issue in psychology: that concerning the relationship between thought and emotion. The most clearly articulated theory in Gray's neuropsychology of emotion concerns the nature of anxiety. Essentially, what is proposed is a cognitive theory of anxiety in which the information processing capacity of the septo-hippocampal system, monoaminergically modulated, is the key factor. According to Gray, states of anxiety may be identified with the activities of this system through the function it serves as a 'comparator', continuously checking predicted against actual events. The theory accommodates both individual differences along the personality dimension of 'trait anxiety' and variations in the form taken by pathological anxiety states. As Gray notes, the fact that many of the structures involved in the neuropsychology of anxiety are also implicated in the neuropathology of Alzheimer's disease itself suggests an intimate bond between cognition and affect.

The innovative neuroanatomical tracing methods so fruitfully exploited by Goldman-Rakic in her studies of the macaque monkey give unique insights into the construction and logic of the networks of communication linking anatomically disparate brain regions. Without the traffic of information through these networks even simple acts of perception and memory could not be sustained. Mapping the interconnectivity of parietal ('perceptual'), limbic ('mnestic'), and prefrontal ('executive') zones, Goldman-Rakic gives flesh to a system by which representational knowledge can be used to guide behaviour. For Goldman-Rakic, the main focus of interest is the principal sulcal area of the prefrontal cortex which is shown to be crucially involved in representational memory. It is suggested that analysis of the functional organization of the principal sulcus may create an important inroad into an understanding of prefrontal function in general. The information processing activity of the principal sulcus is under the modulatory influence of brainstem monoamine projections, and it is proposed that neurochemical disruption of prefrontal functions could account for some of the features of a range of neurological and psychiatric disorders including schizophrenia, Parkinson's disease and Korsakoff's syndrome.

Many of the issues raised in this chapter are brought to the fore by the pressing need for an adequate neuropsychological analysis of Alzheimer's disease, and indeed other degenerative disorders of the brain. It is clear that progress here depends as much on a sophisticated awareness of current cognitive theory as on an appreciation of the anatomy and neurochemistry of the neural systems subserving cognition. For theoretical neuropsychology much stands to be gained from an explicit consideration of the neurochemical contribution to cognitive function. More significantly, perhaps, there is the prospect of tangible benefit for clinical research.

REFERENCES

Atkinson, R.C. and Shiffrin, R.M. (1968). Human memory: a proposed system and its control processes. In *The Psychology of Learning and Motivation* (ed. K.W. Spence and J.T. Spence), pp. 89–195. Academic Press, New York.

Baddeley, A.D. (1982). Implications of neuropsychological evidence for theories of normal memory. *Philosophical Transactions of the Royal Society of London* **B298**, 59–72.

—— and Warrington, E.K. (1970). Amnesia and the distinction between long- and short-term memory. *Journal of Verbal Learning and Verbal Behaviour* **9**, 176–89.

Broks, P., Preston, G.C., Traub, M., Poppleton, P., Ward, C., and Stahl, S.M. (1987). Cognitive effects of scopolamine. (In preparation).

Filley, C.M., Kelly, J., and Heaton, R.K. (1986). Neuropsychologic features of early- and late-onset Alzheimer's disease. *Archives of Neurology* **43**, 574–6.

Gray, J.A. (1977). Drug effects on fear and frustration: possible limbic site of action of minor tranquilizers. In *Handbook of Psychopharmacology*, Vol. 8 (ed. L.L. Iversen, S.D. Iversen, and S.H. Snyder), pp. 433–526. Plenum, New York.

Luria, A.R. (1966). *Higher Cortical Functions in Man*. Tavistock, London.

Milner, B. (1966). Amnesia following operation on the temporal lobes. In *Amnesia* (ed. C.W.M. Whitty and O.L. Zangwill), pp. 109–33. Butterworth, London.

—— (1971). Interhemispheric differences in the localization of psychological processes in man. *British Medical Bulletin* **27**, 272–7.

Norman, D.A. (1968). Toward a theory of memory and attention. *Psychological Review* **75**, 522–36.

Overton, D. (1984). State dependent learning and drug discriminations. In *Handbook of Psychopharmacology*, Vol. 18 (ed. L.L. Iversen, S.D. Iversen, and S.H. Snyder) pp. 59–127. Plenum, New York.

Posner, M.I., Walker, J.A., Friedrich, F.J., and Rafal, R.D. (1984). Effects of parietal injury on covert orienting of attention. *Journal of Neuroscience* **4**, 1863–74.

Rossor, M.N., Iversen, L.L., Reynolds, G.P., Mountjoy, C.Q., and Roth, M. (1984). Neurochemical characteristics of early and late onset types of Alzheimer's disease. *British Medical Journal* **288**, 961–4.

Saffran, E.M. and Marin, O.S. (1975). Immediate memory for word lists and sentences in a patient with deficient auditory short-term memory. *Brain and Language* **2**, 420–33.

Shallice, T. (1979). Neuropsychological research and the fractionation of memory systems. In *Perspectives on Memory Research* (ed. L.G. Nilsson), pp. 257–77. Erlbaum, Hillsdale, N.J.

—— and Warrington, E.K. (1970). Independent functioning of verbal memory stores: a neuropsychological study. *Quarterly Journal of Experimental Psychology* **22**, 261–73.

Warrinton, E.K., Logue, V., and Pratt, R.T.C. (1971). The anatomical localization of selective impairment of auditory verbal short-term memory. *Neuropsychologia* **9**, 377–87.

Waugh, N.C. and Norman, D.A. (1965). Primary memory. *Psychological Review* 72, 89-104.

Weiskrantz, L. (1968). Some traps and pontifications. In *Analysis of Behavioural Change* (ed. L. Weiskrantz), pp. 415-29. Harper and Row, New York.

Part II

NEUROCHEMICAL APPROACHES TO COGNITIVE DISORDERS

12

The role of arginine vasopressin and other neuropeptides in brain-body integration

MICHEL LE MOAL, ROSE-MARIE BLUTHE,
ROBERT DANTZER, FLOYD E. BLOOM, AND
GEORGE F. KOOB

INTRODUCTION

Throughout all of its parts the body probably communicates by means of messengers or regulatory molecules and by this principle influences behaviour. All molecular systems have a source and site of action, autocrine, paracrine, or endocrine (Krieger 1983). The strategy for investigating how these molecular systems affect behaviour is based on their design and their source or site of action (Martinez 1983). In general, the set of receptors upon which peptides act are present both centrally and peripherally. Moreover, it is generally assumed that behaviour, cognition, and memory are dependent upon central processes and thus, if peptides are able to modulate such processes, a central site of action is likely to be involved. The question that remains is whether peripherally administered peptides, as is the case with many drugs, are accessible to the brain. This question is made even more complicated because some parts of the brain can be considered to lie outside the blood-brain barrier and are therefore more accessible to peripherally administered peptide.

Many molecules influence or control behaviour without crossing the blood-brain barrier. The body is not only the target of the brain for responses, adaptation, homeostasis, and behaviour, it is also the source of information for the brain. The body modifications, changes in internal organs and behaviour influence the brain in many features, one of them being the central nervous system equilibrium. Integrative behavioral neurobiology is concerned with the brain-body dialogue in both fundamental research and psychopathology, including psychosomatic medicine.

Neuropeptides: do they integrate body and brain? The question is central to this chapter (Iversen 1981). More and more data enable us to be affirmative, even if the mechanisms involved are still poorly understood. Neuropeptides, such as arginine vasopressin, integrate body, and brain cognitive processes, such as learning and memory (Le Moal *et al.* 1981; Koob *et al.* 1981).

NEUROPEPTIDES AND BEHAVIOUR: GENERAL CONSIDERATIONS

Structural considerations: the blood-brain barrier

The blood-brain barrier and peptides

The factors that seem to account for the existence of the blood-brain barrier depend for a large part on the morphology of brain capillaries and adjoining structures, made up of one layer of endothelial cells, 1 μm thick, surrounded by an amorphous mucopolysaccharide matrix called the basement membrane which is 50 μm thick (Feldman and Quenzer 1984). The endothelial layer of general capillaries has small clefts either between adjacent cells or where the cells circle upon themselves. There are also fenestrae which represent openings of 9–15 nm wide in capillaries—in many ways similar to those found in the kidney—where only small molecules can pass. In other tissues, capillaries have pinocytotic vesicles that envelop and transport larger molecules through the capillary wall; however, in brain capillaries these intercellular clefts are absent for one reason: the adjoining edges are fused, forming very tight junctions. Fenestrae are also absent and pinocytotic vesicles are very rare. On the other hand, mitochondria are more numerous in brain capillary cells. Surrounding these brain capillaries we can identify astrocytic processess or glial feet which cover 85 per cent of the basement membrane and which fill the gap between the endothelial cells and the glial feet. The fate of molecules in the body's circulation depends on their size. For molecules with a high molecular weight, half of them leave the blood and enter the extracellular fluid with a half-life of about 10–30 seconds, while the larger of these molecules of albumin type (60 000 MW and more) remain in the blood much longer with a half-life of several hours. Because of the absence of clefts and fenestrae, small water soluble molecules (MW less than 2000) cannot pass easily from the capillaries into the brain extracellular fluid. To reach the brain's extracellular fluid, such molecules must be actively transported by a carrier mediated process which may limit the types of structures so transported. Although the blood-brain barrier is selectively permeable to water soluble or ionized molecules, it does not greatly impede small lipophilic or non-ionized molecules. Crossing this blood-brain barrier is thus a chemical and metabolic adventure. The details of how molecules which have the ability to reach very limited parts of the brain are also able to interact with deeper regions, distant from these partly open windows in the barrier remains to be discovered. These regions have an interface role between the brain and the periphery. They represent areas from which neuroendocrine (hypothalamic) or physiological integrated functions can respond to changes in the periphery in order to control and modify, by feed-back, this functioning of peripheral organs.

Even if a biological barrier does exist between the periphery and the brain (except perhaps at the level of some small areas in the base of the third ventricule, in the area postrema region, at the choroid plexi, hypothalamus-median eminence, pineal gland), that does not mean that the barrier is totally impermeable. Transfer may occur selectively, after intracellular transport. The blood-brain barrier may also be permeable in the other direction, from the brain to the periphery. When peptides are centrally injected, some diffuse out into the periphery. That is the case for at least some synthetic peptides designed for that purpose. One of the best examples is the case of lipophilic structural analogs of arginine vasopressin which are antagonists either to the V_1 or to the V_2 receptor (Manning and Sawyer 1984), and (AVP) itself, which is not lipophilic. In this respect changes in the effects of the peptide administration may occur if the peripheral administration is given in the form of infusion or bolus injection, whatever side of the blood-brain barrier is treated.

The conclusion that peptides do not enter the parenchyma of brain area protected by the blood-brain barrier has been supported by numerous studies (Cornford *et al.* 1978; Zaildi and Heller 1979; Reppert *et al.* 1981; Ang and Jenkins 1982; Ermish *et al.* 1982, 1985; Wood 1982; Meisenberg and Simmons 1983; Pardridge 1983; Stegner *et al.* 1983). It has been suggested that some peptides are able to change the permeability of the barrier and that minute amounts are able to fill the extracellular brain space (Goldman and Murphy 1981). Although the data are somewhat controversial, it is however, generally accepted that there is an accumulation in the blood-brain barrier free areas, which is independent of the concentration of the injected peptide (Ermish *et al.* 1983). Importantly, at least for AVP, it has been demonstrated that the peptide has a rhythmical pattern in cerebrospinal fluid totally independent of the concentration of the hormone in blood (Reppert *et al.* 1981).

Arginine vasopressin and the blood-brain barrier

The doses required to achieve the behavioural effects described after either peripheral or central administration of the peptide are large compared to normal plasma levels of the hormone (De Wied *et al.* 1975; De Wied 1980; Le Moal *et al.* 1981; Koob *et al.* 1981, Deyo *et al.* 1986). It is interesting to note that the behavioural effects seen after subcutaneous administration of AVP have been obtained also after an osmotic physiological challenge with administration of hypertonic saline; in this case the rise in the level of endogenous neuropeptide was within the range of the 'physiological levels' and only due to the release of AVP in the blood (Koob *et al.* 1985). Moreover, whereas most physiologists have noted that the peptide does not cross the blood-brain barrier, two groups have reported an uptake of labelled AVP into brain tissue (Kastin *et al*. 1976, 1979, 1981; Banks and Kastin 1983; Mens *et al*. 1983). The possibility remains that the high doses used pharmacologically (e.g. by peripheral administration) to produce behavioural effects are necessary because, only a small percentage of the blood-borne hormone is able to enter and interact with receptors in the brain parenchyma.

One of the most interesting studies in this field has recently been published and the method and results raise some interesting points (Deyo *et al*. 1986). In these experiments, the accumulation of (AVP) in brain areas inside the blood-brain barrier (such as the thalamus-hypothalamus, amygdala with the overlying temporal cortex, hippocampus, and cerebral cortex), and in areas outside the blood-brain barrier (such as the median eminence of the hypo-thalamus and the area postrema), was measured after subcutaneous injection of the hormone. The authors showed that: (i) the plasma concentrations of AVP peaked at 5 minutes after subcutaneous injection and decreased in a biphasic manner over the next 115 min; (ii) the concentration of AVP in brain tissue samples peaked at 20 minutes after the subcutaneous injection of AVP; the decrease of AVP in the areas protected by the blood-brain barrier followed the time course seen for plasma; (iii) the concentration of AVP in unprotected brain areas also peaked at 20 minutes, but then declined at rates that differed from other brain areas and plasma; and (iv) the concentration of AVP in the plasma and in most brain areas depended on the dose admi-nistered, while those in the median eminence and in the area postrema did not. It was interesting to see the evolution of AVP levels during physiological challenges. Water deprivation for 24 and 48 hours significantly elevated both the plasma AVP concentration, and the concentration of AVP in the hypo-thalamus and in the amygdala-temporal cortex samples. The increases in AVP after water deprivation were limited to these two regions and were quantitatively much lower than after peripheral administration. However, the most important results were the following: when the brains of anaes-thetized rats were perfused free of blood, there were no changes in regional brain AVP content after subcutaneous treatment with 5000 ng/kg of AVP (the dose classically used in behavioural experiments and which improves learning), except for the median eminence. These data suggest that cir-culating AVP does not enter the parenchyma of brain areas protected by the blood-brain barrier in sufficient quantities to be detected by the assay and that the penetration of AVP into areas not protected by the blood-barrier may be due to differences in the specialized cellular components located there.

In conclusion, these results provide evidence that, even when elevated by 40-fold, AVP in the blood stream does not enter the parenchyma of brain tissue protected by the blood brain barrier. They also support the contention that AVP measured in dissected brain regions resides chiefly in the blood vessels. However, it appears that there are no significant differences in the concentrations of AVP in the median eminence of the hypothalamus, hypothalamus-thalamus, the amygdala-temporal cortex, the area postrema, and the remaining cerebral cortex, at 20 min after the subcutaneous injection of AVP doses (s.c.) of 50, 500, or 5000 ng/kg. After a 5000 ng dose, the difference between AVP concentration in the hypothalamus before and after perfusion cannot be due to the amount present in a few microlitres of plasma volume in this region. These data suggest the presence of accumulation mechanisms, perhaps along the walls of the blood vessels, that would account for the high levels 20 min after subcutaneous administration and for the ease with which AVP was washed out. Only further work will be able to clarify where and how circulatory AVP is bound and released from the brain vessels. More generally it could clarify the interaction of AVP—as for other peptides—and its fate in these areas of specialized blood vessel structure, for instance area postrema and median eminence. However, it is interesting to note that while saline perfusion reduces AVP to control levels in area postrema, the concentration is only reduced to 50 per cent in the median eminence. Specialized non-neural cells, tanocytes, exist in both regions and have been proposed to endocytose peptides from the extracellular space (Scott *et al.* 1974; Joseph *et al.* 1975; Uemura *et al.* 1975; Thompson 1982).

Functional considerations

Strategy and method

The relationships between peptides and behaviour raise some considerations which we would like to mention (see also Le Moal *et al.* 1984). The first is of a causal nature. Most of the studies can be classified in one of two ways. One approach, ascending and integrative, starts at the cellular level and proceeds to more complex levels of organization: systems, circuits, structure, organs. This approach tries to discover which parts of the central nervous system are necessary for a given behaviour or response. The other approach can be labelled as descending or reductionistic. It starts from the behaviour and tries to localize the structures, circuitries and cells involved, and beyond to the cellular machinery or the molecular genetics implicated. For example, experiments are now published which relate a given behavioural response to a given change in m-RNA synthesis. In practice, both of these causal approaches are frequently used for the same purpose: to fill the gap between a response and the cellular physiology, and finally for localization.

The second type of consideration is methodological. We generally study the change in a neural basal activity or in the turnover of a molecule and the concomitant change in a behaviour or response, and finally infer the 'role' of the molecule or of the neural system in the organization of the given behaviour. This approach is of the correlational type. Another approach is more functional and has similarities with the classic endocrinological method: for a given endocrine organ or a brain component, its function is assessed by activation or removal and secondarily by reversal of the deficits by extracts, transplants, or grafted cells. Problems arise when this method is applied in the context of behavioural research because this paradigm involves the assumption that a behaviour is a discrete phenomenon and that the lesioned brain component, the removed organ, the molecule or the peptide is required and sufficient for the execution of the response.

The third consideration is troublesome for behavioural neurobiology: what is behaviour and what are the relationships between accounts of what an organism does, introspects, or thinks, and what its nervous system does? The analysis of this last question is not within the scope of this chapter.

Time-dependency and U-shaped responses

Time-dependency (McGaugh 1966) is a central concept for every one interested in studying peptide effects on memory or learning. Its demonstration ensures, to some extent, that the effects are specific for the given learning process and not due to some proactive effect of the peptide, i.e. on performance, attention, arousal, and other non-specific and non-associative actions (McGaugh 1983). Studies based essentially on animal models have provided conceptual frameworks where different types of memory such as sensory registration, storage, consolidation, and retrieval can be categorized (Heise 1981). The same considerations have been discussed for pharmacological effects on memory (Hunter *et al.* 1977) when the peptide is administered before acquisition, or after training, or before retention. Treatments administered before acquisition or retention are easily confounded by non-specific effects so that facilitating amnesic treatments immediately after training, when consolidation processes are susceptible to alteration (McGaugh 1966) according to a retrograde gradient (Martinez 1983) remains the most popular approach. However, experimental analysis from retrograde and anterograde gradients, leads to the conclusion that the demonstration of time dependency may not be evidence of an endogenous storage process. In the case of pre-training treatment there is no experience to store, and it seems that each treatment, i.e. each peptide, appears to produce its own unique time gradient (Chorover 1976; Martinez 1983). From these considerations the question arises as to why an organism should possess internal mechanisms that would be expressed as a learning and memory gradient. One tentative answer is that this reflects the role of endogenous peptidergic and hormonal mechanisms that influence these neurophysiological functions in a way that exhibits strict contiguity; the learning experience and the peptide response have to occur within certain temporal limits for the molecule to influence the process (Martinez 1983). It is assumed that the peptide exaggerates its normal endogenous function and the analog antagonist would do the opposite, and in so doing may reveal mechanisms or structures important in learning or memory. Unfortunately, the concept that a greater concentration of an agonist facilitates a process may not be valid and may not explain the ubiquitous nature of inverted U-shaped dose effect curves. Typically, the peptides alter learning and memory measures in a dose-dependent monotonic manner up to some optimum dose beyond which the effectiveness of the molecule decreases in a similar monotonic function. Surprisingly, the decline is frequently not correlated with overt or detectable side-effects which could explain the results, and has been obtained after administration of physiologically—but not behaviourally—inactive peptide fragments. This phenomenon is paradoxical and does not fit the classical relationship between a drug and the magnitude of the response, i.e. a hyperbolic function when dose is plotted on a linear scale (Gilman *et al*. 1980). If the peptide acts on systems with a finite number of receptors, the effect of the molecule should be proportional to receptor occupancy with a maximum for the maximum of sites. In the case of an excess of the agonist the receptor may show fatigue or tachyphylaxis and finally produces less effect (Day 1979). One could also imagine an inhibition at the synapse level from recurrent collaterals which, when activated, induce an opposite effect on behaviour. Finally, it is also possible that the different systems on which the peptide acts have different affinities or thresholds, and compete for a final response at the higher doses. It is interesting to remember that the relationship between performance and arousal has a U-shaped function: poor performance at low levels of arousal as at high levels, the best performance being generated at some optimal arousal state in-between. This point has been recently discussed in detail for the behavioural action of vasopressin (Moal *et al*. 1984; Saghal 1984). The U-shaped function raises interesting considerations for the role of intervening variables in learning, and for the differentiation between learning and memory.

Hormones and a peripheral theory of learning

The juxtaposition of the words 'peptides' and 'behaviour' raises questions which are difficult to answer: do peptides influence complex cognitive processes—such as learning and memory—and if so how? (McGaugh and Martinez 1981; Martinez 1983). The answer to the first question has been reviewed in countless chapters and reviews, and is affirmative. More interesting and controversial is the second question. It is of interest to refer to the hormonal and peptidergic environment which is correlated with a particular significant event in the context of an adaptive response. Whether a particular environmental event and a given adaptive behaviour are correlated with a set of hormonal responses, or whether these hormonal-peptidergic responses are non-specific with regard to the various events is a matter of debate. It has been suggested that the peptide response serves to establish in one way or another the importance of the event, and modify some unknown brain processes for better storage of the information relating to the event (Gold and McGaugh 1977; Leshner *et al*. 1981; Martinez 1983). According to Martinez (1983) this hypothesis provides a physiological explanation for the so-called flashbulb or strongly stored memories (Brown and Kulik 1975) and, in analogy with the James–Lange theory of emotion, Martinez suggests a James–Lange view of memory. We meet a bear, run, and remember in vivid details many aspects of the encounter; in other words, the bear encounter provokes a massive autonomic hormonal and peptidergic response which, in consequence, modifies brain processes and information storage in terms of space, time, odours, etc., for significant aspects of the encounter. The event is marked for the individual. This extreme example is at the end of a continuum of modulations of learning. Depending on the state of the general activation more subtle modulations are produced by peptidergic and hormonal systems for increasing or decreasing the strength of the storage. A classic example of such modulation is given by the experiments of Gold and Van Buskirk (1976*a*) which showed that according to the training foot-shock and the dose used, the stress hormone ACTH either enhanced or impaired the training performance in an avoidance paradigm. The time dependency parameter allows such possibilities of modulation when the hormonal-peptidergic internal levels are modified after or before the training and lead to retrograde or anterograde amnesia gradients (Haycock et al. 1977; Karpiak and Rapport 1979). Strangely enough, such an influence of the peripheral levels of a hormone upon learning has been demonstrated in deeply anaesthetized rats. In this preparation it is possible to associate a noise with a leg shock according to a Pavlovian conditioning paradigm only if epinephrine was injected into the animal at the time of training, verifying that this adrenal hormone has not reduced the level of the anaesthesia (Weinberger *et al*. 1984).

NEUROPEPTIDES AND BRAIN-BODY INTEGRATIONS IN LEARNING PROCESSES

A rapid survey of the literature: peripheral actions of hormones or peptides and learning

One way to distinguish between central and peripheral actions of peptides is to compare the central versus peripheral doses needed to influence a given behavioural response such as a learning process. For instance, Bohus and De Wied (1981) inferred that ACTH 4–10 delayed extinction of avoidance behaviour through a direct action on the brain following peripheral administration, because a 100-fold lower concentration was needed for the same purpose following central administration. Similar examples were obtained with arginine vasopressin; central versus peripheral administration with a dose-range of 1 (intraventricular) to 1000 ng (subcutaneous). Sometimes, the central and peripheral administration lead to opposite effects such as for the gut peptides. Whatever the interpretation, the use of a strategy based only on dose comparison does not distinguish between parallel systems; one peripheral—for instance, the autonomic nervous system—and another central which may operate similarly, but by different mechanisms. Some experiments suggest this possibility. Melegini *et al.* (1978) tested the hypothesis that peripherally administered diethyldithiocarbamate (DDC) produced learning impairment through an action on central catecholamine pathways. The problem of localization is less complex in the case of catecholamines because peripheral and central pathways are well characterized. In fact, DDC inhibits the synthesis of epinephrine and norepinephrine in the periphery (including the adrenal medulla) as well as in the brain. DDC injected intraperitoneally produces amnesia in a passive avoidance learning paradigm, and this defect is reversed by intracerebral replacement of the amine. Surprisingly, norepinephrine, which does not cross the blood-brain barrier, when injected peripherally also attenuated the amnesia induced by DDC, but the effective dose was between 5 and 500 times greater than the affective central dose. Whatever the interpretation, the data suggest a peripheral mechanism of action of norepinephrine, at least in the lower dose range.

Met-enkephalin administered in rats intraperitoneally impairs acquisition of an active avoidance response; this peptide does not penetrate easily into the brain and that has led to the suggestion that the effect is due to a peripheral mechanism (Rigter *et al.* 1980*a*, *b*). This behavioural action was blocked by the opiate antagonist naloxone, but naloxone easily enters the brain and its action does not differentiate between peripheral or central mechanisms. In the case of opioid peptides, fortunately, a quaternary form of naloxone, known as methylnaloxonium (naloxonium) does not penetrate the brain and also antagonizes the enkephalin actions (for discussion see Martinez 1983). The result suggests a peripheral action on opioid receptors 'somewhere' outside the brain. In another experiment (Martinez *et al.* 1984*a*, *b*), mice were injected before training either with enkephalin or with the combination enkephalin + naloxonium; only the first group was impaired in the acquisition of a one-way active avoidance response. Some authors have proposed a central hypothesis of peptide behavioural action because similar effects were obtained after local intracerebral or after peripheral injections, and many examples exist of precise and clear cut—if not always specific—peptide action after intracerebral infusion. In other words, if an effect on learning and memory produced with a peripheral injection could be reproduced by a more local administration within a specific region of the brain, that was the proof that both effects were produced through an action at the same site, the given brain region. The authors generally infuse the peptide in a region supposed to have a role in learning, in memory or for the behaviour studied. This strategy is in itself a matter of debate, so controversial is the precise localization of a function in the central nervous system. Gallagher and Kapp (1978) have reported that naloxone injected intracerebrally, within the amygdala region, enhanced retention of a passive avoidance response. Later Messing *et al.* (1979) found that the same drug administered i.p. produced the same effect. There were two possible interpretations: first to pretend that the naloxone effect was either or both peripheral and central regardless of the route of administration; and secondly, that it was not useful to distinguish between parallel central and peripheral mechanisms, because the unique site of action for the retention of this particular aversive response was amygdala. Following these results McGaugh and Martinez (1981) and Martinez (1983) showed that naloxonium i.p. had the same effect as naloxone injected centrally. In other words, the antagonism of peripheral opioid receptor systems produced qualitatively the same action as that of naloxone. It is therefore possible to affect learning through peripheral opioid receptors.

The role of peripheral adrenergic systems in mediating drug effects on conditioning has been supported by several experiments. Removal of the adrenal medulla blocks the enhancing action of amphetamine and nicotine in avoidance learning (Orsingher and Fulginiti 1971; Martinez *et al.* 1980). In these experiments, the normal activating effect of nicotine is mediated by its action on the cholinergic receptors located in splanchnic nerve terminals and in adrenal chromaffin cells, and secondly by the release of catecholamines. Conversely, catecholamines administered systemically alter avoidance learning (Gold and Van Buskirk 1975, 1976*b*). Epinephrine can be considered as a learning modulatory hormone (Gold and McCarty 1981; McCarty and Gold 1981) in the sense that a good retention of an avoidance response is correlated with a higher plasma level of epinephrine. The result fits well with the idea that the hormonal response reflects the importance of the event and establishes it.

In conclusion, it is likely that catecholamine hormones and opioid peptides, such as enkephalin, which do not cross the blood-brain barrier affect avoidance conditioning through peripheral mechanisms when injected at the periphery. The integrity of the adrenal medulla is necessary for enkephalin action (Martinez and Rigter 1982) whereas naloxonium or an antiserum directed against leu-enkephalin, which by themselves enhance acquisition, attenuate the effects of enkephalin (Martinez *et al.* 1984 *a, b*).

Peptides and the brain-body dialogue: the gut peptides

Another example is given by the so-called gut peptides. Some peptides such as somatostatin, thyrotropin-releasing factor, substance P . . . were first discovered in the brain and then extracted from the gut while other peptides, such as neurotensin and cholecystokinin, were first found in the gut and subsequently in the brain. During the last few years a growing number of peptides have been discovered on both sides of the blood-brain barrier, in the gut and in brain tissues (Krieger and Martin 1981). This phenomenon was likely to have important implications for brain-gut interactions and has given rise to the concept of a brain-gut axis (Dockray 1982; Pappas *et al.* 1985). Moreover, it has been demonstrated that these peptides are secreted down the vagus to the gut (Dockray 1982). An elegant series of experiments from Taché's and Pappas' groups has provided the basis for a hypothesis to explain the physiological significance of this remarkable phenomenon. These authors have examined the central and peripheral actions of three peptides on gastric function: cholecystokinin octapeptide (CCK-8), bombesin, and somatostatin, which exist in both the gut and brain. Intravenous infusion of CCK-8, in doses of 50, 100, and 200 pmol/kg/hr, caused 28, 38, and 52 per cent inhibition, respectively, of the rate of gastric emptying of a liquid meal in dogs. By contrast, the injection of 32, 64, and 128 pmol/kg into the lateral cerebral ventricle of these dogs accelerated gastric emptying by 6, 26, and 32 per cent, respectively. Bombesin, which stimulated gastric acid secretion in a dose-dependent manner, but which has no effect on the submaximal acid response to pentagastrin when administered peripherally, inhibited in a dose-dependent manner the submaximal response to pentagastrin when given centrally, with a maximal inhibition of 66 $\pm$ 5 per cent, at a dose of bombesin of 180 pmol/kg. Similarly, somatostatin-14 caused graded inhibition of pentagastrin-stimulated acid response when it was given centrally. Maximal inhibition of 51 per cent of the pentagastrin response occurred with a peripheral dose of somatostatin of 800 pmol/kg/hr. By contrast, maximal augmentation of the pentagastrin response of 78 per cent occurred when a dose of 400 pmol/kg of the peptide was injected into the lateral ventricle. These three peptides have an opposite role in the regulation of gastric function when administered peripherally or centrally. For the authors these observations reflect a more general phenomenon common to the neuropeptides of the brain-gut axis: the activation of the peptides in the brain might be a method of turning off gut function in the interdigestive period and, conversely, turning off the central activity of peptides might serve as a means of stimulating gut function (Pappas *et al.* 1985). Here again much has to be done to demonstrate the pathways mediating the central effects. However, at least for bombesin, the sympathetic pathway is proposed to explain how the central action of bombesin is relayed to the stomach (Taché *et al.* 1982). It seems that gut neuropeptides acting centrally can influence vagal afferent inputs and that they may modulate efferent vagal and sympathetic activity (Ewart and Wingate 1983), the vagus nerve being in this case considered as a 'peptidergic highway connecting the brain and the gut' (Pappas *et al.* 1985).

Another example of a role for a neuropeptide in the brain periphery relationships is provided by corticotropin-releasing factor (CRF). Intravenous CRF inhibits basal pentagastrin-stimulated gastric acid secretion in rats and dogs in a dose-dependent, long-lasting, and reversible manner while growth hormone-releasing factor is without effect. This action is not blocked by naloxone, indomethacin, or by hypophysectomy, or by adrenalectomy; it does not involve a decrease in gastrin release and is not dependent upon prostaglandin synthesis. Again this action of CRF seems to significantly require the integrity of the nerve vagus, along with other mechanisms (Taché *et al.* 1984). Moreover, CRF injected into the cisterna magna or the lateral hypothalamus also inhibited basal and pentagastrin-stimulated gastric acid secretion in rats bearing pylorus ligation or gastric fistula (Taché *et al.* 1983). This suppression was not observed after infusion of CRF within the frontal cortex. It is interesting to note that the inhibitory effect of CRF was blocked by vagotomy and adrenalectomy, but not by hypophysectomy or naloxone treatment, which confirms that CRF acts from the brain on the body through modulation of the autonomic nervous system, i.e. via vagal and adrenal mechanisms, and not through hypophysiotropic effects. These results are of great interest because they might provide a biological link between the hypothalamus and the cognitive-emotional systems acting upon the organs as an expression of emotional and psychosomatic disturbances. These findings that CRF mimics the autonomic and endocrine response to stress, and that it alters gastric secretion in a manner similar to that of various stressors suggest that CRF may have a role in the pathophysiologic gastric response to stressors (Taché *et al.* 1983).

Peripherally injected arginine vasopressin improves learning by means of peripheral mechanisms

Arginine vasopressin and behaviour: controversies

The primary physiological action of arginine-vasopressin is to conserve body water by acting on the kidney (for review see Sawyer 1964). The pathological loss of vasopressin secreting neurons from the hypothalamus results in a pronounced and chronic diuresis, a condition known as diabetes insipidus (Hays 1980). Vasopressin also has potent pressor actions and although this pressor effect requires significantly higher doses than the anti-diuretic one, it is physiologically significant during hypovolemic or hypotensive crises (Zerbe *et al.* 1982). Vasopressin also has a corticotropin-releasing factor like effect (Gillies and Lowry 1979). The neuropeptide is present not only in the hypothalamo-neuro-hypophysial system, but also in neural pathways that project to various and specific areas of the central nervous system (Buijs 1978; Sofroniew 1980). Axon terminals are found in the proximity of cell clusters believed to participate in autonomic functions such as the nucleus of the solitary tract-dorsal motor nucleus of the vagus complex or sympathetic preganglionic neurons in the intermediolateral column of the spinal cord. Terminals are also found in regions given to regulate adaptive behaviours and learning processes such as prosencephalic limbic structures. Physiological and anatomical studies have prompted an interest in the possible central role of AVP for modulating autonomic functions. AVP released from the neuro-pituitary acts at the periphery on various organs through receptors belonging to the smooth muscle-type (V_1) whose intracellular signal is dependent upon the hydrolysis of membrane inositol phospholipids, or to the kidney-type (V_2) associated with adenylate cyclase (Jard 1981, 1983). Arginine vasopressin-like immunoreactivity has also been found in the superior central ganglion where functional V_1 receptors have been detected, suggesting that AVP plays a role in peripheral autonomic function (Kiraly *et al.* 1986).

In early behavioural work De Wied and associates (1965) found hypophy-sectomized rats to be deficient in a number of behavioural situations, especially the acquisition and extinction of aversively motivated tasks. These deficiencies were reversed by administration of a raw pituitary extract, pitressin, and, in later work, by lysine vasopressin (LVP) in microgram amounts injected subcutaneously (Bohus *et al.* 1973). Vasopressin also reversed the behavioural deficits observed in Brattleboro strain rats with con-genital diabetes insipidus (Bohus *et al* 1975; De Wied *et al.* 1975). Further-more, LVP injected subcutaneously delayed extinction in an active avoi-dance task in intact animals (De Wied 1971) and in passive (inhibitory) avoi-dance (Ader and De Wied 1972) as did intracerebroventricular injection of nanogram quantities of AVP (De Wied 1976). These same investigators reported that AVP enhanced retention of the passive avoidance response when injected subcutaneously either just after the training test (foot-shock) or just before the retention test, but not at times in between (Bohus *et al.* 1978), suggesting that AVP enhanced both consolidation and retrieval of memory.

Evidence to support the hypothesis that vasopressin produces its beha-vioural effects independently of its classical renal or pressor effects came from the work of De Wied and associates with different analogs of vaso-pressin. For example, desglycinamide-lysine vasopressin (DG-AVP), a vasopressin analog with minimal pressor and renal activity, reverses the behavioural deficits associated with diabetes insipidus (De Wied *et al.* 1975) and has effects which are similar to those of vasopressin in tests with active avoidance. DG-AVP or DG-LVP also counteract disruptions of consolida-tion induced with diethyldithiocarbamate, pentylenetretrazol, puromycin, and CO_2-induced amnesia (Lande *et al.* 1972; Rigter *et al.* 1974; Bookin and Pfeifer 1977; Asin 1980). The relative behavioural potencies of these analogs suggests that the ring structure of vasopressin may be most important for the 'consolidation' of acquired avoidance responses, while the C-terminal appears to be more important for reversing the effects of amnesia treatments (Van Ree *et al.* 1978).

With few exceptions, the data regarding a role for vasopressin in memory came from studies employing aversively motivated tasks. Obviously, this limits the nature of the behavioural conclusions that can be drawn from such data. If vasopressin does have 'memory' enhancing properties, this effect should also be demonstrable using positively motivated tasks. Rats trained to discriminate sides of a T-maze using a sex reward showed a facilitation of retention with DG-LVP (Bohus 1977). Rats receiving vasopressin during acquisition training in a black-white discrimination T-maze task showed a prolongation of extinction, but only on the side using the black discriminative stimulus (Hostetter *et al.* 1977). DG-AVP has been shown to facilitate the acquisition of an autoshaping response and to prolong extinction of this response (Messing and Sparber 1983). In one of the only observations to date of an effect of the peptide in an appetitively-motivated task, post-training administration of AVP can facilitate subsequent test performance in a one trial water-finding task (Ettenberg *et al.* 1983*a*, *b*). More recently Packard and Ettenberg (1985) used an elevated eight-arm radial maze and tested the effects of AVP administration on the spatial learning abilities of food-deprived rats. Following 18 days of reinforced training, each animal was briefly exposed to the maze with no food available in any of the eight food-cups; immediately after this preliminary trial animals were injected with a simple subcutaneous dose of either saline, AVP, or DG-AVP. Additional extinction trials were conducted at 2, 4, 6, and 8 hr post-injection. The results showed that AVP potentiates this radial maze extinction behaviour while DG-AVP produced behavioural results opposite to those predicted by a memory facilitation hypothesis. Incidentally, these results also suggest that peripheral endocrinological responses may be necessary to demonstrate memory-enhancing effects following peripherally administered AVP.

Recent findings with vasopressin antagonists have reopened some questions raised by this earlier work of De Wied and associates (for review see Le Moal *et al.* 1984). A pressor antagonist analog of arginine vasopressin, 1-deaminopenicillamine, 2-(0-methyl)tyrosine-AVP [dPTyr-(Me)AVP] which prevented the AVP pressor response (Bankowski *et al.* 1978), blocked the effects of both subcutaneously (Le Moal *et al.* 1981) and intraventricularly (Koob *et al.* 1985) injected AVP on prolongation of extinction of active avoidance. The antagonist had little effect on its own except at high doses where it produced a facilitation of extinction of active avoidance (Koob *et al.* 1981). This antagonist also blocked the effects of AVP on passive avoidance, but at somewhat higher doses (Lebrun *et al.* 1984). Again these effects indicate either that signals from peripheral visceral sources play an important role in the subsequent behavioural changes or that the receptors responsible for AVP pressor effects are similar to those in the central nervous system leading to its behavioural action. These observations have prompted a significant effort to distinguish between these two hypotheses. In two recent studies De Wied and associates have shown that a new analog of vasopressin (pGlu4,Cyt6)-AVP-4-8,-(AVP4-8), is much more active than AVP itself in facilitating passive avoidance when injected peripherally post-training (Burbach *et al.* 1983; De Wied *et al.* 1984). This analog has no pressor activity and these effects can be reversed by a potent AVP pressor antagonist (De Wied *et al.* 1984), even when the antagonist is injected in nanogram amounts intracerebroventricularly. Similar results were obtained using AVP itself (De Wied *et al.* 1984), leading De Wied and associates to conclude that both peripherally and centrally derived AVP act on CNS receptors.

Another area of controversy in vasopressin research centres on the divergent results obtained using Brattleboro rats, devoid of AVP both in the CNS and pituitary. Whereas the early reports claimed that Brattleboro rats were impaired in retention of active and passive avoidance (De Wied *et al.* 1975; Bohus *et al.* 1975), several studies have demonstrated that non-Utrecht Brattleboro rats are unimpaired in a variety of learning tests (Celestian *et al.* 1975; Bailey and Weiss 1979; Brito *et al.* 1980; Carey and Miller 1982). Indeed, in other laboratories Brattleboro rats often show longer retention latencies for shock avoidance, i.e. better memory than non-Brattleboro rats, and some controversy exists as to the proper control groups in both camps (for point/counterpoint, see Gash and Thomas 1983, 1984; De Wied 1984*a, b*).

Clinical studies with vasopressin and vasopressin analogs devoid of endocrine or physiological effects such as dDAVP have shown mixed results in various situations, in normal subjects or depressed, cognitively impaired, in memory disorders. Some authors have reported improvements (Legros *et al.* 1978; Oliveros *et al.* 1978; Anderson *et al.* 1979; Weingartner *et al.* 1981; Beckwith *et al.* 1982; Laczi *et al.* 1983). Other studies have not been so positive (Blake *et al.* 1978; Koch-Hendriksen and Nielsen 1981; Jenkins *et al.* 1981; Fewtrell *et al.* 1982).

Arginine vasopressin, interoceptive signals, and facilitation of stimulus associations

Work in our laboratories has been devoted to assessing the peripheral effects of this hormone and more generally to determining if interoceptive signals perceived by the subject may induce a state of increased arousal which could facilitate stimulus association and hence affect subsequent behaviour. Using an antagonist to reverse the pressor effects of AVP (Le Moal *et al.* 1981, 1984) we were able to block the effects on memory. The same results were obtained by injecting the antagonist intracerebroventricularly, but only at doses which reversed the peripheral effects of AVP injected subcutaneously

(Lebrun *et al.* 1985). The antagonists are highly lipophilic and easily cross the blood-brain barrier (Le Moal *et al.* 1982). In brief (i) AVP injected peripherally improves learning only at doses which induce physiological effects, (ii) when these physiological effects, i.e. systemic blood pressure increase, are blocked by a V_1 antagonist, the behavioural action of AVP is also blocked, (iii) central injections of AVP at doses which improve learning do not alter systemic blood pressure and do not produce aversive stimulus properties, but these effects are blocked either by small doses of a V_1 antagonist when injected centrally or high doses of the same antagonist when injected peripherally suggesting that the receptor mediating the central effects is functionally similar to the AVP V_1 receptor, (iv) AVP administered centrally acts by central mechanisms, unknown at the present time and AVP administered peripherally acts by peripheral mechanisms related to its physiological effects (Le Moal *et al.* 1982, 1984, 1985; Ettenberg 1984; Ettenberg *et al.* 1983*a*, *b*; Lebrun *et al.* 1984, 1985; De Wied *et al.* 1984; Bluthe *et al.* 1985*a*, *b*; Koob *et al.* 1985, 1986).

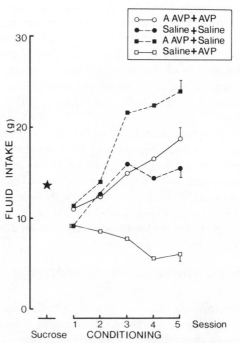

FIG. 12.1.　Effects of dPTyr(Me) AVP on conditioned taste aversion to AVP. The figure shows mean fluid intake during the last 2 days of sucrose presentation(star) and over repeated conditioning trials for groups treated with 10 μg/kg AVP or saline preceded by an injection of 50 μg/kg dPTyr(Me) AVP (AAVP) or saline, after milk presentation in a single bottle test. Each point is the mean of five rats. The standard error of each group mean is given for the last test session.

It is essential to investigate mechanisms by which post-training administration of the neuropeptide modifies behaviour; increases resistance of conditioned avoidance, improves retention of appetitive learning, and at the same time has aversive or arousal properties. The specificity of the aversive stimulus properties of AVP has been investigated extensively in our laboratory (Bluthe *et al.* 1985*a*, *b*; Dantzer *et al.* 1982), mainly by using the conditioned taste aversion (CTA) paradigm. When an animal is given the opportunity to ingest a solution with a particular taste followed by the administration of a toxic agent (e.g. ionizing radiation or lithium chloride), it will subsequently show an aversion to the initially consumed solution. This CTA has been initially conceptualized as a form of classical conditioning in which some type of malaise or gastrointestinal distress serves as an unconditioned stimulus. The compelling nature of this phenomenon has generated considerable interest both for behavioural analysis and in pharmacology. Many

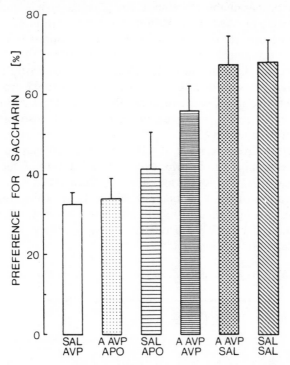

FIG. 12.2. Effects of dPTyr(Me) AVP on conditioned taste aversion to apomorphine or to AVP. The figure shows means + SEM percent preferences for saccharin in a two-bottle test session with five rats per group. Animals were injected with saline (SAL), 10 µg/kg AVP (AVP) or 0.5 mg/kg apomorphine (APO), preceded by saline or 50 µg/kg dPTyr(Me) AVP (AAVP), following saccharin consumption on days 1 and 3 of the experiment and the test session was run on day 5, by presenting two drinking bottles, one containing the saccharin solution and the other water.

agents have now been reported to be effective in eliciting a taste aversion. This is true for most psychotropic drugs, including self-administered drugs. The fact that taste aversion can be obtained with such a wide variety of agents has been interpreted as indicating that overt illness is not a requisite for the aversion to occur. Alternative interpretations emphasize the importance of factors such as aversiveness or novelty induced by the drugs which are responsible for CTA. This model is useful for the study of the stimulus complex produced by molecules which control behaviour, and if AVP is able to produce visceral afferent signals placing the animal in a state of altered arousal, a more direct way of studying such effects and the way they control behaviour would be to look for possible aversive stimulus properties of AVP. In a first set of experiments rats injected subcutaneously with AVP following presentation of a milk solution developed a marked aversion to the taste of this solution (Fig. 12.1). Incidently, dDAVP, devoid of pressor activity, but still antidiuretic was unable to induce CTA. These aversive stimulus properties of AVP were blocked by the V_1 antagonist. These results clearly demonstrate that AVP induced CTA by interacting with vasopressor-like receptors, not with the antidiuretic activity of AVP (no effect with dDAVP). In another set of experiments, by using the same techniques we demonstrated that CTA induced by vasopressin was blocked by prior exposure to the peptide, but not to another aversive agent, such as apomorphine (Fig. 12.2). A prior exposure to another hypertensive agent, angiotensin II (AII), also blocked CTA induced by AVP and this effect was fully reciprocal since prior exposure to AVP blocked the aversive effect of AII (Table 12.1). We also demonstrated that the protection offered by prior exposure to AII was not due to an endogenous release of AVP since the aversive properties of AII were not blocked by administration of the V_1 antagonist (Table 12.2). Taken together these results suggest strongly that the interoceptive cues which are responsible for the CTA induced by AVP are related to the hypertensive action of the neuropeptide.

TABLE 12.1. *Effects of prior administration of AII on conditioned taste aversion to AVP*

Pretreatment	Conditioning		
	SAL	AVP	AII
SAL	86.1 ± 2.58	30.4 ± 2.80	31.0 ± 6.19*
AVP	81.4 ± 6.47	51.7 ± 3.51*†	55.6 ± 8.70*†
AII	76.1 ± 5.77	62.8 ± 4.79*†	79.8 ± 4.53†

Results are expressed as mean (± SEM) percentage preferences for saccharin in two two-bottle test sessions with four rats per group. † indicates significant difference from saline conditioning, $P < 0.05$ (Newman–Keuls test). Single asterisk indicates significant difference between groups within a given conditioning treatment, $P < 0.05$ (Newman–Keuls test).

TABLE 12.2. *Effects of AAVP on conditioned taste aversion to AII*

Treatment	Percentage preference for saccharin
SAL + SAL	80.2 ± 2.10
SAL + AII	57.3 ± 7.63*
AAVP + SAL	77.2 ± 2.69
AAVP + AII	53.4 ± 4.69*

Results are expressed as mean (± SEM) percentage preferences for saccharin in a two-bottle test session with seven rats per group. Asterisk indicates significant difference from other means at $P < 0.01$.

The conclusion that AVP improves memory by mechanisms involving peripheral arousing actions (Le Moal *et al.* 1984; Koob *et al.* 1985) has also been discussed by other authors (Saghal 1984; Ettenberg 1984). The aversive actions of AVP have been tested by Ettenberg (1984) with two behavioural assays: (i) a CTA, and the results are similar to those described above, (ii) a conditioned place test in which rats learned to avoid a distinctive environment associated with AVP administration. Ettenberg also used appetitive learning paradigms to study memory action of AVP. Here the memory effects of AVP were observed in a one-trial food-finding task where non-deprived rats were briefly exposed to a large open-field that contained an alcove in which a high-incentive food reward (sweetened milk) was freely available. AVP injected immediately upon removal from the apparatus produced faster latencies to relocate the alcove (compared to vehicle controls) when tested 48 hr later. Finally, both the memory and aversive responses to AVP were prevented in a dose-dependent manner by immediate pretreatment with intracerebroventricular infusions of the pressor V_1 antagonist dPTyr (Me) AVP, but very large doses were required to block the neuropeptide behavioural action. These results suggest that the critical site of action was far from the ventricles. The doses of antagonist used were those necessary to block the vasopressor effect of AVP (Lebrun *et al.* 1985).

Arginine vasopressin administered peripherally facilitates social memory

The neurohormone AVP affects learning and consequently memory by means of identified peripheral signals. As reviewed earlier in this chapter, for the doses used, the peripherally injected peptide does not cross the blood-brain barrier. As mentioned above most of the tests used so far were based on aversively motivated paradigms. We have recently developed a social memory test, based on recognition of conspecific identity, a form of memory very similar to factual memory in humans. In rodents, individual recognition is based mainly upon chemosensory cues. Sexually-experienced males can easily discriminate between the odour of a stranger and that of a cagemate, as well as between the odour of two different cagemates (Carr *et al.* 1976). The length of time during which such social recognition remains in the absence of the congener may be used as an index of the memory for this particular stimulus. The value of this index is not yet known for most forms of social memory. In a recent study, however, Thor and Holloway (1982) observed that adult male rats were able to discriminate between individual juveniles and that the memory they were forming for a particular individual was short-lasting. This memory was subsequently found to be based on the olfactory characteristics of the stimulus animal (Sawyer *et al.* 1984). We first examined whether this form of memory could be modulated by retroactive inhibition or facilitation induced by presentations of appropriate stimulus animals interposed between the first and second exposure to the original juvenile. The results showed that adult male rats spend a great amount of time investigating novel juveniles. In contrast, rats re-exposed to the same juvenile 30 min after the initial exposure display little investigatory behaviour; if the re-exposure occurs 2 hr later, the juvenile is thoroughly investigated. These results can be

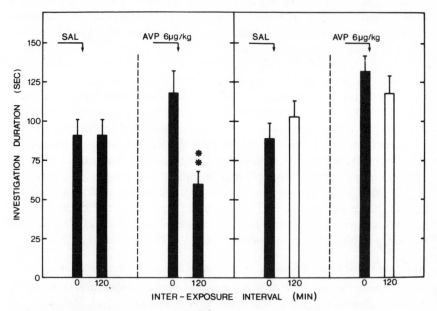

FIG. 12.3.   Retroactive facilitating effects of AVP on social memory. Means and SEM of social investigation times by adult male rats (*n* = 10) on first and second exposure to the same juvenile stimulus (left panel) or to a different juvenile (right panel). In each case, 6 μg/kg AVP was injected immediately after the first exposure. **P < 0.01 compared to time 0.

interpreted to mean that rats form a transient memory for a particular juvenile. We have also demonstrated that memory was enhanced when the initial exposure to the juvenile was followed by another exposure to the same juvenile (retroactive facilitation) and impaired when exposure to the original juvenile was followed by exposure to another juvenile (retroactive interference). Having validated this memory model, the second objective was to determine whether social memory could be modulated by subcutaneously administered AVP. As expected AVP had retroactive facilitating effects on social memory (Fig. 12.3) and these effects were blocked by the vasopressor antagonist dPTyr(Me)AVP (Fig. 12.4). These results suggest that AVP has a prepotent role in modulating the mnemonic processing of chemosensory information associated with social interactions (Dantzer *et al.* 1987).

Data accumulated over the last 5 years led us to propose that AVP influences behaviour, and undoubtly learning, by peripheral mechanisms. These mechanisms are complex, but different in nature from those acting when the peptide is injected centrally. Here we could discuss again what we called a homology of function (Le Moal 1985). A given neuropeptide can influence a given function in a different or complementary manner by central and/or peripheral mechanisms achieving better adaptation of the organism. Given its widespread central and peripheral distribution and action, AVP is

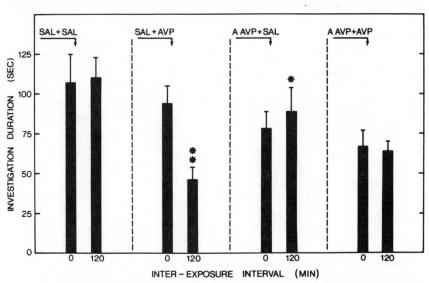

FIG. 12.4.   Effects of 30 μg/kg dPTyr(Me) AVP on retroactive facilitating effects of 6 μg/kg AVP on social memory. Means and SEM of social investigation times by adult male rats (n = 7) on first and second exposure to the same juvenile stimulus. Each group is identified by two terms describing the two treatment administered at 5-minute intervals after the first exposure [AAVP = dPTyr(Me) AVP]. *$P < 0.05$; **$P < 0.01$ compared to time 0.

a neuropeptide on which more and more studies are now concentrated. All the studies mentioned are of a pharmacological nature. However, recent studies have attempted to answer more physiological questions. Given the fact that AVP is released from the pituitary in response to physiological challenges it was interesting to establish whether situations which increased peripheral levels of AVP were able to influence behaviour and more specifically learning situations. In other words the objective was to evaluate the relationship between the behavioural effects of exogenously administered AVP and the behavioural functions of endogenously released AVP. In the same experiment we demonstrated that a potent peripheral osmotic stimulus, an intraperitoneal injection of hypertonic saline, at doses known to release AVP both centrally and peripherally, produced behavioural effects similar to those of exogenously administered AVP. These results were obtained by using an active avoidance task (Koob *et al.* 1985) or a passive inhibitory avoidance paradigm (Lebrun *et al.* 1985) In each case the facilitation of learning was reversed by pretreatment with the V_1 antagonist dPTyr (Me) AVP, and only at doses which blocked the increase in blood pressure suggesting, firstly that endogenously released AVP may also produce behavioural effects and, secondly, that peripheral mechanisms are involved.

CONCLUDING REMARKS

The complexity of the relationships between molecules and behaviour has led an increased number of neurobiologists to believe that progress can only be made by initiating research at the cellular or at the molecular level, or by studying the role of biological molecules in simple preparations such as *Aplysia* through simple 'behavioural models' provided by such preparations. We do not deny a need for these approaches nor the enormous impact of cellular and molecular neurobiology on our culture and everyday scientific life, nor do we deny that we are expecting a lot from the exploding molecular genetics in behavioural neurosciences, neurology, and psychiatry. Neurobiology is prone to reductionism. Each level of organization has its own logic and complexity, and the development of cognitive neuropsychology is a good example that much has to be done to understand integrative processes. Integrative neurobiology must combine, on one hand, top-down and bottom-up approaches and, on the other hand, the languages of the body. We think not because we have a mind or a soul, but ontogenetically and at the origin also because we have a body. Languages suggest information and syllables, words, sentences, syntaxes. Hormones and neuropeptides contribute to the necessary interactions between 'periphery', internal organs, muscles, etc., and brain (Iversen 1981); they constitute important information components.

Ideally, integrative behavioural neurobiology should have first some kind of holistic theory based on all the possible sources of information available on the peptide under investigation. The theory is then mainly tested by correlational studies aimed at finding out: (i) whether specific behavioural processes are related to the peptide; (ii) which role, as a code in a language, has the given peptide; (iii) the receiver and the origin, why and how. If there are some relationships, experimental activities manipulating the level of one factor while holding the other constant should be undertaken. All this can only be achieved if the constructs on which the initial theory is built (e.g. behavioural strategies, active *v.* passive attitudes, etc.) are described at a highly operational level, for example, in terms of information processing (e.g. selective attention, use of exteroceptive *v.* interoceptive information) or of organization of basic motor patterns (e.g. defensive or offensive postures). It is only in this way that description of the functions of neuropeptides at the behavioural level can join together with other approaches (Dantzer *et al.* 1986). From a biological perspective it is worth speculating why such molecules are present on both sides of the blood barrier (homology of function ?) and whether the role of these neuropeptides in behaviour or cognitive processes may somehow be related to their classical hormonal action. Such speculations may help us to enlarge the concept of homeostasis including the regulations of body, organs, systems, and of central states, the behaviour being an essential component of these regulations.

REFERENCES

Ader, R. and De Wied, D. (1972) D. Effects of lysine vasopressin on passive avoidance learning. *Psychonomic Science* **29**, 46–8.

Anderson, L.T., David, R., Bonnet, K., and Dancis, J. (1979). Passive avoidance learning in Lesch–Nyhan disease: effect of 1-Desamino-8-Arginine-vasopressin. *Life Sciences* **24**, 905–10.

Ang, V.T.Y. and Jenkins, J.S. (1982). Blood-cerebrospinal fluid barrier to arginine-vasopressin, demopressin and desglycinamide arginine-vasopressin in the dog. *Journal of Endocrinology* **93**, 319–25.

Asin, K.E. (1980). Lysine vasopressin attenuation of diethylidithio-carbamate-induced amnesia. *Pharmacology, Biochemistry, and Behaviour* **12**, 343–6.

Bailey, W.H. and Weiss, J.M. Evaluation of a 'memory deficit' in vasopressin-deficient rats. *Brain Research* **162**, 174–8.

Bankowski, K., Manning, M., Haldar, J., and Sawyer, W.H. (1978). Design of potent antagonists of the vasopressor response to arginine-vasopressin. *Journal of Medical Chemistry* **21**, 850–3.

Banks, W.A. and Kastin, A.J. (1983). CSF-plasma relationships for DSIP and some other neuropeptides. *Pharmacology, Biochemistry and Behaviour* **19**, 1037–40.

Beckwith, B.E., Petros, T., Kanaan-Beckwith, S., Couk, D.I., and Hauge, R.J. (1982). Vasopressin analog (DDAVP) facilitates concept learning in human males. *Peptides* **3**, 627–30.

Blake, D.R., Dodd, M.J., and Evans, J.G. (1978). Vasopressin in amnesia. *Lancet* **i**, 608.

Bluthé, R.M., Dantzer, R., and Le Moal, M. (1985*a*). Peripheral injections of vasopressin control behaviour by way of interoceptive signals for hypertension. *Behavioural Brain Research* **18**, 31–9.

——, ——, R., Mormède, P., and Le Moal, M. (1985*b*). Specificity of aversive stimulus properties of vasopressin. *Psychopharmacology* **87**, 238–41.

Bohus, B. (1977). Effect of desglycinamide-lysine vasopressin (DG-LVP) on sexually motivated T-maze behavior of the male rat. *Hormones and Behaviour* **8**, 52–61.

—— and De Wied, D. (1981). Actions of ACTH- and MSH-like peptides on learning, performance and retention. In *Endogenous Peptides and Learning and Memory Processes* (ed. J.L. Martinez Jr., R.A. Jensen, R.B. Messing, H. Rigter, and J.L. McGaugh), pp. 59–77. Academic, New York.

——, Gispen, W.H., and De Wied, D. (1973). Effect of lysine vasopressin and ACTH-$_{4-10}$ on conditioned avoidance behavior of hypophysectomized rats. *Neuroendocrinology* **11**, 137–43.

——, Van Wimersma-Greidamus, T., and De Wied, D. (1975). Behavioral and endocrine responses of rats with hereditary hypothalamic diabetes insipidus (Brattleboro Strain). *Physiological Behaviour* **14**, 609–15.

——, Urban, I., Van Wimersma-Greidmus, T.B., and De Wied, D. (1978). Opposite effects of oxytocin and vasopressin on avoidance behavior and hippocampal theta rhythm in the rat. *Neuropharmacology* **17**, 239–47.

Bookin, H.B. and Pfeifer, W.D. (1977). Effect of lysine vasopressin on pentylenetetrazol-induced retrograde amnesia in rats. *Pharmacology, Biochemistry, and Behaviour* **7**, 51–4.

Brito, G.N.O., Thomas, G.J., Gingold, S.I., and Gash, D.M. (1980). Behavioural characteristics of vasopressin-deficient rats (Brattleboro strain). *Brain Research Bulletin* **6**, 71–5.

Brown, R. and Kulik, J. (1975). Flashbulb memories. *Cognition* **5**, 73–99.

Buijs, R.M. (1978). Intra- and extra hypothalamic vasopressin and oxytocin pathways in the rat: pathways to the limbic system, medulla oblongata and spinal cord. *Cell Tissue Research* **192**, 423–35.

Burbach, J.P.H., Kovacs, G.L., De Wied, D., Van Nispen, J.W., and Greven, H.M. (1983). A major metabolite of arginine vasopressin in the brain is a highly potent neuropeptide. *Science* **221**, 1310–12.

Carey, R.J. and Miller, M. (1982). Absence of learning and memory deficits in the vasopressin deficient rat (Brattleboro strain). *Behavioural Brain Research* **6**, 1–13.

Carr, W.J., Yee, L., Gable, D., and Marasco, E. (1976). Olfactory recognition of conspecifics in domestic Norway rats. *Journal of Comparative Psychology* **90**, 821–8.

Celestian, J.F., Carey, R.J., and Miller, M. (1975) Unimpaired maintenance of a conditioned avoidance response in the rat with diabetes insipidus. *Physiological Behaviour* **15**, 707–11.

Chorover, S. (1976). An experimental critique of 'consolidation studies' and an alternative 'model-systems' approach to the biophysiology of memory. In *Neural Mechanisms of Learning and Memory* (ed. M.R. Rosenzweig and E.L. Bennett), pp. 561–82. MIT Press, Cambridge.

Cornford, E.M., Braun, L.D., Crane, P.D., and Oldendry, W.H. (1978). Blood-brain barrier restriction of peptides and the low uptake of enkephalins. *Endocrinology* **103**, 1297–303.

Dantzer, R., *et al.* (1982). Aversive stimulus properties of peripherally injected arginine vasopressin. In *Drug Discrimination: Applications in CNS Pharmacology* (ed. F.C. Colpaert and J.L. Slangen), pp. 305–14. Elsevier Biomedical Press/Janssen Research Foundation, Amsterdam.

——, Tazi, A., Bluthé, R.M., Koob G.F., and Le Moal, M. (1986). Neuropeptides and Behavior. In *Ion Channels Membranes* (ed. R.D. Keynes and J.M. Ritchie), pp. 347–62. Alan R. Liss Inc., New York.

——, Bluthé, R.M., Koob, G.F., and Le Moal, M. (1987). Modulation of social memory in male rats by neurohypophyseal peptides. *Psychopharmacology* **91**, 363–8.

Day, M.D. (1979). *Autonomic Pharmacology. Experimental and Clinical Aspects.* Churchill Livingston, New York.

De Wied, D. (1965). The influence of the posterior and intermediate lobe of the pituitary and pituitary peptides on the maintenance of a conditioned avoidance response in rats. *International Journal of Neuropharmacology* **4**, 157–67.

—— (1971). Long term effect of vasopressin on the maintenance of a conditioned avoidance response in rats. *Nature* **232**, 58–60.

—— (1976). Behavioral effects of intraventricularly administered vasopressin and vasopressin fragments. *Life Sciences* **19**, 685–90.

—— (1980). Pituitary-adrenal system hormones and behavior. In *Selye's Guide to Stress Research* (ed. H. Selye), Vol. 1 pp. 252–79. Van Nostrand Reinhold Co., New York.

—— (1984a). The importance of vasopressin in memory. *Trends in Neuroscience* **7**, 62–4.

—— (1984b). The importance of vasopressin in memory. *Trends in Neuroscience* **7**, 109.

——, Bohus, B., and Van Wimersma Greidamus, T.J.B. (1975). Memory deficit in rats with hereditary diabetes insipidus. *Brain Research* **85**, 152–6.

——, Gaffori, O., Van Ree, J.M., and De Jong, W. (1984). Central target for the behavioural effects of vasopressin neuropeptides. *Nature* **308**, 276–8.

Deyo, S.N., Shoemaker, W.J., Ettenberg, A., Bloom, F.E., and Koob, G.F. (1986). Subcutaneous administration of behaviorally effective doses of arginine-vasopressin change brain AVP content only in median eminence. *Neuroendocrinology* **42**, 260–6.

Dockray, G.J. (1982). The physiology of cholecystokinin in brain and gut. *British Medical Bulletin* **38**, 253–8.

Ermish, A., Landgraf, K., Heinold, G., and Stuba, G. (1982). Vasopressin, blood-barrier and memory. In *Neuronal plasticity and Memory Formation* (ed. C. Marsan and H. Matthies), pp. 147–52. Raven Press, New York.

——, Rinhle, H.J., Neubert, K., Hartrdt, B., and Lardgraf, R. (1983). On the blood-brain barrier to peptides (^{3}H) B-casomorphin-5 uptake by eighteen brain regions *in vivo*. *Journal of Neurochemistry* **41**, 1229–33.

——, ——, Landgraf, R., and Hess, J. (1985). Blood-brain barrier and peptides. *Journal of Cerebral Blood Flow and Metabolism* **5**, 350–2.

Ettenberg, A. (1984). Intracerebroventricular application of a vasopressin antagonist peptide prevents the behavioral action of vasopressin. *Behavioural Brain Research* **14**, 201–11.

——, Le Moal, M., Koob, G.F., and Bloom, F.E. (1983*a*). Vasopressin potentiation in performance of a learned appetitive task: reversal by a pressor antagonist analog of vasopressin. *Pharmacology Biochemistry and Behaviour* **18**, 645–7.

——, Van der Kooy, D., Le Moal, M., Koob, G.F., and Bloom, F.E. (1983*b*). Can aversive properties of (peripherally-injected) vasopressin account for its putative role in memory? *Behavioural Brain Research* **7**, 331–50.

Ewart, W.R. and Wingate, D.L. (1983). Cholecystokinin octapeptide and gastric mechanoreceptor activity in rat brain. *American Journal of Physiology* **244**, G613–7.

Feldman, R.S. and Quenzer, L.F. (1984). *Fundamentals of Neuropsychopharmacology*. Sinauer Associates, Sunderland, MA.

Fewtrell, W.D., House, A.O., Jamie, P.F., Oates, M.R., and Cooper, J.E. (1982). Effects of vasopressin on memory and new learning in a brain-injured population. *Psychological Medicine* **12**, 423–5.

Gallagher, M. and Kapp, B.S. (1978). Manipulation of opiate activity in the amygdala alters memory processes. *Life Sciences* **23**, 1973–8.

Gash, D.M. and Thomas, G.J. (1983). What is the importance of vasopressin in memory processes? *Trends in Neuroscience* **6**, 197–8.

—— and —— (1984). Reply from Don M. Gash and Garth J. Thomas. *Trends in Neuroscience* **7**, 64–5.

Gillies G. and Lowry, P.J. (1979). Corticotrophin releasing factor may be modulated vasopressin. *Nature* **278**, 463–5.

Gilman, A.G., Mayer, S., and Melman, K.L. (1980). Pharmacodynamics: Mechanisms of drug action and the relationship between drug concentrations and effect. In *Goodman and Gilman's The Pharmacological Basis of Therapeutics* (ed. A. Goodman Gilman, L.S. Goodman, and A. Gilman), pp. 28–39. Macmillan, New York.

Gold, P.E. and McCarty, R. (1981). Plasma catecholamines: changes after footshock and seizure-producing frontal cortex stimulation. *Behavioral and Neural Biology* **31**, 247–60.

—— and McGaugh, J.L. (1977). Hormones and memory. In *Neuropeptide Influences on the Brain and Behavior* (ed. L.H. Miller, C.A. Sandman, and A.J. Kastin), pp. 127–43. Raven Press, New York.

—— and Van Buskirk, R.B. (1975). Facilitation of time-dependent memory processes with post-trial epinephrine injections. *Behavioral Biology* **13**, 145–53.

—— and —— (1976*a*) Enhancement and impairment of memory processes with post-trial injections of adrenocorticotrophic hormone. *Behavioral Biology* **16**, 387–400.

—— and —— (1976*b*). Effects of post-trial hormone injections on memory processes. *Hormones and Behavior* **7**, 509–17.

Golman, H. and Murphy, S. (1981). An analog of ACTH/MSH 4–9 ORG-2766 reduces permeability of the blood-brain barrier. *Pharmacology, Biochemistry and Behaviour* **14**, 845–8.

Haycock, J.W., Van Buskirk, R., and McGaugh, J.L. Effects of catecholaminergic drugs upon memory storage processes in mice. *Behavioral Biology* **20**, 281–310.

Hays, R.M. (1980). Agents affecting the renal conservation of water. In *The Pharmacological Basis of Therapeutics* (ed. A.G. Gilman, L.S. Goodman, and A. Gilman), pp. 916–34. Macmillan, New York.

Heise, G.A. (1981). Learning and memory facilitators: experimental definition and current status. *Trends in Pharmacological Sciences* 2, 158–160.

Hostetter, G., Jubb, S.L., and Kozlowski, G.P. (1977). Vasopressin affects the behavior of rats in a positively-rewarded discrimination task. *Life Sciences* 21, 1323–8.

Hunter, B., Zornetzer, S.F., Jarvik, M.E., and McGaugh, J.L. (1977). Modulation of learning and memory: effects of drugs influencing neurotransmitters. In *Handbook of Psychopharmacology* Vol .18 (ed. L.L. Iversen, S.D. Iversen, and S.H. Snyder), pp. 531–77. Plenum Press, New York.

Iversen, S.A. (1981). Neuropeptides: do they integrate body and brain? *Nature* 291, 454.

Jard, S. (1981). Les isorécepteurs de la vasopressine dans le foie et les reins: relation entre fixation d'hormone et réponse biologique. *Journal of Physiology (Paris)* 77, 621–8.

Jard, S. (1983). Vasopressin isoreceptors in mammals: relation to cyclic AMP-independent transduction mechanisms. *Current Topics in Membrane Transportation* 18, 225–85.

Jenkins, J.S., Mather, H.M., Coughlan, A.K., and Jenkins, D.G. (1981). Desmopressin and desglycinamide vasopressin in post traumatic amnesia. *Lancet* i, 39.

Joseph, S.A., Sorrentino, S., and Sundberg, D.K. (1975). Releasing hormones, LRF and TRF, in cerebrospinal fluid of the third ventricle. In *Brain-endocrine Interactions-II. The Ventricular System in Neuroendocrine Mechanisms* (ed. K.M. Knigge, D.E. Scott, R.M. Kobayashi and S. Ishii), pp. 306–17. Karger, Basel.

Karpiak, S.E. and Rapport, M.M. (1979). Inhibition of consolidation and retrieval stages of passive-avoidance learning by antibodies to gangliosides. *Behavioral and Neural Biology* 27, 146–56.

Kastin, A.J., Nissen, C., Schalley, A.W., and Coy, D.H. (1976). Blood-brain barrier, half-time disappearance, and brain distribution for labelled enkephalin and a potent analog. *Brain Research Bulletin* 1, 583–9.

——, ——, ——, and —— (1979). Additional evidence that small amounts of a peptide can cross the blood-brain barrier. *Pharmacology, Biochemistry and Behaviour* 11, 317–19.

——, ——, and Coy, D.H. Permeability of blood-brain barrier to DSIP peptides. *Pharmacology, Biochemistry and Behaviour* 15, 955–9.

Kiraly, M., Audigier, S., Tribollet, E., Barberis, C., Dolivo, M., and Dreifuss, J.J. (1986). Biochemical and electrophysiological evidence of functional vasopressin receptors in the rat superior cervical ganglion. *Proceedings of the National Academy of Sciences, USA* 83, 5335–9.

Koch-Hendriksen, N. and Nielsen, H. (1981). Vasopressin in post traumatic amnesia. *Lancet* ii, 38–9.

Koob, G.F., *et al.* (1981). Arginine vasopressin and a vasopressin antagonist peptide: opposite effects in extinction of active avoidance in rats. *Regulatory Peptides* 2, 153–64.

Koob, G.F. *et al.* (1985). Use of arginine vasopressin antagonists in elucidating the mechanism of action for the behavioral effects of arginine vasopressin. In *Vasopressin* (ed. R.W. Schreir), pp. 197–201. Raven Press, New York.

Koob, G.F., Dantzer, R., Bluthé R.M., Lebrun, C., Bloom, F.E., and Le Moal, M. (1986). Central injections of arginine vasopressin prolong extinction of active avoidance. *Peptides* 7, 213–18.

Krieger, D.T. (1983). Brain peptides: What, where, and why? *Science* 222, 975–85.

—— and Martin, J.B. (1981) Brain peptides. *New England Journal of Medicine* 304, 876–85.

Laczi, F. *et al.* (1983). Effects of desglycinamide-arginine vasopressin (DG-AVP) on memory processes in diabetes insipidus patients and non diabetic subjects. *Acta Endocrinologica* **102**, 205-12.

Lande, S., Flexner, T.B., and Flexner, L.L. (1972). Effects of corticotrophin and desglycinamide lysine vasopressin on suppression of memory by puromycin. *Proceedings of the National Academy of Sciences USA* **69**, 558-60.

Lebrun, C.J., Rigter, H., Martinez, J.L. Jr., Koob, G.F., Le Moal, M., and Bloom, F.E. (1984). Antagonism of effects of vasopressin (AVP) on inhibitory avoidance by a vasopressor antagonist peptide [dPTyr (Me)AVP]. *Life Sciences* **35**, 1505-12.

——, Le Moal, M., Koob G.F., and Bloom F.E. (1985) Vasopressin pressor antagonist injected centrally reverses peripheral behavioral effects of vasopressin but only at doses that reverse increases in blood pressure. *Regulatory Peptides* **11**, 173-81.

Legros, J.J. *et al.* (1978). Influence of vasopressin on learning and memory. *Lancet* i, 41-2.

Le Moal, M. (1985). Neuropeptides et comportement: quelques réflexions méthodologiques et critiques. *Annals of Endocrinology* **46**, 55-60.

—— *et al.* (1981). Vasopressor receptor antagonist prevents behavioural effects of vasopressin. *Nature* **291**, 491-3.

——, Koob, G.F., Mormède, P., Dantzer, R., and Bloom, F.E. (1982). Vasopressin pressor antagonist reverses central behavioral effects of vasopressin. *Neuroscience Abstracts* **8**, 368.

——, *et al.* (1984). Behavioural effects of peripheral administration of arginine vasopressin: a review of our search for a mode of action and a hypothesis. *Psychoneuroendocrinology* **9**, 319-42.

Leshner, A.I., Merkle, D.A., and Mixon, J.F. (1981). Pituitary adrenocortical effects on learning and memory in social situations. In *Endogenous Peptides and Learning and Memory Processes* (ed. J.L. Martinez, Jr, R.A. Jensen, R.B. Messing, H. Rigter, and J.L. McGaugh), pp. 159-79. Academic, New York.

Manning, M. and Sawyer, W.H. (1984). Design and uses of selective agonistic and antagonistic analogs of the neuropeptides oxytocin and vasopressin. *Trends in Neuroscience* **7**, 6-9.

Martinez, J.L. (1983). Endogenous modulators of learning and memory. In *Theory in Psychopharmacology,* Vol. 2 (ed. S. Cooper), pp. 47-74. Academic Press, London.

Martinez, J.L., Jr and Rigter, H. (1982). Enkephalin actions on avoidance conditioning may be related to adrenal medullary function. *Behavioural Brain Research* **6**, 289-99.

—— *et al.* (1980). Attenuation of amphetamine-induced enhancement of learning by adrenal demedullation. *Brain Research* **195**, 433-43.

——, Conner, P., Dana, R.C., Chavkin, C., Bloom, F.E., and De Graff, J. (1984*a*). Endogenous modulation of peripheral Leu-enkephalin systems affects avoidance conditioning. *Neuroscience Abstracts* **10**, 176.

——, Olson, K., and Hilston, C. (1984*b*). Opposite effects of Met-enkephalin and Leu-enkephalin on a discriminated shock-escape task. *Behavioral Neuroscience* **98**, 487-95.

McCarty, R. and Gold, P.E. (1981). Plasma catecholamines: effects of footshock level and hormonal modulators of memory storage. *Hormones and Behavior* **15**, 168-82.

McGaugh, J.L. (1966). Time-dependent processes in memory storage. *Science* **153**, 1351-8.

—— (1983). Hormonal influences on memory storage. *Journal of the American Psychology Association* **38**, 161-74.

—— and Martinez, J.L., Jr (1981) Learning modulatory hormones: an introduction to endogenous peptides and learning and memory processes. In *Endogenous Peptides and Learning and Memory Processes* (ed. J.L. Martinez Jr, A. Jensen, R.B. Messing, H. Rigter, and J.L. McGaugh), pp. 1-3. Academic, New York.

Meisenberg, G. and Simmons, W.H. (1983). Peptides and the blood-brain barrier. *Life Sciences* **32**, 2611-23.

Meligini, J.A., Ledergerber, S.A., and McGaugh, J.L. (1978). Norepinephrine attenuation of amnesia produced by diethyldithiocarbamate. *Brain Research* **149**, 155-64.

Mens, W.B.J., Witter, A., and Van Wimersma-Greidamus, T.J.B. (1983). Penetration of neurohypophysial hormones from plasma into cerebrospinal fluid (CSF): half-times of disappearance of these neuropeptides from CSF. *Brain Research* **262**, 143-9.

Messing, R.B. and Sparber, S.B. (1983). Des-gly-vasopressin improves acquisition and slows extinction of autoshaped behavior. *European Journal of Pharmacology* **89**, 43-51.

—— Messing, R.B. *et al.* (1979). Naloxone enhancement of memory. *Behavioral and Neural Biology* **27**, 266-75.

Oliveros, J.C. *et al.* (1978). Vasopressin in amnesia. *Lancet* **i**, 42.

Orsingher, C.A. and Fulginiti, S. (1971) Effects of alpha-methyl tyrosine and adrenergic blocking agents on the facilitation action of amphetamine and nicotine on learning in rats. *Psychopharmacologia (Berlin)* **19**, 231-40.

Packard, M.G. and Ettenberg, A. (1985). Effects of peripherally injected vasopressin and des-glycinamide vasopressin on the extinction of a spatial learning task in rats. *Regulatory Peptides* **11**, 51-63.

Pappas, T.N., Taché, Y., and Debas, H.T. (1985). Opposing central and peripheral actions of brain-gut peptides: a basis for regulation of gastric function. *Surgery* **98**, 183-90.

Pardridge, W.M. (1983). Neuropeptides and the blood-brain barrier. *Annual Review of Physiology* **45**, 73-83.

Reppert, S.M., Artman, H.G., Swaminanthan, S., and Fisher, D.A. (1981). Vasopressin exhibits a rhythmic pattern in cerebrospinal fluid but not in blood. *Science* **213**, 1256-7.

Rigter, H., Van Riezen, H., and De Wied, D. (1974). The effects of ACTH and vasopressin-analogues on CO_2-induced retrograde amnesia in rats. *Physiological Behaviour* **13**, 381-8.

—— *et al.* (1980*a*). Enkephalins interfere with acquisition of an active avoidance response. *Life Sciences* **26**, 337-45.

—— *et al.* (1980*b*). Enkephalin and fear-motivated behavior. *Proceedings of the National Academy of Sciences USA* **77**, 3729-32.

Saghal, A. (1984). A critique of the vasopressin memory hypothesis. *Psychopharmacology* **83**, 215-28.

Sawyer, W.H. (1964). Vertebrate neurohypophysial principles. *Endocrinology* **75**, 981-90.

Sawyer, T.F., Henhgehold, A.K., and Perez, W.A., (1984) Chemosensory and hormonal mediation of social memory in male rats. *Behavioural Neuroscience* **98**, 908-13.

Scott, D.E., Krobisch, D.G., Knigge, K.M., and Kozlowksi, G.P. (1974). *In vitro* analysis of the cellular localization of luteinizing hormone releasing factor (LRF) in the basal hypothalamus of the rat. *Cell and Tissue Research* **149**, 371-8.

Sofroniew, M.V. (1980). Projections from vasopressin, oxytocin and neurophysin neurons to neural targets in the rat and human. *Journal of Histochemistry and Cytochemistry* **28**, 475-8.

Stegner, H., Artman, H.G., Leake, R.D., and Fisher, D.A. (1983). Does DDAVP (1. Deamino-8-D-Arginine-vasopressin) cross the blood-CSF barrier? *Neuroendocrinology* **37**, 262-5.

Taché, Y., Brown, M., and Collu, R. (1982). Central nervous system actions of bombesin-like peptides. In *Brain Peptides and Hormones* (ed. R. Collu *et al.*), pp. 183-96. Raven Press, New York.

——, Goto, Y., Gunion, M.W., Vale, W., River, J., and Brown, M. (1983). Inhibition of gastric acid secretion in rats by intracerebral injection of corticotropin-releasing factor. *Science* **222**, 935-7.

——, ——, M., Rivier, J., and Debas, H. (1984). Inhibition of gastric acid secretion in rats and in dogs by corticotropin-releasing factor. *Gastroenterology* **86**, 281-6.

Thompson, S.A. (1982). Localization of immunoreactive prolactin in ependyma and circumventricular organs of rat brain. *Cell and Tissue Research* **225**, 79-94.

Thor, D.H. and Holloway, W.R. (1982). Social memory of the male laboratory rat. *Journal of Comparative Physiology and Psychology* **96**, 1000-6.

Uemura, H., Asai, T., Nozoki, M., and Kobayashi, H. (1975). Ependymal absorption of luteinizing hormone-releasing hormone injected into the third ventricle of the rat. *Cell and Tissue Research* **160**, 443-52.

Van Ree, J.M., Bohus, B., Versteeg, D.H., and De Wied, D. (1978). Neurohypophyseal principles and memory processes. *Biochemical Pharmacology* **27**, 1793-800.

Weinberger, N.M., Gold, P.E., and Sterneberg, D.B. (1984). Epinephrine enables Pavlovian fear conditioning under anesthesia. *Science* **223**, 605-7.

Weingartner, H. *et al.* (1981). Effects of vasopressin on human memory functions. *Science* **211**, 601-3.

Wood, J.H. (1982). Neuroendocrinology of cerebrospinal fluid: peptides, steroids and other hormones. *Neurosurgery* **11**, 293-305.

Zaildi, S.M.A. and Heller, H. (1979). Can neurohypophysial hormones cross the blood-cerebrospinal fluid barrier? *Journal of Endocrinology* **60**, 195-6.

Zerbe, R.L., Bayorth, M.A., and Feuerstein, G. (1982). Vasopressin: an essential pressor factor for blood pressure recovery following hemorrhage. *Peptides* **3**, 509-14.

13

The neurochemistry of cortical dementias

M. ROSSOR

INTRODUCTION

A discussion of the neurochemistry of cortical dementias implies first that the distinction between cortical and subcortical dementias is valid, and second that a description of the pathology in terms of the neurotransmitter systems involved contributes to the understanding of the cognitive impairment. These assumptions need to be considered before describing the available neurochemical data on the cortical dementias.

Albert and colleagues first used the term subcortical dementia to describe the cognitive impairment in progressive supranuclear palsy (Albert *et al.* 1974). They drew attention in this disease to a slowness of thought processes, apathy, poor memory, and an impairment of manipulation of acquired knowledge and, by contrast, a lack of aphasia, apraxia, and agnosia. The term subcortical dementia reflected the prominent brain stem pathology in progressive supranuclear palsy. This was compared with Alzheimer's disease in which there is dramatic cortical pathology, and in which aphasia, apraxia, and agnosia are prominent. Independently, McHugh and Folstein (1975) described features in Huntington's disease resembling those described by Albert *et al.* (1974), again commenting on the lack of aphasia, apraxia, and agnosia, and suggesting that this pattern of cognitive impairment related to subcortical pathology. The concept of a subcortical dementia has since been expanded to include the cognitive impairment which may be found in Parkinson's disease, Wilson's disease, spinocerebellar degeneration, basal ganglia calcification, and the lacunar state. It has also been suggested that the 'pseudodementia' associated with depression can also be considered in this category (for review see Cummings and Benson 1984). A number of criticisms have been made of the overall concept and have been reviewed recently by Whitehouse (1986). A semantic criticism has been that the term subcortical implies a clear anatomical and pathological distinction from cortical dementia. This is pertinent to the observed changes in subcortical structures in Alzheimer's disease, such as within the locus coeruleus and nucleus basalis (see below), and to the impairment of subcortical systems projecting to the cortex in the subcortical dementias. Damage to an ascending

233

projection to cortex might mimic dysfunction within that terminal field in the cortex. Thus, reduced cortical glucose utilization can be demonstrated in animals with lesions of the nucleus basalis (London *et al*. 1984), and Positron Emission Tomography (PET) scanning in progressive supranuclear palsy reveals hypometabolism within the frontal cortex (D'Antona *et al*. 1985). Indeed, the abnormalities in some subcortical dementias resemble intrinsic cortical frontal lobe disease. The main question, which as yet remains unanswered, is whether the differences that can be seen between cortical and subcortical dementias are greater than those that exist within, for example, the spectrum, of Alzheimer's disease. Despite the criticisms, however, the distinction can be useful clinically and can provide a framework for the discussion of the neurochemical pathology.

The belief that a description of the neurotransmitter abnormalities found in the cortical dementias is useful derives much from the model of Parkinson's disease, in which the akinesia can be usefully described in terms of the striatal dopamine deficit arising from damage to the nigrostriatal projection neurons. A similar approach to cognitive impairment offers a new conceptual framework (Drachman 1978) which has important implications for therapeutic strategies. It would be predicted that therapy based upon such neurotransmitter abnormalities would be of most value in those diseases in which there is widespread, but selective damage to specific populations of neurons. It is less likely to be useful where cognitive impairment arises from discrete anatomical infarcts or tumours. Present data suggest that Alzheimer's disease involves selective degeneration of cortical and sub-cortical neurons, and may be usefully described in terms of neurotransmitter involvement. Multiinfarct dementia may be less usefully described in these terms, and it is unknown whether focal cortical degenerations can be similarly approached. The latter comprise a poorly defined group of patients who present with dysphasia, apraxia, or agnosia and may have focal atrophy on CT scan. Some develop a dementia associated with Alzheimer pathology, but some are reported not to progress (Mesulam 1982). It is not known whether specific neurons are involved within the areas of focal degeneration nor what determines the anatomical vulnerability within these areas. Neurochemical markers do not display prominent changes between cortical areas, although minor differences are observed such as higher concentrations of somatostatin within association cortices (Hayashi and Oshima 1986). It is possible that such neurochemical heterogeneity within cerebral cortex may contribute to the regional vulnerability that can be observed.

Alzheimer's disease can be viewed as the prototypical cortical dementia. It is also the dementia best studied neurochemically and as such forms the main basis of this chapter. The neurochemical abnormalities found in the cortex are discussed first, followed by the subcortical changes. Brief reference is made to other cortical dementias and finally the approach to histopathological and clinical correlates is discussed.

ALZHEIMER'S DISEASE

Although Alzheimer's disease is considered the prototypical cortical dementia characterized by the development of aphasia, apraxia, and agnosias, the clinical course may be very varied. Memory impairment can be a prominent feature well into the course of the disease with late development of 'cortical features'. The presence of other features such as akinesia reflects the heterogeneity that can be found in the disease (Mayeux *et al.* 1985).

Neuropathologically, Alzheimer's disease is characterized by widespread neocortical senile plaques and neurofibrillary tnalges which may also be found in normal old age within the hippocampus (for review see R.H. Perry 1986). Senile plaques consist of three components, namely dystrophic neuronal processes, glial processes, and a central amyloid core, although the proportion varies during the development of a plaque. Neurofibrillary tangles represent an abnormal collection of protein intraneuronally and can be visualized at the electron miscroscope level as consisting of paired helical filaments with a diameter of 10 nm. Neurofibrillary tangles and senile plaques are found throughout the neocortex, with the exception of primary sensory and motor cortices. Neurofibrillary tangles are found particularly in the association areas and in the early stages of the disease have a predilection for layer 3 pyramidal neurons. The frequent finding of cerebral atrophy in Alzheimer's disease suggests the presence of cortical neuronal loss, although the extent is controversial. Despite some early reports of normal cell counts, there is general agreement that neuronal loss does occur over and above that found in normal ageing, although it may not be demonstrable in the elderly Alzheimer patient (Mountjoy *et al.* 1983). Cell loss is greatest from the temporal lobes, and attrition of neuronal processes may also be seen and precede perikaryal loss.

Intrinsic cortical neurotransmitters

Neurotransmitters considered as intrinsic to the cerebral cortex are those utilized by neurons whose cell bodies lie within the cortex and project either as local circuit neurons, corticocortical association fibres or as efferent projection systems. Knowledge of the neurotransmitter circuitry within the cerebral cortex is still rudimentary, but considerable advances have been made recently by the use of immunohistochemistry with anterograde and retrograde tracer techniques. A broad classification of cerebral cortical neurons has been derived from their appearance after Golgi silver staining. The Golgi technique stains only a small number of neurons within a section and thus provides very detailed morphological information. The two main classes of cortical cells are the pyramidal and non-pyramidal neurons. Pyramidal cells have a characteristic cone shape with an ascending apical dendrite and descending axon; they are the principal efferent neurons. Various subclasses of non-pyramidal cells have been described on the basis of the appearance of the cell body and dendrites, and include spiny and non-spiny, and multipolar and bipolar. The use of immunohistochemistry has permitted the location of specific neurotransmitters, particularly peptides, to morphologically defined populations of neurons (for reviews see Emson and Lindvall 1986; Peters and Jones 1984).

Gamma aminobutyric acid (GABA)

The majority of non-pyramidal interneurons stain with antisera against GABA or the associated enzyme glutamic acid decarboxylase (GAD). Thus, GABA constitutes a major cortical inhibitory transmitter system and it has been estimated that perhaps as many as a third of synapses within the brain utilize GABA. This ubiquity makes it an obvious candidate for involvement in any intrinsic cortical pathology in Alzheimer's disease. However, there are problems in assessing the integrity of this system, particularly at autopsy. The activity of glutamic acid decarboxylase has been found to be low in cerebral cortex, but the activity can be non-specifically depressed by prolonged terminal illness and in biopsy samples, as opposed to autopsy samples, is found to be normal (Spillane *et al.* 1977). Measurements of GABA itself can avoid the influence of the agonal state (Spokes *et al.* 1979), although concentrations of the amino acid may be difficult to interpret due to the presence of a glial uptake system and a small metabolic pool. Nevertheless, concentrations of GABA are reduced (Rosser *et al.* 1982; Ellison *et al.* 1986; Sasaki *et al.* 1986) and low synaptosomal GABA uptake also indicates damage to GABA neurons (Hardy *et al.* 1986*a*). The reduction in GABA concentration is about 30 per cent and is maximal in the temporal lobe. The variability in the published studies of GABA markers may relate to diferences in the age of the cases analysed since reduced GABA concentrations may be confined to younger cases with sparing in the very old (Rossor *et al.* 1984). By contrast, cortical GABA receptors are normal (Cross *et al.* 1984). The presence of a GABA deficit might be predicted from the observed neuropeptide abnormalities since there is evidence of peptide co-existence within GABA neurons (see below).

Excitatory amino acids

The amino acids glutamate and aspartate may be utilized as cortical excitatory amino acids. Both amino acids have potent excitatory effects on cortical cells, both can be released on depolarization and there exists a high affinity uptake system. Glutamate and possibly aspartate are believed to be transmitters of pyramidal cells, which are preferentially affected by neuro-fibrillary tangle formation in Alzheimer's disease. Direct measurements of glutamate concentration are very difficult to interpret since glutamate is also involved in intermediary metabolism and one cannot reliably define the neurotransmitter pool. Glutamate concentrations are generally reduced when measured at autopsy (Arai *et al.* 1984*a*; Sasaki *et al.* 1986), but more reliable may be measurements of glutamate uptake and release. Glutamate release from synaptosomes obtained at biopsy is normal, but it is not established that only the neurotransmitter pool is released under these conditions (Smith *et al.* 1983).

Using D-aspartate the glutamate uptake site can also be measured in synaptosomes obtained from rapid autopsy samples and a recent report indicates that this is reduced particularly in temporal lobe (Hardy *et al.* 1986*b*). The post-synaptic glutamate receptor is also reduced throughout the cerebral cortex (Greenamyre *et al.* 1985).

Cortical neuropeptides

A large number of peptides have been described which are localized to neurons, are released on depolarization, and are believed to have a role in slow chemical signalling in CNS. Four peptides in particular are found in relatively high concentration in cerebral cortex and can be visualized within cortical neuron perikarya, namely cholecystokinin (CCK) vasoactive intestinal polypeptide (VIP), somatostatin (SRIF), and neuropeptide-Y (NPY). Within the cortex these peptides are found predominantly within the non-spiny non-pyramidal interneurons. As well as staining for these various peptides, many non-pyramidal neurons also stain with antisera against GAD. Double staining demonstrates that many of these neurons are GABAergic, but also contain a co-existent peptide (Hendry *et al.* 1984). Moreover, more than one peptide, for example somatostatin and neuropeptide Y, may co-exist within a single cortical neuron.

Cholecystokinin and vasoactive intestinal polypeptide

These two cortical neuropeptides are discussed together since the available data indicate no significant alteration in their concentration in Alzheimer's disease (Ferrier *et al.* 1983; Rossor and Iversen 1986). Arai *et al.* (1984*a*, *b*, *c*) reported a small reduction in VIP concentration in angular and insular cortex, but these were the only areas amongst twenty examined. The normal concentrations of these two peptides are of particular interest for two reasons. Firstly, this may be adduced as evidence of relative sparing of some neurons and the associated neurotransmitters. Such an interpretation needs to be made with some caution since tissue shrinkage consequent upon the cell loss may mask reduction in a chemical marker when expressed as a concentration. Nevertheless, the finding of normal or even increased concentrations [see corticotrophic-releasing factor (CRF) receptors below] may be more important observations than widespread neurotransmitter losses. The second point of interest in relation to the normal concentrations of VIP is the reported co-existence between this peptide and acetylcholine. Immuno-staining in the rat sensorimotor cortex with antisera against VIP and the bio-synthetic enzyme for acetylcholine, choline acetyltransferase (ChAT), demonstrates co-existence within a population of non-pyramidal bipolar neurons (Eckenstein and Baughman 1984). Such co-existence is not found within the ascending cholinergic projection. If a similar co-existence is demonstrated within intrinsic cortical cells in man, then the normal VIP concentrations in Alzheimer's disease would indicate sparing of these cells with the cortical cholinergic deficit confined to the ascending projection (see below).

Somatostatin and neuropeptide-Y

Of the many peptides measured in cerebral cortex, somatostatin is the only one to show a consistent reduction in Alzheimer's disease (Davies *et al.* 1980; Rossor *et al.* 1980; Ferrier 1983; Arai *et al.* 1984c; Beal *et al.* 1985). The reduction in somatostatin is maximal in the temporal lobe and may be confined to temporal cortex in older cases (Rossor *et al.* 1984). Somatostatin receptor binding in cortex has also been measured, and the total number of receptors found to fall in proportion to the reduction in somatostatin (Beal *et al.* 1985).

A proportion of somatostatin neurons in human cerebral cortex also stain with antisera against neuropeptide Y (Vincent *et al.* 1982) and in view of this reported co-existence, a commensurate loss of neuropeptide-Y reactivity might be expected. Two studies, however, have found no alteration in neuro-peptide-Y concentration (Allen *et al.* 1984; Rossor *et al.* 1986). This might imply alterations in turnover rather than neuronal disintegration or alter-natively that co-existence is found only in a subset of somatostatin neurons. More recently, however, Beal *et al.* (1986), found a reduction in neuropeptide Y concentration in cerebral cortex, and the presence of degenerating neurons which are immunoreactive for neuropeptide-Y in Alzheimer's disease (Chan-Palay *et al.* 1985) makes it likely that at least osme neuropeptide-Y containing neurons are involved.

Corticotropin-releasing factor (CRF)

Less is known about the anatomical localization of CRF immunoreactivity in the cerebral cortex, but immunostaining of intrinsic neurons has been reported in animal studies. It is possible that there is an additional contribu-tion to cortex from co-existence within the ascending cholinergic system. Two recent studies found a reduction in CRF concentrations (Bissette *et al.* 1985; De Souza *et al.* 1986) and the latter also reported a reciprocal increase in CRF receptor binding suggestive of denervation supersensitivity. This implies that the post-synaptic cells are intact and functional, again providing evidence of selective vulnerability within the cortex.

Subcortical systems in Alzheimer's disease

Although the concept of Alzheimer's disease as a cortical dementia directs attention to the intrinsic cortical neurotransmitter systems, abnormalities are found within subcortical structures and in particular involve the diffuse systems which project to cerebral cortex. Damage to the cholinergic projection from the basal forebrain to cerebral cortex is a consistent feature and as one of the first abnormalities to be described has attracted considerable attention. To what extent the observed changes in the ascending cholinergic and monoamine projections are secondary to primary pathology within the cortex, or are co-incidentally, but independently affected is not yet established. Observed retrograde changes in the cholinergic nuclei following cortical lesions indicate that subcortical loss can occur as a secondary feature of damage to the cerebral cortex (Sofroniew *et al.* 1983), although the presence of large numbers of tangles within subcortical nuclei has been adduced as evidence of coincidental involvement.

Cholinergic system

Reduced activity of choline acetyltransferase (ChAT) (a reliable marker of cholinergic neurons) within the cerebral cortex in Alzheimer's disease was one of the first neurochemical abnormalities described, and has been amply confirmed in many studies (for review see Perry, E.K. 1986). The cholinergic abnormality was quickly interpreted in the light of the observed effects on cognition in animals of disrupting cholinergic function and led to the 'cholinergic hypothesis' of Alzheimer's disease (for review see Collerton 1986). The loss of ChAT activity is maximal in the temporal lobe, although it is found throughout the cerebral cortex including the primary and motor sensory areas (Rossor *et al.* 1982). Synthesis of acetylcholine is reduced in biopsy samples, demonstrating that the reduced enzyme activity is of functional importance. Histochemical, tracer, and lesion studies in a variety of animal species, including primates, have established that the majority of ChAT activity is located within terminals of the extrinsic projection from basal forebrain (Emson and Lindvall 1986). A small population of intrinsic cholinergic cortical cells can be demonstrated, some of which co-exist with VIP (see above), but their existence and relative importance in human cortex is not established. The origin of the cholinergic projection is a diffuse cluster of nuclei extending from the medial septal nucleus and diagonal band of Broca to the nucleus basalis of Meynert (Mesulam *et al.* 1984).

The loss of enzyme activity in cerebral cortex in Alzheimer's disease can be related to degeneration of this ascending projection from the basal forebrain (Whitehouse *et al.* 1982).

The discovery of a cholinergic abnormality immediately raised hopes of transmitter replacement therapy, particularly if the post-synaptic receptor is intact. However, the status of the cholinergic receptor has been controversial. Early reports of normal concentrations of muscarinic receptors defined by [³H] quinuclidinyl benzilate (QNB), were disputed by other authors who found modest reductions, particularly in the temporal lobe. These discrepancies may have been explained by the work of Mash *et al.* (1985) who have defined the muscarinic receptor binding in terms of M1 and M2 receptors. From lesions of the cholinergic projection in animals it is believed that M1 receptors are post-synaptic and M2 receptors are located pre-synaptically on the cholinergic terminal. In Alzheimer's disease the number of M1 receptors is unchanged, but M2 receptors are reduced by about 25 per cent which is reflected in the overall reduction in [³H] QNB binding. More recently, it has been possible to measure nicotinic receptors using direct agonist binding and these are also reduced (Whitehouse *et al.* 1986). The location of nicotinic receptors is not definitely established, but a proportion are likely to be presynaptic.

In addition to the cortical cholinergic deficit, reduced ChAT activity is also found in a variety of subcortical nuclei which do not project to the cerebral cortex. Cell loss and neurofibrillary tangle formation is prominent in the amygdala and ChAT activity is often as low as that found in temporal cortex. Other areas include the caudate nucleus and medial thalamus with sparing of the putamen and ventrolateral thalamus (Rossor *et al.* 1982). The significance of these subcortical changes to the cognitive deficit is unknown but the development of profound memory disturbance with midline thalamic lesions is of interest.

Monoamine and serotonin systems

There is now good evidence for damage to the noradrenergic projection from the locus coeruleus to the cerebral cortex and to the cortical serotonin pro-

TABLE 13.1. *Neurochemical/histological correlates in Alzheimer's disease*

Pathological feature	Neurotransmitter
Cell loss	
Nucleus basalis	Acetylcholine
	?S2 receptors/M2 receptors
Locus coeruleus	Noradrenaline
Raphe nucleus	Serotonin
Cerebral cortex	?Somatostatin/GABA
Neurofibrillary tangles	Somatostatin
	?Glutamate
Senile plaques	Non-specific. Staining for GAD, ChAT, tyrosine hydroxylase, VIP, CCK, NPY, somatostatin

S2 = serotonin 2
M2 = muscarinic 2

jection from raphe nuclei. Concentrations of both noradrenaline and of 5-hydroxytryptamine are reduced in cerebral cortex (Adolfsson *et al.* 1979; Arai *et al.* 1984*b*), and noradrenaline and serotonin uptake are both reduced in cerebral biopsy samples (Benton *et al.* 1982). Adrenoceptors are normal in the cerebral cortex but serotonin S2 receptors, defined using ketanserin binding, are reduced by about 50 per cent in frontal cortex (Reynolds *et al.* 1984). The location of the S2 receptor is not clearly established, but there is preliminary evidence for a location on cholinergic terminals which suggests that the S2 receptor loss may be a further reflection of the damage to the ascending cholinergic projection. In contrast to the noradrenaline and serotonin abnormalities, concentrations of dopamine are normal in cerebral cortex (Arai *et al.* 1984*b*) providing further evidence of selectivity of the neurochemical changes (Table 13.1).

OTHER CORTICAL DEMENTIAS

A number of diseases are discussed together here because of the paucity of information about the neurochemical changes. The term multi-infarct dementia covers a variety of pathologies, and although the lacunar state may be usefully viewed as a subcortical dementia (Cummings and Benson 1984), the dementia arising from multiple cortical infarcts may be more appropriately considered as a cortical dementia. Cases with multiple cortical infarcts have been examined, but usually as a control group in studies of Alzheimer's disease. Reductions in ChAT activity and, to a lesser extent, GABA markers have been reported (Rossor *et al.* 1982), but it is likely that this reflects focal tissue loss, rather than selective involvement of these sytems. No data are available on other neurochemical markers which might permit an assessment of neurochemical specificity within the locus of pathology. By contrast to Alzheimer's disease, Pick's disease is rare. Histologically, there is prominent intrinsic cortical damage with cell loss and intraneuronal inclusions particularly in the frontal and temporal lobes. Despite some reports of reduced cell counts in the nucleus basalis, ChAT activities are generally normal, but muscarinic receptors are reduced (Yates *et al.* 1980). Another disease which might be considered as a cortical dementia is Down's syndrome. Patients who survive into their fifth decade develop a superadded dementia and although there is little information on the clinical features, it is associated with the development of neurofibrillary tangles throughout the cortex. Reduced ChAT activity and noradrenaline concentrations accompany the histopathological changes (Yates *et al.* 1983).

INTERPRETATION OF NEUROCHEMICAL CHANGES IN CORTICAL DEMENTIAS

There are many problems attendant upon the measurement of chemical markers in degenerative disease, particularly at autopsy (for review see Rossor 1986). Many non-specific factors such as agonal state, autopsy delay, and drug therapy may influence neurochemical determinations and render interpretation very difficult. Moreover, it may be impossible to determine whether an alteration in concentration is due to neuronal disintegration or to an alteration in turnover. It is tempting to relate the loss of a chemical marker in a degenerative disease to neuronal damage, but this may not necessarily be valid. An additional problem is that with tissue shrinkage the concentration of a neurotransmitter may remain unchanged and yet there may still be an overall loss. The combination of neurochemistry with immunohisto-chemistry and classical histopathology has enabled some of these problems to be resolved, and the observed changes to be related more precisely to damage to a particular neuronal population. Once these changes have been established it may be possible to define more precisely the functional role of such systems and to relate the neurotransmitter abnormalities to the clinical deficit.

Histopathological correlations

The development of immunohistochemical techniques suitable for staining post-mortem tissue has provided an impetus to determining the relationship of neurochemical changes to the specific features of Alzheimer's disease, namely neurofibrillary tangles, senile plaques, and cell loss. Neurofibrillary tangles may be found in a number of subcortical nuclei, such as the locus coeruleus, raphe nucleus, and nucleus basalis, all of which have established neurotransmitter identities, namely noradrenergic, serotonergic, and cholinergic. However, the nature of the tangle-bearing cells within the cortex is much less secure. Roberts *et al.* (1985) reported somatostatin, but not VIP or CCK immunoreactivity within tangle-bearing cells of the cortex. This would be important evidence that tangle formation selectively involves specific biochemically defined neurons, but it is unlikely to be unique to somatostatin neurons. In view of the predilection for tangles to form within pyramidal neurons early in the disease, it is likely that the glutamate neurons are also affected.

Immunostaining in human autopsy tissue tends to visualize neuronal processes more readily than perikarya and so it has been much easier to study senile plaques. It is now apparent that a large number of neurotransmitter systems contribute to plaque formation with staining of dystrophic neurites within the plaque using a variety of antisera. Thus, of the intrinsic markers,

CCK, VIP, NPY, somatostatin, and GAD immunostaining have all been demonstrated (Price *et al.* 1986). In addition to the involvement of intrinsic neurons, the terminals of the ascending cholinergic and noradrenergic projections also contribute to plaque formation. Senile plaques stain strongly for acetylcholinesterase and, more recently, specific anti-ChAT immunostaining of neurites in the senile plaques of aged monkeys has been demonstrated. It is notable that some peptides, e.g. CCK and VIP, can be demonstrated within plaques, but do not alter in overall tissue concentration.

Cell loss can be readily demonstrated in some of the subcortical nuclei such as nucleus basalis and locus coeruleus, providing correlates of the observed cholinergic and noradrenergic deficits. The neurotransmitter identity of those cells lost in cerebral cortex, however, is unknown. Degenerating somatostatin- and NPY-positive neurons have been reported (Joynt and McNeill 1984; Chan-Palay *et al.* 1985), but it is likely that other systems are involved. From the biochemical data glutamate pyramidal neurons and non-pyramidal GABA/somatostatin neurons are the most likely candidates.

Clinical biochemical correlates

There is very little known about the role of specific cortical transmitters, as opposed to specific cortical areas, in cognitive function. There is a lack of pharmacological agents for the manipulation of these systems. In animals cysteamine, which depletes brain somatostatin, impairs passive avoidance tasks (Bakhit and Swerdlow 1986) and conversely intracerebroventricular injection of the peptide attenuates electroconvulsive shock amnesia (Vecsei *et al.* 1984). A specific pattern of cognitive impairment which can be attributed to a somatostatin deficit has not, however, been established. GABA antagonists cause seizures, and although these occur in Alzheimer's disease they are not a prominent feature. It is not clear what the effect of chronic mild impairment of GABA function may have on cognition.

More information is available on the effects of manipulation of the subcortical projections, particularly the cholinergic and noradrenergic system (for review see Chapter 14; Collerton 1986). As discussed by Agid many of these systems are damaged in subcortical dementia, and it is tempting to seek the clinical features of Alzheimer's disease which might be shared by the subcortical dementias and thus related to specific damage to the ascending projections. A similar approach has been made by Fuld (1984), who utilized those features of cognitive impairment which can be seen following cholinergic blockade to predict which patients might have Alzheimer's disease. Although the search for clinical features in Alzheimer's disease which can be attributed to a cholinergic deficit is an attractive one, the effect of cholinergic impairment superimposed upon cortical damage may be quite different from that found in subcortical dementia alone. Related to this is the possibility that the clinical deficit arising from damage to an input system is dependent upon the area of cerebral cortex innervated.

For example, frontal lobe symptomatology can be seen in those subcortical dementias where damage to projections to the frontal lobe are found. It is not known whether selective damage to the ascending cholinergic projection to Broca's area would result in an aphasia. The fact that damage to the ascending cholinergic projection in Alzheimer's disease is widespread, and includes the primary motor and sensory cortex, but does not result in motor weakness or sensory disturbance, suggests that additional cortical pathology is required. Although these problems remain unsolved the distinction between cortical and subcortical dementia is currently valuable to the clinician and a neurochemical approach to diffuse degenerative disease offers an exciting new framework for future research.

REFERENCES

Adolfsson, R., Gottfries, C.G., Roos, B.E., and Winblad, B. (1979). Changes in the brain catecholamines in patients with dementia of Alzheimer type. *British Journal of Psychiatry* 135, 216–23.

Albert, M.L., Feldman, R.G., and Willis A.L. (1974). The 'subcortical dementia' of progressive supranuclear palsy. *Journal of Neurology, Neurosurgery and Psychiatry* 37, 121–30.

Allen, J.M. *et al.* (1984). Elevation of neuropeptide Y (NPY) in substantia innominata in Alzheimer's type dementia. *Journal of Neurological Science* 64, 325–31.

Arai, H., Kobayashi, K., Ichimiya, Y., Kosaka, K., and Iizuka, R. (1984a). A preliminary study of free amino acids in the postmortem temporal cortex from Alzheimer-type dementia patients. *Neurobiology of Ageing* 5, 319–21.

——, H., Kosaka, K., and Iizuka, T. (1984b). Changes of biogenic amines and their metabolites in postmortem brains from patients with Alzheimer-type dementia. *Journal of Neurochemistry* 43, 388–93.

——, Moroji, T., and Kosaka, K. (1984c). Somatostatin and vasoactive intestinal polypeptide in postmortem brains from patients with Alzheimer-type dementia. *Neuroscience Letters* 52, 73–8.

Bakhit, C. and Swerdlow, N. (1986). Behavioural changes following central injection of cysteamine in rats. *Brain Research* 365, 159–63.

Beal, M.F., Mazurek, M.F., Tran, V.T., Chattha, G., Bird, E.D., and Martin, J.B. (1985). Reduced numbers of somatostatin receptors in the cerebral cortex in Alzheimer's disease. *Science* 229, 289–91.

——, ——, Chattha, G.K., Svendsen, C.N., Bird, E.D., and Martin, J.BC. (1986). Neuropeptide Y immunoreactivity is reduced in cerebral cortex in Alzheimer's disease. *Annals of Neurology* (in press).

Benton, J.S. *et al.* (1982). Alzheimer's disease as a disorder of isodendritic core. *Lancet* i, 456.

Bissette, G., Reynolds, G.P., Kilts, C.D., Widerlov, E., and Nemeroff, C.B. (1985). Corticotropin-releasing factor-like immunoreactivity in senile dementia of the Alzheimer type. *Journal of the American Medical Association* 254, 3067–9.

Chan-Palay, V., Lang, W., Allen, Y.S., Haesler, V., and Polak, J.M. (1985). Cortical neurons immunoreactive with antisera against neuropeptide Y are altered in Alzheimer's type dementia. *Journal of Comparative Neurology* **238**, 390–400.

Collerton, D. (1986). Cholinergic function and intellectual decline in Alzheimer's disease. *Neuroscience* **19**, 1–28.

Cross, A.J. *et al.* (1984). Studies on neurotransmitter receptor systems in neocortex and hippocampus in senile dementia of the Alzheimer type. *Journal of Neurological Science* **64**, 109–17.

Cummings, J.L. and Benson, D.F. (1984). Subcortical dementia. Review of an Emerging Concept. *Archives of Neurology* **41**, 874–9.

D'Antona, R. *et al.* (1985). Subcortical dementia. Frontal cortex hypometabolism detected by positron tomography in patients with progressive supranuclear palsy. *Brain* **108**, 785–99.

Davies, P., Katzman, R., and Terry, R.D. (1980). Reduced somatostatin-like immunoreactivity in cerebral cortex from cases of Alzheimer's disease and Alzheimer senile dementia. *Nature* **288**, 279–80.

De Souza, E.B., Whitehouse, P.J., Kuhar, M.J., Price, D.L., and Vale, W.W. (1986). Alzheimer's disease: reciprocal changes in corticotropin-releasing factor (CRF)-like immunoreactivity and CRF receptors in cerebral cortex. *Nature* **39**, 593–600.

Drachman, D.A. (1978). Central cholinergic systems and memory. In *Psychopharmacology: a Generation of Progress* (ed. M.A. Lipton, A. Dimascio, and K.F. Killan), pp. 651–2. Raven Press, New York.

Eckenstein, F. and Baughman, R.W. (1984). Two types of cholinergic innervation in cortex, one co-localised with vasoactive intestinal polypeptide. *Nature* **309**, 153–5.

Ellison, D.W., Beal, M.F., Mazurek, M.F., Bird, E.D., and Martin, J.B. (1986). A post-mortem study of amino acid neurotransmitters in Alzheimer's disease. *Annals of Neurology* (in press).

Emson, P.C. and Lindvall, O. (1986). Neuroanatomical aspects of neurotransmitters affected in Alzheimer's disease. *British Medical Bulletin* **42**(i), 57–62.

Ferrier, I.N. *et al.* (1983). Neuropeptides in Alzheimer-type dementia. *Journal of Neurological Science* **62**, 159–70.

Fuld, P.A. (1984). Test profile of cholinergic dysfunction and of Alzheimer-type dementia. *Journal of Clinical Neuropychology* **6**, 380–92.

Greenamyre, J., Penney, J., Young, A., D'Amato, C., Hicks, S., and Shoulson, I. (1985). Alterations in L-glutamate binding in Alzheimer's and Huntington's disease. *Science* **227**, 1496–9.

Hardy, J. *et al.* (1986*a*). A disorder of cortical GABAergic innervation in Alzheimer's disease. *Neuroscience Letters* (in press).

—— *et al.* (1986*b*). Region-specific loss of glutamate innervation in Alzheimer's disease. *Neuroscience Letters* (in press).

Hayashi, M. and Oshima, K. (1986). Neuropeptides in cerebral cortex of macaque monkey (Macaca fuscata fuscata). Regional distribution and ontogeny. *Brain Research* **364**, 360–8.

Hendry, S.H.C., Jones, E.G., De Felipe, J., Schemechel, D., Brandon, C., and Emson, P.C. (1984). Neuropeptide-containing neurons of the cerebral cortex are also GABAergic. *Proceedings of the National Academy of Sciences USA*, **81**, 6526–30.

Joynt, R.J. and McNeill, T.H. (1984). Disease-correlated morphologic changes of somatostatin-like-containing neurons of the cortex in senile dementia of the Alzheimer type. *Neurology* **34** (Suppl. 1), 120.

London, E.D. *et al.* (1984). Decreased cortical glucose utilisation after ibotenate lesion of the rat ventromedial globus pallidus. *Journal of Cerebral Blood Flow and Metabolism* **4**, 381–90.

Mash, D.C., Flynn, D.D., and Potter, L.T. (1985). Loss of M_2 muscarinic receptors in the cerebral cortex in Alzheimer's disease and an experimental cholinergic denervation. *Science* **228**, 1115–17.

Mayeux, R., Stern, Y., and Spanton, S. (1985). Heterogeneity in dementia of the Alzheimer-type: evidence of subgroups. *Neurology*, **35**, 453–61.

McHugh P.R. and Folstein, M.F. (1975). Psychiatric syndromes of Huntington's chorea: A clinical and phenomenologic study. In *Psychiatric Aspects of Neurologic Disease* (ed. D.F. Benson and D. Blumer), pp. 267–85. Grune and Stratton Inc. New York.

Mesulam, M.M. (1982). Slowly progressive aphasia without generalised dementia. *Annals of Neurology* **11**, 592–8.

——, Mufson, E.J., Levey, A.I., and Wainer, E.H. (1984). Atlas of cholinergic neurons in the forebrain and upper brainstem of the macaque based on monoclonal choline acetyltransferase immunohistochemistry and acetycholinesterase histochemistry. *Neuroscience* **12**, 669–86.

Mountjoy, C.Q., Roth, M., Evans, N.J.R., and. Evans, H.M. (1983). Cortical neuronal counts in normal elderly controls and demented patients. *Neurobiology of Ageing* **4**, 1–11.

Perry, E.K. (1986). The cholinergic hypothesis—ten years on. *British Medical Bulletin* **42**(i), 63–9.

Perry, R.H. (1986). Recent advances in neuropathology. *British Medical Bulletin* **42**(i), 34–41.

Peters, A. and Jones, E.G. (ed.) (1984). *Classification of Cortical Neurons in Cerebral Cortex Vol 1. Cellular Components of the Cerebral Cortex.* Plenum Press, New York.

Price, D.L. *et al.* (1986). Alzheimer's disease: A multi-system disorder. In *Neuropeptides in Neurologic and Psychiatric Disease* (ed. J.B. Martin and J.D. Barchas), pp. 209–14. Raven Press, New York.

Reynolds, G.P., Arnold, L., Rossor, M.N., Iversen, L.L., Mountjoy, C.Q., and Roth, M. (1984). Reduced binding of (^{3}H) ketanserin to cortical 5-HT$_2$ receptors in senile dementia of the Alzheimer type. *Neuroscience Letters* **44**, 47–51.

Roberts, G.W., Crow, T.J., and Polak, J.M. (1985). Location of neuronal tangles in somatostatin neurons in Alzheimer's disease. *Nature* **314**, 92–4.

Rossor, M.N., Emson, P.C., Mountjoy, C.Q., Roth, M., and Iversen, L.L. (1980). Reduced amounts of immunoreactive somatostatin in the temporal cortex in senile dementia of Alzheimer type. *Neuroscience Letters* **20**, 373–7.

——, Garrett, N.J., Johnson, A.L., Mountjoy, C.Q., Roth, M., and Iversen, L.L. (1982). A post-mortem study of the cholinergic and GABA systems in senile dementia. *Brain* **105**, 313–30.

——, Iversen, L.L., Reynolds, G., Mountjoy, C.Q., and Roth, M. (1984). Early and late onset types of Alzheimer's disease are neurochemically distinct. *British Medical Journal* **288**, 961–4.

—— and —— (1986). Non-cholinergic neurotransmitter abnormalities in Alzheimer's disease. *British Medical Bulletin* **42**(i), 70–4.

——, Emson, P.C., Dawbarn, D., Mountjoy, C.Q., and Roth, M. (1986). Neuropeptides and Dementia. *Progressive Brain Research* **66**, 143–59.

—— (1986). Post-mortem neurochemistry of human brain. *Progress in Brain Research* **65**, 167–75.

Sasaki, H., Muramoto, O., Kanazawa, I., Arai, H., Kosaka, K., and Iizuke, R. (1986). Regional distribution of amino acid transmitters in post-mortem brains of presenile and senile dementia of Alzheimer type. *Annals of Neurology* **19**, 263–9.

Smith, C.C.T., Bowen, D.M., Sims, N.R., Neary, D., and Davison, A.N. (1983). Amino acid release from biopsy samples of temporal neocortex from patients with Alzheimer's disease. *Brain Research* **264**, 138–41.

Sofroniew, M.V., Pearson, R.C.A., Eckenstein, F., Cuello, A.C., and Powell, T.P.S. (1983). Retrograde changes in cholinergic neurons in the basal forebrain of the rat following cortical damage. *Brain Research* **289**, 370–4.

Spillane, J.A., White, P., Goodhardt, M.J., Flack, R.H.A., Bowen, D.M., and Davison, A.N. (1977). Selective vulnerability of neurons in organic dementia. *Nature* **266**, 558–9.

Spokes, E.G.S., Garrett, N.J., and Iversen, L.L. (1979). Differential effects of agonal status on measurements of GABA and glutamate decarboxylase in human post-mortem brain tissue from control and Huntington's chorea subjects. *Journal of Neurochemistry* **33**, 773–8.

Vecsei, L. *et al.* (1984). Comparative studies with somatostatin and cysteamine in different behavioural tests in rats. *Pharmacology, Biochemistry, and Behaviour* **21**, 833–7.

Vincent, S.R. *et al.* (1982). Neuropeptide coexistence in human cortical neurons. *Nature* **298**, 65–7.

Whitehouse, P.J. (1986). The concept of subcortical and cortical dementia: another look. *Annals of Neurology* **19**, 1–6.

——, Price, D.L., Struble, R.G., Coyle, J.I., and DeLong, M.R. (1982). Alzheimer's disease and senile dementia—loss of neurons in the basal forebrain, *Science* **215**, 1237–9.

—— *et al.* (1986). Nicotinic acetylcholine binding sites in Alzheimer's disease. *Brain Research* **371**, 146–51.

Yates, C.M., Simpson, J., Maloney, A.F.J., and Gordon, A. (1980). Neurochemical observations in a case of Pick's disease. *Journal of Neurological Science* **48**, 257–63.

Yates, C.M. *et al.* (1983). Catecholamines and cholinergic enzymes in pre-senile and senile Alzheimer-type dementia and Down's syndrome. *Brain Research* **280**, 119–26.

14

Anatomoclinical and biochemical concepts of subcortical dementia

YVES AGID, MERLE RUBERG, BRUNO DUBOIS,
AND BERNARD PILLON

INTRODUCTION

The most widely accepted criteria for the diagnosis of dementia are those proposed by the *Diagnostic and Statistical Manual of Mental Disorders,* 3rd edn (DSM III) (American Psychiatric Association 1980), which include: intellectual deterioration sufficiently severe to interfere with social or occupational functions; memory impairment; impaired abstract thinking and/or judgement; aphasia, apraxia, or agnosia; difficulty with constructional tasks; personality changes; but clear consciousness. These criteria apply quite well to Alzheimer's disease, defined as 'cortical dementia' because the histopathological lesions in the brains of these patients lie essentially in the cortex and hippocampus. They apply less well, however, to 'subcortical dementia' where cell loss occurs in deep grey matter rather than in the cerebral cortex (see reviews in Albert 1978; Butters *et al*. 1978; Mayeux 1981; Benson 1983; Cummings and Benson 1984; Agid *et al*. 1986*a*; Freedman and Albert 1986; Whitehouse 1986; Ruberg and Agid 1987). As an anatomoclinical entity, subcortical dementia is distinguished by two essential characteristics: (a) memory disorders associated with frontal lobe-like deficits, including bradyphrenia or slowing of mental processes, and sometimes with depression; (b) absence of aphasia, apraxia, and agnosia.

Although the clinical picture of subcortical dementia can be caused by tumours, particularly in the floor and walls of the third ventricle (Williams and Pennybaker 1954), or vascular accidents, notably lacunar states (Cummings and Benson 1984), it is most frequently observed in degenerative diseases of the central nervous system in which progressive lesions of the basal ganglia cause movement disorders. The three neurodegenerative diseases classically evoked as subcortical dementias are Huntington's chorea with lesions in the striatum, particularly the caudate nucleus (see Butters, this volume), Parkinson's disease with severe neuronal loss in the substantia nigra (Hornykiewicz 1966) and progressive supranuclear palsy with severe

248

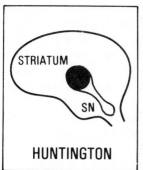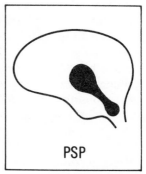

FIG. 14.1. Subcortical dementia: subcortical lesions.

neuronal loss in the striatum and substantia nigra, associated with degeneration of other structures in the basal ganglia, upper brainstem, and cerebellum (Steele *et al*. 1964) (Fig. 14.1). In these cases, because neuronal death is progressive, intellectual deterioration is slow and inexorable.

The following discussion of the concept of subcortical dementia will focus on two anatomoclinical paradigms: Parkinson's disease and progressive supranuclear palsy. The putative lesions that can be detected by biochemical assay of post-mortem brain tissue from patients with these diseases will be presented along with histopathological data, and hypotheses concerning the contribution of the lesions to cognitive impairment will be discussed.

'SUBCORTICAL DEMENTIA' IN PARKINSON'S DISEASE

The prevalence of dementia in Parkinson's disease is difficult to assess, but a figure of 30 per cent is often advanced. The figure might be lower, however, because of erroneous inclusion of patients with non-degenerative or iatrogenic Parkinsonian syndromes, Parkinsonian patients with severe depressive states that cannot be distinguished from dementia, or because the diagnostic criteria for dementia may be inappropriate (Brown and Marsden 1984; Agid *et al*. 1986*b*). However, the prevalence of Parkinsonian dementia might, in fact, be higher as suggested by retrospective clinical and histological analyses of autopsy cases (Dubois *et al*. 1985*a*). The absence of a precise definition of Parkinsonian dementia is the major cause of disagreement.

The cognitive deficits observed in Parkinsonian patients include the following. (a) Memory is impaired, particularly retrieval processes or the functional utilization of memory stores, rather than memory formation (see review in Ruberg and Agid 1987). (b) Visuospatial and perceptual motor functions are disturbed (Bowen *et al*. 1972; Flowers 1978; Boller *et al*. 1984; Stern *et al*. 1984), although the existence of a generalized visuospatial defect in Parkinson's disease is still the subject of controversy (Brown and Marsden 1986). (c) Cognitive and behavioural disorders resembling those found in patients with lesions of the frontal lobe have frequently been reported (see review in Taylor *et al.* 1986). These include decreased verbal fluency (Lees and Smith 1983), a tendency to perseveration (Bowen *et al.* 1975), decreased aptitude for set shifting, whether verbal, figurative or motor (Cools *et al.* 1984), abnormal behaviours like those caused by frontal lobe dysfunction (Pillon *et al.* 1986), inability to maintain repetitive sequences of gestures, or to reverse behaviours in response to audiovisual stimuli (Morel-Maroger 1978), bradyphrenia (Rogers 1986; Pillon *et al.* 1987).

Bradyphrenia or psychic akinesia is often evoked by analogy with akinesia or decreased initiation of movement characteristic of Parkinsonian patients. As the etymology of the word indicates, bradyphrenia (from the greek bradus—slow—and phrenos—thought) might best be defined as cognitive slowing independent of the motor component and has been observed in memory scanning (Wilson *et al.* 1980), auditory discrimination (Hansch *et al.* 1982) and visual recognition (Pillon *et al.* 1987). Bradyphrenia was first defined as a composite syndrome including cognitive and psychiatric elements: decreased attention and capacity for effort, slight memory impairment and slowness of thinking, difficulty in decision making, and reduction of emotional reaction (Naville 1922; Hassler 1953). In this sense, it resembles—or may be a part of—frontal syndrome.

Depression is observed in about 50 per cent of Parkinsonian patients (see review in Mayeux 1981), although its pathophysiological significance is still a matter of controversy. An association between intellectual deterioration and depression in Parkinsonian patients has sometimes been noted (Mayeux *et al.* 1981). It is interesting in this respect that psychomotor slowing is as characteristic of depression as it is of bradyphrenic syndrome (Ruberg and Agid 1987).

The characteristics listed above do not compose a monolithic picture which is identical in all Parkinsonian patients; a spectrum of intellectual disorders can be observed, ranging from a few limited frontal signs (Lees and Smith 1983) to florid dementia with severe memory impairment (Martilla and Rinne 1976). Nor can each of the various symptoms be attributed to a lesion of a specific type of neuron or brain structure in Parkinsonian patients. Not only is the inventory of these lesions far from complete, the neuronal interactions underlying brain functions, even at the lowest level of definition, are too complex to permit this kind of reduction. However, as the term 'subcortical' suggests, Parkinsonian dementia must be considered as an anatomoclinical entity, that is a syndrome of intellectual impairment resulting from subcortical lesions. What are the principal lesions that may alter mental function in patients? For simplicity, the cerebral cortex being essentially intact, we will describe only the major subcortical neuronal systems which are lesioned in

Parkinsonian patients: the nigrostriatal dopaminergic system and the long ascending subcortico-cortical monoaminergic (dopaminergic, noradrenergic, serotoninergic), and cholinergic systems.

Degeneration of the nigrostriatal dopaminergic pathway

Neuronal loss in the substantia nigra and decreased dopamine concentrations (more than 70 per cent) in the substantia nigra and the striatum (Bernheimer *et al.* 1973) are the evidence that the nigrostriatal dopaminergic pathway is severely lesioned in Parkinson's disease. Although the striatal target cells seem to be spared, as indicated by the normal concentrations of choline acetyltransferase (index of cholinergic neurons), glutamic acid decarboxylase (index of GABAergic neurons), and D1 and D2 dopamine receptors (see review in Agid *et al.* 1987), denervation of these cells affects output from the striatum to the cerebral cortex via the striato-pallido-thalamo (ventrolateral nucleus)-cortical (motor cortex) and the nigro (pars reticulata)-thalamo

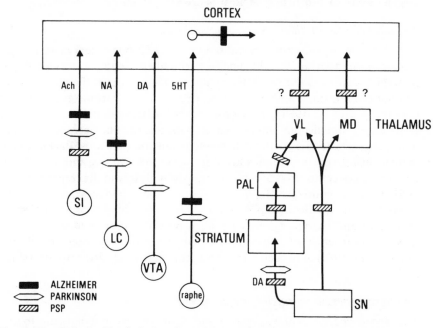

FIG. 14.2. Principal brain lesions in patients with Alzheimer's disease, Parkinson's disease and progressive supranuclear palsy. The black, white, and hatched bars indicate lesioned neuronal systems in Alzheimer's disease, Parkinson's disease, and progressive supranuclear palsy, respectively. ACh, acetylcholine; NA, noradrenaline; DA, dopamine; 5HT, serotonin; SI, substantia innominata; LC, locus coeruleus; VTA, ventral tegmental area; SN, substantia nigra; PAL, globus pallidus; VL, ventrolateral thalamic nucleus; MD, mediodorsal thalamic nucleus.

(ventrolateral nucleus)-cortical (premotor and prefrontal cortex) systems (Fig. 14.2).

There are two reasons for thinking that decreased dopaminergic transmission in the striatum might be responsible for intellectual changes in Parkinsonian patients. (a) A significant association has been found between bradykinesia (a symptom known to result from degeneration of the dopaminergic nigrostriatal system) and impaired performance on tests of spatial orientation, memory and visuospatial reasoning (Marttila and Rinne 1976; Mortimer *et al*. 1982). (b) Intellectual changes reminiscent of frontal lobe dysfunction have been detected in patients with MPTP-induced Parkinsonism (Stern and Langston 1985). Since only the dopaminergic neurons in the substantia nigra seem to be severely affected by MPTP, this lesion may be responsible for the moderate frontal lobe-like symptomatology observed in intoxicated patients, and for similar symptoms in patients with idiopathic Parkinsonism.

Degeneration of subcortico-cortical neuronal systems

The mesocorticolimbic dopaminergic pathway

A decrease in dopamine concentrations of about 50–60 per cent in neocortex, hippocampus, amygdala, and nucleus accumbens, and the loss of dopaminergic neurons in the ventrotegmental area suggest that the mesocortical and mesolimbic dopaminergic neurons degenerate in Parkinsonian patients (Javoy-Agid *et al*. 1984). Two observations suggest that this lesion may contribute to intellectual impairment in patients. (a) The decrease in dopamine concentrations is greater in demented than non-demented Parkinsonian patients in both the prefrontal and entorhinal cortex (Ruberg and Agid 1987). Dopamine concentrations are, however, extremely low in the cerebral cortex and assays are, consequently, quite imprecise; these data need, therefore, to be confirmed. (b) Behavioural abnormalities attributed to dysfunction of prefrontal and limbic regions have been observed in animal studies after selective destruction of dopaminergic neurons in the ventrotegmental area (Simon *et al*. 1980) or of their terminals in the prefrontal cortex (Brozowski *et al*. 1979).

The noradrenergic coeruleo-cortical system

The noradrenergic pathway projecting from the locus coeruleus to the cerebral cortex is partially lesioned in Parkinsonian patients as indicated by cell loss in the locus coeruleus and decreased noradrenaline concentrations (at least 50 per cent) in most cortical areas, particularly in the amygdala, the hippocampus and the frontal cortex (Scatton *et al*. 1983).

Lesions of the locus coeruleus in the rat cause deficiencies in selective attention, impairing learning and memory, and disorganizing complex behaviours (Iversen 1984; Ogren *et al*. 1984), effects which are also observed after lesions of the frontal cortex. Cognitive deficiencies, particularly attentional disorders, might therefore be expected in Parkinsonian patients with these lesions. Two types of data support this hypothesis. (a) Reaction times on continuous performance tasks have been observed to correlate positively with cerebrospinal fluid 3-methoxy-4-hydroxy-phenylene glycol (MHPG) concentrations, although the concentrations of the metabolite were normal (Stern *et al*. 1984). (b) Noradrenaline concentrations have been shown to be significantly lower in demented than in non-demented Parkinsonian patients in both the locus coeruleus (Cash *et al*. 1987) and in the entorhinal cortex, although not in the frontal cortex or hippocampus (Scatton *et al*. 1983).

Noradrenergic hypoactivity may also be implicated in the depressive states observed in patients, in accordance with the noradrenergic hypothesis of depression (van Praag 1982). The tricyclic antidepressants, which increase synaptic uptake of the amine and beta-adrenergic agonists (Lecrubier *et al*. 1980) are effective in the treatment of depression. The former have proven efficacious in depressed Parkinsonian patients as well (Strang 1976).

Serotoninergic raphé-cortical neurons

The ascending serotoninergic neurons are partially lesioned in Parkinsonian patients, judging from the decreases in serotonin concentrations observed in the striato-pallidal complex (Bernheimer *et al*. 1961) and in certain cortical areas (Scatton *et al*. 1983). The decrease is greatest in the hippocampus and frontal cortex, less pronounced in the cingular and entorhinal cortex, but non-significant in the amygdala and nucleus accumbens (Scatton *et al*. 1983). The density of imipramine binding sites, some of which are localized on serotoninergic nerve terminals, is reduced in the putamen and frontal cortex of Parkinsonian patients (Raisman *et al*. 1986).

Serotonin has been shown to play a role in learning behaviour in animals (Green and Heal 1985). It may therefore be hypothesized that the central serotoninergic deficiency is responsible for some cognitive disorders in patients. This deficiency may also be implicated in depression in Parkinsonian patients in accordance with the serotonin hypothesis of affective disorders (van Praag 1982). (a) 5-HIAA concentrations, the main metabolite of serotonin, are lower in the cerebrospinal fluid of depressed Parkinsonian patients than in those who are not depressed (Mayeux *et al*. 1984). (b) Imipramine-like drugs which are inhibitors of serotonin uptake have antidepressant activity in these as well as other depressed patients.

Cholinergic septo-hippocampal and innominato-cortical cholinergic systems

Choline acetyltransferase activity, indicative of cholinergic innervation, is reduced in the neocortex and the hippocampus, but not in the basal ganglia of Parkinsonian patients (Ruberg *et al.* 1982; Perry *et al.* 1983). This cholinergic deficiency most likely results from degeneration of neurons originating in the substantia innominata, where reduced choline acetyltransferase activity (Dubois *et al.* 1983) and cell loss (Whitehouse *et al.* 1983) have been detected, and in the septum. The lower the activity of the enzyme in the various cortical areas, the more the patients were found to be demented. This suggests that intellectual deterioration may, at least partly, result from degeneration of the innominato-cortical and septo-hippocampal neurons. Choline acetyltransferase activity was also decreased, although to a lesser extent, in the frontal cortex and hippocampus of non-demented patients, indicating that the onset of degeneration of subcortico-cortical cholinergic neurons may precede the deterioration of acetylcholine-dependent cognitive functions (Dubois *et al.* 1983).

Data from experiments using animal models and pharmacological treatment of patients support the hypothesis that cholinergic lesions can cause intellectual deterioration. (a) Following lesions of the basal forebrain (including the substantia innominata) in rats, which resulted in a cortical cholinergic deficiency, behavioural anomalies indicative of cognitive impairment are observed (Table 14.1). (b) The performance of Parkinsonian patients on memory tests is found to be below control levels before pharmacological intervention, and is further impaired in tests of intermediate (delayed recall and recall with interference), but not immediate memory, by chronic administration of anti-cholinergic drugs (Sadeh *et al.* 1982). (c) In another set of experiments, the hypothesis that a presymptomatic cholinergic lesion exists in non-demented Parkinsonians was substantiated. Administration of subthreshold doses of anticholinergics to non-demented Parkinsonian patients impairs performance on tests of visual memory, whereas the performance of control subjects is not affected (Dubois *et al.* 1987). (d) That there is a relationship between dementia in Parkinsonian patients and the cholinergic deficiency is also suggested by the analysis of confusional episodes in Parkinsonian patients receiving anticholinergic medication: confusional episodes (which may be considered acute and severe memory disorders) are essentially observed in demented rather than non-demented Parkinsonian patients; among the former, they are observed in practically all those patients who are treated with anticholinergic drugs (De Smet *et al.* 1982).

Role of subcortical, subcortico-cortical, and cortical lesions in the cognitive disorders of Parkinsonian patients

Massive lesions of the dopaminergic nigrostriatal pathway and partial destruction (about 50 per cent) of the long ascending dopaminergic,

TABLE 14.1. *Principal behavioural disturbances observed after lesions of the ventral tegmental area (A10) and the nucleus basalis of Meynert*

Behaviour	Tasks	Lesions	
		A10	Nucleus basalis
Spontaneous activity	Circular corridor	Le Moal *et al.* (1969) (HF)	Dubois *et al.* (1985*b*) (HF-IBO)
	Activity cage		Wishaw *et al.* (1985) (IBO)
Exploratory behaviour	Four-hole box	Gaffori *et al.* (1980) (HF)	Dubois *et al.* (1985*b*) (HF-IBO)
	Eight-hole box		
Hoarding behaviour		Stinus *et al.* (1978) (HF)	Dubois *et al.* (1985*b*) (HF-IBO)
Spatial learning	T-maze	Simon *et al.* (1980) (6-OHDA)	Hepler *et al.* (1985) (IBO)
	Radial maze		Dubois *et al.* (1985*b*) (HF-IBO)
			Hepler *et al.* (1985) (IBO)
	Morris water maze		Wishaw *et al.* (1985) (IBO)
Avoidance learning	Passive avoidance task	Galey *et al.* (1977) (6-OHDA)	Flicker *et al.* (1983) (IBO)
			Friedman *et al.* (1983) (KA)
			Hepler *et al.* (1985) (IBO)

Abbreviations: HF, radiofrequency lesion; IBO, ibotenic acid lesion; KA, Kainic acid lesion; 6-OHDA, 6-hydroxydopamine lesion.

noradrenergic, serotoninergic, and cholinergic systems projecting to the cortex may be implicated directly or indirectly (Fig. 14.2) in the genesis of cognitive symptomatology in Parkinsonian patients, in addition to the effects of local cortical pathology observed in some patients.

The nigrostriatal dopaminergic system is entirely subcortical, but lesions of these neurons may affect cortical function indirectly, as indicated by the 'frontal syndrome' observed in MPTP-intoxicated patients. This syndrome remains moderate, however, and can only be detected in neuropsychological tests involving complex tasks (Stern and Langston 1985). The indirect effect of the nigrostriatal lesion on cortical function may be mediated via the pallido-thalamo-cortical system to the frontal (premotor) cortex or via the substantia nigra pars reticulata and its thalamo-cortical (prefrontal) output (Fig. 14.2). In patients with idiopathic Parkinson's disease, unlike those with MPTP-induced Parkinsonism, an additional subcortico-cortical dopaminergic lesion affecting the mesocorticolimbic neurons, should aggravate cognitive impairment by a direct effect on cortical function, as seen in animal studies.

If the central dopaminergic deficiency plays a primordial role in intellectual impairment of Parkinsonian patients, restoration of dopaminergic transmission in the brain by L-dopa should correct these cognitive defects. This is not the case, however, except in early stages of the disease (Ricklan *et al*. 1976), either when evaluated by standard neuropsychological tests (Brown *et al*. 1984) or by measurements of cognitive slowing (Pillon *et al*. 1987). At most, a discrete awaking effect is observed (Marsh *et al*. 1971).

Lesions of the other subcortico-cortical pathways also contribute to deafferentation of the neocortex (prefrontal, in particular) and limbic regions including the hippocampus. The widely distributed monoaminergic systems, which are thought to play a modulatory role, should probably be distinguished, from the ascending cholinergic systems which project in a topographically defined manner to delimited areas of the cerebral cortex (Price and Stern 1985), and which may play a more specific role in memory functions, at least in the hippocampus.

Distinguishing the contributions of the subcortical and subcortico-cortical lesions to the symptoms of intellectual deterioration in Parkinsonian patients is difficult, however, since alteration of neurons in the cerebral cortex is suspected in a certain number of cases (see Ruberg and Agid 1987). (a) Alzheimer-like histological changes (neurofibrillary tangles and senile plaques) are found in the cerebral cortex of patients, especially in those with severe dementia (Boller *et al*. 1980). (b) This observation, and the selective decrease in somatostatin concentrations in the cortex of demented subjects only (Epelbaum *et al*. 1983) indicate not only that Parkinsonian intellectual deterioration may possibly be associated with neuronal loss in the cortex when dementia is severe, but also that dementia in Parkinson's and

Alzheimer's disease may have a similar pathophysiology.

In addition to the difficulty in assigning functions to precise pathways that are lesioned, the characteristics of intellectual deterioration (i.e. the results of neuropsychological studies) are difficult to interpret in Parkinson's disease. One reason is that the patients tend to be studied as if they form a homogeneous cohort. However, several subgroups can be distinguished, as a function of the duration of the disease, its severity and the nature of the motor symptoms. Two factors seem particularly relevent: age and non-dopaminergic lesions. Severe dementia is indeed more common in older patients (Sweet *et al.* 1976; Martilla and Rinne 1976), perhaps because neuronal dysfunction due to aging amplifies the consequences of neuronal loss characteristic of Parkinsonian patients. The intellectual disorders in Parkinson's disease also result from destruction of other neuronal systems besides those containing dopamine. In support of this hypothesis, the degree of intellectual deterioration has been found to be significantly correlated with the Parkinsonian score assessed when patients are treated with L-dopa, considered to reflect non-dopaminergic brain lesions (Esteguy *et al.* 1985), but not with that part of the Parkinsonian score that can be improved by L-dopa, symptoms due to dopaminergic dysfunction (B. Pillon, B. Dubois, and Y. Agid, unpublished results).

It may be that those cognitive disorders in Parkinsonian patient which resemble clinical frontal lobe symptoms, bradyphrenia in particular, evolve slowly and progressively in all patients, as a result of dysfunction of subcortico-cortical pathways and the dopaminergic nigrostriatal pathway. The portion of mental disorders that correspond to true 'subcortical dementia', may be observed selectively at the beginning of the disease (Lees and Smith 1983). The severe destruction of certain systems of subcortical

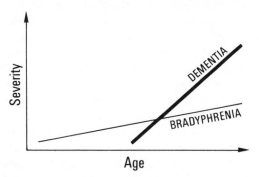

FIG. 14.3. Hypothetical evolution of cognitive impairment in Parkinson's disease. Bradyphrenia and frontal lobe-like symptomatology, i.e. subcortical dementia in the true sense of the word, could progressively worsen, but to variable degrees, in most if not all patients. Dementia, mainly characterized by memory disorders resulting from cortical lesion occurs only in a small number of patients.

origin (particularly cholinergic) and the appearance of histological and bio-chemical Alzheimer-like changes in the cerebral cortex would contribute to the severe dementia observed in a small number of patients, notably aged subjects (Agid *et al.* 1984) (Fig. 14.3). However, regardless of whether there is continuity or discontinuity in the evolution of cognitive disorders in Parkinson's disease, many of the neuronal systems which play a role in their genesis have yet to be identified. Whether the evolution of cognitive disorders in Parkinson's disease is continuous or discontinuous remains, however, to be determined; indeed, even the neuronal systems which play a role in their genesis are still far from known.

SUBCORTICAL DEMENTIA IN PROGRESSIVE SUPRANUCLEAR PALSY

The best known example of subcortical dementia is progressive supranuclear palsy. The formulation of the hypothesis that 'subcortical dementia' con-stitutes a clinical entity was based on the behavioural abnormalities observed in this disease (memory loss, impaired manipulation of acquired knowledge, personality changes marked by apathy or inertia, general slowness of thought processes), in view of the absence of evident cortical pathology in post-mortem brain tissue. The term was later changed to 'fronto-subcortical dementia' (Albert 1978) as the similarities between the cognitive and beha-vioural disorders in this disease and the effects of frontal lobe lesions were recognized. This apparent similarity has recently been substantiated by neuropsychological testing (Maher *et al.* 1987; Cambier *et al.* 1985; Pillon *et al.* 1986). The 'frontal syndrome' in patients with progressive supranuclear palsy can be observed in early stages of the disease, but develops into a major frontal motor syndrome after several years of evolution. Physiological evidence of hypometabolism in the frontal cortex of patients with progressive supranuclear palsy has been provided by PET-scan analysis of glucose utilization using 18-fluorodeoxyglucose (D'Antona *et al.* 1985). Despite clinical and metabolic evidence of frontal lobe impairment, no lesions can be detected in this structure. Indeed, neurofibrillary tangles in post-mortem brain tissue from patients with this disease, a rough indication of the severity of neuronal loss in various brain structures, are found mainly in the sub-stantia nigra (origin of the nigrostriatal dopaminergic neurons), the pallido-subthalamic complex, the pretectal area, the superior colliculi, striatum, dentate nucleus, thalamus, etc. (see review in Agid *et al.* 1986*a*). In summary, then, this disease is characterized by severe lesions of the basal ganglia and upper brain stem, contrasting with the integrity of the cere-bral cortex. Are the biochemical data concordant with histopathological observations?

Biochemical deficiencies in post-mortem brains from patients with progressive supranuclear palsy

The nigrostriatal dopaminergic pathway is severely damaged

Decreases in dopamine concentrations in the striatum and substantia nigra of the order of 90 per cent indicate that the nigrostriatal system is massively lesioned (Bokobza *et al.* 1984), as in Parkinson's disease. Unlike the latter, however, progressive supranuclear palsy is also characterized by degeneration of striatal target cells as well, indicated by decreased choline acetyltransferase and glutamic acid decarboxylase activity (Ruberg *et al.* 1985), and D2 dopamine receptor density (Ruberg *et al.* 1985; Baron *et al.* 1985).

Long ascending subcortico-cortical neurons are for the most part intact

Assays of dopamine, noradrenaline, and serotonin concentrations in cortical regions show that the long ascending mesocorticolimbic dopaminergic, coeruleo-cortical noradrenergic and raphé-cortical serotoninergic pathways are relatively untouched in patients with progressive supranuclear palsy (Kish *et al.* 1985; Ruberg *et al.* 1985). The cholinergic innominato-cortical and septo-hippocampal systems are altered, however, but to a much lesser degree than in Parkinson's or Alzheimer's disease. Small decreases in choline acetyltransferase activity were observed in some neocortical regions (but not the temporal cortex), in contrast to the basal ganglia where decreases reached almost 50 per cent (Agid *et al.* 1986*a*). This result is compatible with neuropathological observations of relatively moderate neuronal loss in the substantia innominata (Tagliavini *et al.* 1983).

The cerebral cortex is generally spared

Neurofibrillary tangles, an indication of neuronal pathology, are notably absent from major cortical areas such as the temporal and occipital cortex as well as the amygdala. However, in more than 50 per cent of cases, neurofibrillary tangles have been observed in the frontal cortex and the hippocampus, although they are often much less numerous than in patients with Alzheimer's or Parkinson's disease, even in the early stages. The integrity of the cortex has been confirmed by the biochemical assays that have been performed up to now. GABAergic interneurons in the cortex would seem to be intact, since no significant decreases have been found in cortical glutamic acid decarboxylase activity in either the frontal, cingular or temporal cortex, or in the amygdala, although a decrease in the activity of the enzyme of about 40 per cent was found in the hippocampus (Ruberg *et al.* 1985). Certain peptidergic neurons in the cortex, for example those containing met-enkephalin, leu-enkephalin, cholecystokinin-8, and substance P may also be judged intact, since the concentrations of these peptides have been found to be normal. The same is true for somatostatin-14 concentrations in the cortex of patients with progressive supranuclear palsy (Agid *et al*. 1986*a*), as opposed to intellectually impaired Parkinsonian subjects and patients with Alzheimer's disease.

Anatomo-biochemical substrate for frontal lobe-like symptomatology in progressive supranuclear palsy

The possibility that the frontal lobe symptomatology characteristic of progressive supranuclear palsy results from prefrontal deafferentation due to suppression of the output from the basal ganglia and from the thalamus must be seriously considered, since instrumental and mnemonic deficits observed in these patients are not likely to result from lesions in the temporo-parietal-occipital cortex which seems to be anatomically intact. These disorders may, however, result from a dysfunction of the basal ganglia or from the deactivation of posterior associative areas due to prefrontal hypoactivity (Fig. 14.4). The lesions of the long ascending monoaminergic and cholinergic pathways would seem to be too limited to cause the severe disorders observed, but those of the basal ganglia or thalamus (Fig. 14.2) are possible candidates. (a) Output from the substantia nigra to the caudate nucleus and putamen

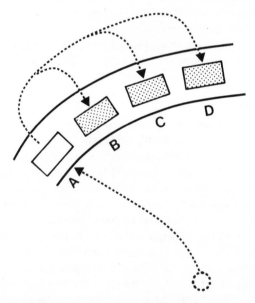

FIG. 14.4. Cortical deafferentation in progressive supranuclear palsy. In progressive supranuclear palsy, the various cognitive deficits which are not of 'frontal' origin may result from deactivation of associative cortex (cognitive programmes B, C, D) as a result of prefrontal hypoactivity (cognitive programme A) which itself results from deafferentation due to subcortical lesions (basal ganglia and upper brain stem).

(dopaminergic) and to the ventrolateral thalamus (GABAergic ?) is damaged. (b) The fact that the striatum and the pallidum are sites of severe neuronal loss, detected both histologically and biochemically, strongly suggests that the striato-pallidal and pallido-thalamic systems are damaged. (c) Cell loss in thalamic nuclei projecting to the prefrontal cortex has also been observed (Steele *et al.* 1964), although the precise nuclei and the neuro-transmitters involved have not yet been identified.

Other lesions in patients with progressive supranuclear palsy may aggravate the output deficiency from the basal ganglia, for example that of the cholinergic pedunculopontine nucleus (Zweig *et al.* 1985) which projects to the substantia nigra, the pallidum and the subthalamic nucleus.

DISCUSSION

The anatomoclinical and biochemical data on Parkinson's disease, pro-gressive supranuclear palsy, and Alzheimer's disease indicate that at least three types of neuronal lesions are involved in the pathophysiology of dementias (Fig. 14.5): (a) Degeneration of long ascending subcortico-cortical systems (cholinergic, noradrenergic, and dopaminergic) found to a limited extent in patients with progressive supranuclear palsy, somewhat more in patients with Alzheimer's disease, but most of all in patients with Parkinson's disease; (b) lesions of subcortical structures which disrupt communication with the cerebral cortex through the thalamus, most characteristic of progressive supranuclear palsy, although denervation of the striatum in Parkinsonian patients may also have an effect on cortically

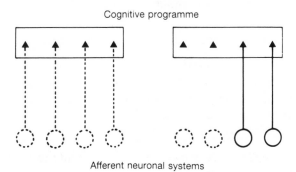

Cognitive programme

Afferent neuronal systems

FIG. 14.5. Cortical deafferentation and cognitive disorders. In subcortical dementia, where the cerebral cortex is generally spared, cognitive disorders (i.e. dysfunction of cognitive programmes) may occur either when the afferent subcortico-cortical systems are partially damaged, for example, partial lesion of each of the four afferent pathways (left), or severe lesion of only some of the afferent pathways (right).

directed output from these structures; (c) loss of intracortical neurons, as suggested by the presence of histopathological stigmata and decreased somatostatin concentrations characteristic of Alzheimer's disease, but also found to a limited extent in Parkinson's disease. What these lesions really do to patients is still largely, though not entirely, in the realm of speculation. However, even if these anatomo-biochemical data are still incomplete, they support some hypotheses concerning the pathophysiology of cognitive disorders in the dementias.

Cortical deafferentation or demodulation in subcortical dementia

If it is assumed that cognitive functions depend on the integrity of neuronal circuits (presumably cortical) programmed for the execution of these functions, cognitive disorders can occur in two ways: either where neurons in which the programme 'resides' are lost or in the process of degeneration, or where the afferent input has degenerated or is deficent. Alzheimer's disease is illustrative of the former, the subcortical dementias of the latter. There are, in effect, two types of afferent input to the cortex that may be disrupted in these diseases. The afferents may be either command specific, as is perhaps the case with the thalamic relays in progressive supranuclear palsy, or the cholinergic subcortico-cortical neurons in Parkinson's disease, or they may modulate cortical activity as, for example, the ascending monoaminergic pathways in Parkinson's disease. The distinction may be of importance since substitutive treatments can be envisaged as long as the cognitive programmes are themselves intact. This poses the question of whether dysfunction of one of these subcortico-cortical systems may by itself cause a cognitive disorder, and if so, beyond what threshold.

Threshold of denervation

The notion that behaviour is modified until a certain level of denervation is reached is clearly illustrated in Parkinson's disease, where the principal motor signs appear only when about 70 per cent of the dopaminergic nigro-striatal neurons have degenerated (Bernheimer *et al.* 1973; Agid and Blin 1987). Is the same true for cognitive disorders?

Cognitive disorders may appear beyond a threshold of destruction of a single system of neurons

Insofar as the long ascending subcortico-cortical pathways are concerned, it has been demonstrated, for example, that destruction of the cholinergic inno-minato-cortical pathway causes various behavioural disorders which seem to result from cognitive dysfunction (Table 14.1). In Parkinson's disease, a severe destruction of the innominato-cortical cholinergic pathway is observed in demented patients suggesting that the destruction of cholinergic afferents to the cortex contributes to intellectual deficiencies (Ruberg *et al.* 1982). However, a significant decrease in choline acetyltransferase activity of about 30 per cent is observed in cortical regions of patients in the absence of intellectual deterioration (Dubois *et al.* 1983). In this case, neuronal degeneration is present but insufficient to cause cognitive disorders. Evidence that this is the case has been obtained with non-demented Parkinsonian patients whose performance on a test of visual memory was impaired by administration of a subthreshold dose of scopolamine whereas control subjects were not affected (Dubois *et al.* 1987), thus confirming the existence of an underlying alteration of central cholinergic transmission. Two phases in the degeneration of cholinergic input to the cortex may therefore be distinguished: in the first phase neuronal degeneration is moderate and asymptomatic; in the second phase, neuronal loss becomes sufficient for the first cognitive disorders to appear (Agid *et al.* 1984). At the onset of the disease, despite advancing degeneration, normal cholinergic transmission may be maintained first by hyperactivity of the remaining cholinergic neurons, then by post-synaptic supersensitivity of muscarinic receptors. Beyond a certain threshold of cholinergic denervation, intellectual impairment would occur when these synaptic adjustments are no longer sufficient to compensate for neuronal loss. At this stage, when Parkinsonian patients are intellectually impaired, anticholinergic drugs can provoke confusional states, probably because pre-existent presynaptic cholinergic denervation is abruptly aggravated through blockade of cortical cholinergic receptors.

Cognitive disorders may appear only when more than one neuronal system is damaged

At least three configurations can be envisaged depending on whether the systems lesioned regulate a given function conjointly, in parallel or inversely, as in the following examples.

Conjoint regulation   In the rat, simultaneous destruction of the dopaminergic innervation of the nucleus accumbens or the striatum, by local microinjection of 6-hydroxydopamine in the mesocorticolimbic or nigrostriatal systems, disrupts the conditioned avoidance response, whereas selective destruction of one or the other of the systems has no effect (Koob *et al.* 1984). Extended to human subjects, this may indicate that disorders of cognitive functions controlled by the mesencephalic dopaminergic systems may be observed in Parkinsonians only when there is sufficient destruction of the dopaminergic mesocorticolimbic system in addition to severe lesion of the nigrostriatal system.

Parallel regulation   This may be the case when both the dopaminergic and cholinergic subcortico-cortical systems are lesioned in Parkinsonian patients. It has been observed that the behavioural anomalies caused in rats by bilateral lesion of the nigrostriatal and mesocorticolimbic dopaminergic systems resemble those which are observed after lesion of the innominato-cortical cholinergic system (Table 14.1). The possibility that cognitive disorders will appear either when a sufficient number of cholinergic neurons alone degenerate, or when the sum of partial destructions of the ensemble of afferent modulating systems (Fig. 14.5) reaches the necessary threshold, should be investigated.

Inverse regulation   This situation has been observed, where destruction of one ascending system (noradrenergic) attenuates or suppresses the effects of lesion of another system (dopaminergic) (Taghzouti *et al.* 1986).

In patients with Parkinson's or Alzheimer's disease, the cortically directed monaminergic and cholinergic systems are variably destroyed, perhaps explaining the diversity of symptoms (memory disorders, cognitive slowing, or depression, etc.) observed among individual subjects.

Cognitive syndromes in various types of dementia and the spectrum of neuronal lesions

Can the symptoms found in the various types of dementia be superimposed on the distribution of subcortical and cortical lesions, some of which have been characterized biochemically? A hypothesis is proposed in Table 14.2. Briefly, the principal syndromes commonly found in dementia are of three types: specific cognitive disorders, frontal lobe-like syndrome, and depression. Each of these syndromes includes a variety of symptoms which may overlap.

Cognitive disorders may be focal, such as aphasia, apraxia, or agnosia. They may involve memory, impairing storage mechanisms, or the use of memory stores. The latter, however, is also in the domain of frontal syndrome which associates bradyphrenia with the loss of various cognitive functions, such as memory impairment, visuospatial disorders, sensorimotor defects, etc., all of which are under the control of the prefrontal cortex (Nauta 1971). Bradyphrenia is, in fact, as characteristic of depression as it is of frontal syndrome, and can be distinguished from, and may be independent of the alterations of mood generally associated with the word depression. Each of these syndromes seems to correspond to a given pathological condition: focal cortical deficits and memory impairment are characteristic of Alzheimer's disease; frontal lobe symptomatology is characteristic of progressive supranuclear palsy and Parkinson's disease, associated in the latter with depression and in some cases with Alzheimer-like cortical pathology.

TABLE 14.2. *Cognitive syndromes in various types of dementia and the neuronal lesions observed*

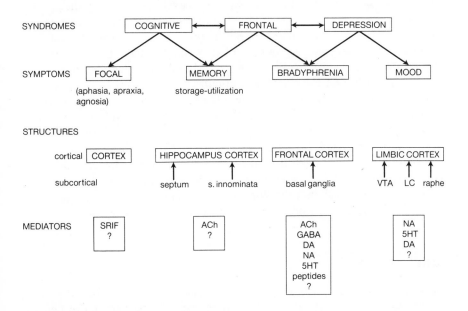

Can the types of lesions observed in patients with cortical and subcortical dementias explain the distribution of symptoms (i.e. the syndromes) which characterize them? Loss of intracortical neurons in the temporo-parieto-occipital areas is thought to be responsible for the focal cognitive deficits found in Alzheimer's disease. The identity of the cells that are lost is not known, aside from those containing neuropeptides, such as somatostatin-14 (see Rossor, this volume), if decreased somatostatin levels indeed result from neuronal loss. Memory disorders may result either from lesions in cortical structures (mainly, but not exclusively the hippocampus), or from degeneration of afferent neuronal systems such as the cholinergic septo-hippocampal and innominato-cortical systems, or the ascending monoaminergic regulatory systems. Bradyphrenia may be caused by lesions of the frontal cortex itself (although a distinction must be made between dysfunction of the dorso-lateral and orbitofrontal lobes), or from degeneration of the basal ganglia or the ascending monoaminergic systems (the very definition of subcortical dementia). The latter may also determine the appearance of depression in patients.

Patients with Alzheimer's disease have severe lesions of intracortical neurons and of the innominato-cortical and septo-hippocampal cholinergic systems, associated with some loss of the ascending monoaminergic neurons. Patients with Parkinson's disease have less cortical pathology, but more substantial lesions of the ascending cholinergic and monoaminergic systems, denervating the frontal cortex, in particular. Patients with progressive supranuclear palsy have basal ganglia lesions massively denervating the frontal cortex, but without the effects of lesion of the long ascending cholinergic and monoaminergic systems. It would seem, then, that the frontal syndrome observed in progressive supranuclear palsy can be distinguished from the frontal-depressive syndrome with associated acetylcholine-dependent memory disorders of Parkinsonian patients.

Although this schema is incomplete with respect to the lesions known (or to be discovered) in these patients, and the analysis of the symptoms summary, it does explain how the types of lesions found in the various dementia syndromes can begin to explain the characteristic constellations of the symptoms. The specific features of cognitive deterioration in each patient or group of patients will depend on the distribution of lesions, their severity, and the order in which they develop in the course of the diseases. This view of things is obviously oversimplified, but we hope not entirely false.

REFERENCES

Agid, Y. and Blin, J. (1987). Nerve cell death in degenerative diseases of the central nervous system: clinical aspects. In *Selective Neuronal Death* (ed. CIBA Foundation), pp. 5–29. John Wiley, New York.

——, Ruberg, M., Dubois, B., and Javoy-Agid, F. (1984). Biochemical substrates of mental disturbances in Parkinson's disease. In *Advances in Neurology, vol. 40: Parkinson-Specific Motor and Mental Disorders* (ed. R.G. Hassler and J.F. Christ), pp. 211–18. Raven Press, New York.

—— et al. (1986a). Progressive supranuclear palsy: anatomoclinical and biochemical considerations. In *Advances in Neurology*, Vol. 45 (ed. D. Yahr and K.J. Bergmann), pp. 194–206. Raven Press, New York.

—— et al. (1986b). Parkinson's disease and dementia. Proceedings of the XIII World Congress of Neurology, Hamburg, Sept. 1–6, 1985. *Clinical Neuropharmacology*.

——, Javoy-Agid, F., and Ruberg, M. (1987). Biochemistry of neurotransmitters in Parkinson's disease. In *Movement Disorders*, Vol. 2 (ed. C.D. Marsden and S. Fahn), pp. 166–230. Butterworth, London.

Albert, M.L. (1978). Subcortical dementia. In *Alzheimer's Disease: Senile Dementia and Related Disorders* (ed. R. Katzman, R.D. Terry, and K. Beck), pp. 173–80. Raven Press, New York.

American Psychiatric Association (1980). *Diagnostic and Statistical Manual of Mental Disorders*, 3rd edition (DSM-III). APA, Washington, DC.

Baron, J.C., Mazière, B., Loc'h, C., Sgouropoulos, P., Bonnet, A.M., and Agid, Y. (1985). Progressive supranuclear palsy: loss of striatal dopamine receptors demonstrated *in vivo* by positron tomography. *Lancet* i, 1163–4.

Benson, D.F. (1983). Subcortical dementia: a clinical approach, In *The Dementias* (ed. R. Mayeux and W.G. Rosen), pp. 185–94. Raven Press, New York.

Bernheimer, H., Birkmayer, W., and Hornykiewicz, O. (1961). Verteilung de-

5-hydroxytryptamins (Serotonin) im Gehirn des Menschen und sein Verhalten bei Patienten mit Parkinson-Syndrom. *Klinische Wochenschrift.* **39**, 1056–9.

——, ——, ——, Jellinger, K., and Seitelberger, F. (1973). Brain dopamine and the syndromes of Parkinson and Huntington. *Journal of Neurological Science* **20**, 415–55.

Bokobza, B., Ruberg, M., Scatton, B., Javoy-Agid, F., and Agid, Y. (1984). ^{3}H-Spiperone binding, dopamine and HVA concentrations in Parkinson's disease and supranuclear palsy. *European Journal of Pharmacology* **99**, 167–75.

Boller, F., Mizutani, T., Roessmann, U., and Gambetti, P. (1980). Parkinson disease, dementia and Alzheimer disease: clinicopathological correlations. *Annals of Neurology* **7**, 329–35.

——, Pasafume, D., Keefe, N.C., Rogers, K., Morrow, L., and Kim, Y. (1984). Visuospatial impairment in Parkinson's disease. *Archives of Neurology* **41**, 485–90.

Bowen, F.P., Hoehn, M.H., and Yahr, M.D. (1972). Parkinsonism: alterations in spatial orientation as determined by a route-walking test. *Neuropsychologia* **10**, 355–61.

——, Kamienny, R.S., Burns, M.M., and Yahr, M.D. (1975). Effects of levodopa treatment on concept formation, *Neurology* **25**, 701–4.

Brown, R.G. and Marsden, C.D. (1984). How common is dementia in Parkinson's disease. *Lancet* **ii**, 1262–5.

—— and —— (1986). Visuospatial function in Parkinson's disease. *Brain* **109**, 987–1002.

——, ——, Quinn, N., and Wyke, M.A. (1984). Alterations in cognitive performance and affect-arousal state during fluctuations in motor function in Parkinson's disease, *Journal of Neurology, Neurosurgery and Psychiatry* **47**, 454–65.

Brozowski, T.J., Brown, R.M., and Goldman, P. (1979). Cognitive deficit caused by regional depletion of dopamine in prefrontal cortex of rhesus monkey. *Science* **205**, 929–31.

Butters, N., Sax, D., Montgomery, K., and Tarlow, S. (1978). Comparison of neuropsychological deficits associated with early and advanced Huntington's disease. *Archives of Neurology* **35**, 585–9.

Cambier, J., Mason, M., Viadern, F., Limodier, J., and Strube, A. (1985). Le syndrome frontal de la maladie de Steele-Richardson-Olszewski. *Revue Neurologique* **141**, 528–36.

Cash, R., Dennis, T., L'Heureux, R., Raisman, R., Javoy-Agid, F., and Scatton, B. (1987). Parkinson's disease and dementia: norepinephrine and dopamine in locus coeruleus. *Neurology* **37**, 42–6.

Cools, A.R., van den Bercken, J., Horstink, M., van Spaendonck, K., and Berger, H. (1984). Cognitive and motor shifting aptitude disorder in Parkinson's disease, *Journal of Neurology, Neurosurgery and Psychiatry* **47**, 443–53.

Cummings, J.L. and Benson, D.F. (1984). Subcortical dementia. *Archives of Neurology* **41**, 874–9.

D'Antona, R. *et al* (1985). Subcortical dementia: frontal cortex hypometabolism detected by positron tomography in patients with progressive supranuclear palsy. *Brain* **108**, 785–99.

De Smet, Y., Ruberg, M., Serdaru, M., Dubois, B., Lhermitte, F., and Agid, Y. (1982). Confusion, dementia and anticholinergics in Parkinson disease. *Journal of Neurology, Neurosurgery and Psychiatry* **45**, 1161–4.

Dubois, B., Ruberg, M., Javoy-Agid, F., Ploska, A., and Agid, Y. (1983). A sub-cortico-cortical cholinergic system is affected in Parkinson's disease. *Brain Research* **288**, 213–18.

——, Hauw, J.J., Ruberg, M., Serdaru, M., Javoy-Agid, F., and Agid, Y. (1985*a*) Démence et maladie de Parkinson: corrélations biochimiques et anatomo-cliniques. *Revue Neurologique* **141**, 184–93.

——, Mayo, W., Agid, Y., LeMoal, M., and Simon, H. (1985*b*). Profound disturbances of spontaneous and learned behaviours following lesion of the nucleus basalis magnocellularis in the rat. *Brain Research* **338**, 249–58.

——, Danzé, F., Pillon, B., Cusimano, J., Lhermitte, F., and Agid, Y. (1987). Cholinergic-dependent cognitive deficits in non-demented Parkinsonian patients. *Annals of Neurology* (in press).

Epelbaum, J., Ruberg, M., Moyse, E., Javoy-Agid, F., Dubois, B., and Agid, Y. (1983). Somatostatin and dementia in Parkinson's disease. *Brain Research* **278**, 376–9.

Esteguy, M., Bonnet, M., Kefalos, J., Lhermitte, F., and Agid, Y. (1985). Le test à la L-DOPA dans la maladie de Parkinson. *Revue Neurologique* **141**, 413–15.

Flicker, C., Dean, R.C., Watkins, D.L., Fisher, S.K., and Bartus, R.T. (1983). Behavioural and neurochemical effects following neurotoxic lesions of a major cholinergic input to the neocortex in the rat. *Pharmacology, Biochemistry, and Behaviour* **18**, 973–81.

Flowers, K. (1978). Lack of prediction in the motor behaviour of Parkinsonism. *Brain* **101**, 35–52.

Freedman, M. and Albert, M. (1986). Subcortical dementia. In *Handbook of Clinical Neurology, Vol. 2: Behavioural Neurology* (ed. P.J. Vinken and G.W. Bruyn), pp. 311–16. North Holland Press, Amsterdam.

Friedman, E., Lerer, B., and Kuster, J. (1983). Loss of cholinergic neurons in the rat neocortex produces deficits in passive avoidance learning. *Pharmacology, Biochemistry, and Behaviour* **19**, 309–12.

Gaffori, O., Le Moal, M., and Stinus, L. (1980). Locomotor hyperactivity and hypo-exploration after lesion of the dopminergic A10 area in the ventral mesencephalic tegmentum (VMT) of rats, 1980. *Behavioural Brain Research* **1**, 313–29.

Galey, D., Simon, H., and Le Moal, M. (1977). Behavioral effects of lesions in the A10 dopaminergic area of the rat, 1977. *Brain Research* **124**, 83–97.

Green, A.R. and Heal, D.J. (1985). The effects of drugs on serotonin-mediated behavioural model. In *Neuropharmacology of Serotonin* (ed. A.R. Green), pp. 326–65. Oxford University Press, Oxford.

Hansch, E.C., Syndulko, K., Cohen, S.N., Goldberg, Z.I., Potvin, A.R., and Tourtellote, W. (1982). Cognition in Parkinson's disease: an event-related potential perspective. *Annals of Neurology* **11**, 599–607.

Hassler, R. (1953). Extrapyramidal-motorische Systeme und Erkrankungen. In *Handbuch der Innerin Medizin*, Vol. 3, pp. 676–904. Springer Verlag, Berlin.

Hepler, D.J., Wenk, G,J., Cribbs, B.J., Olton, D.S., and Coyle, J.T. (1985). Memory impairments following basal forebrain lesions. *Brain Research* **346**, 8–14.

Hornykiewicz, O. (1966). Dopamine (3-hydroxytyramine) and brain function. *Pharmacology Review* **18**, 925–64.

Iversen, S. (1984). Cortical monoamines and behavior. In *Monoamine Innervation of*

Cerebral Cortex (ed. L. Descarries, T.A. Reader, and H.H. Jasper), pp. 321–49. Alan R. Liss, New York.

Javoy-Agid, F. *et al.* (1984). Biochemical neuropathology of Parkinson disease. In *Advances in Neurology, Vol. 40: Parkinson-Specific Motor and Mental Disorders* (ed. R.G. Hassler and J.F. Christ), pp. 189–98. Raven Press, New York.

Kish, S.J., Chang, L.J., Mirchandani, L., Shannak, K., and Hornykiewicz, O. (1985). Progressive supranuclear palsy: relationship between extrapyramidal disturbances, dementia, and brain neurotransmitter markers, *Annals of Neurology* **18**, 530–6.

Koob, G.F., Simon, H., Herman, J.P., and Le Moal, M. (1984). Neuroleptic-like disruption of the conditioned avoidance response requires destruction of both the mesolimbic and nigrostriatal dopamine systems. *Brain Research* **303**, 319–29.

Lecrubier, Y., Puech, A.J., Jouvent, R., Simon, R., Simon, P., and Widlocher, D. (1980). A beta adrenergic stimulant (salbutamol) versus clomipramine in depression: a controlled study. *British Journal of Psychiatry* **136**, 354–8.

Lees, A.J. and Smith, E. (1983). Cognitive deficits in the early stages of Parkinson's disease. *Brain* **106**, 257–70.

Le Moal, M., Cardo, B., and Stinus, L. (1969). Influence of ventral mesencephalic lesions on various spontaneous and conditioned behaviors in the rat. *Physiology and Behavior* **4**, 567–73.

Maher, E.R., Smith, E.M., and Lees, A.J. (1985). Cognitive deficits in the Steele–Richardson–Olszewski syndrome (progressive supranuclear palsy). *Journal of Neurology, Neurosurgery and Psychiatry* **48**, 1234–9.

Marsh, G.G., Markham, C.M., and Ansel, R. (1971). Levodopa's awakening effect on patients with Parkinsonism. *Journal of Neurology, Neurosurgery and Psychiatry* **34**, 209–18.

Marttila R.J. and Rinne, U.K. (1976). Dementia in Parkinson's disease, *Acta Neurologica Scandinavica* **54**, 431–41.

Mayeux, R. (1981). Depression and dementia in Parkinson's disease. In *Movement Disorders* (ed. C.D. Marsden and S. Fahn), pp. 75–95. Butterworth Scientific, London.

Mayeux, R., Stern, Y., Rosen, J., and Leventhal, J. (1981). Depression, intellectual impairment and Parkinson's disease. *Neurology* **31**, 645–50.

Mayeux, R., Stern, Y., Coté, L., and Williams, J.B.W. (1984). Altered serotonin metabolism in depressed patients with Parkinson's disease. *Neurology (Clev.)* **34**, 642–6.

Morel-Maroger, A. (1978). Effects positifs de la L-DOPA sur une symptomatologie 'marginale' observée chez les Parkinsoniens. *L'Encéphale* **IV**, 223–31.

Mortimer, J.A., Pirozzolo, F.J., Hansch, E.C., and Webster, D.D. (1982). Relationship of motor symptoms to intellectual deficits in Parkinson disease. *Neurology* **32**, 133–7.

Nauta, W.J.H. (1971) The problem of the frontal lobe: a reinterpretation. *Journal of Psychiatric Research* **8**, 167–87.

Naville, F. (1922). Les complications des séquelles mentales de l'encéphalite epidemique. *L'Encéphale* **17**, 369–75.

Ogren, S.O., Archer, T., and Ross, C.B. (1984). Norepinephrine in learning and memory—the status of the cognitive deficit. In *Catecholamines: Neuropharmacology and Central Nervous System—Theoretical Aspects* (ed. E. Usdin, A.

Carlsson, A. Dahlstrom, and J. Engel), pp. 285–92. Alan R. Liss, New York.

Perry, R. *et al.* (1983). Cortical cholinergic deficit in mentally impaired Parkinsonian patients. *Lancet* **ii**, 789–90.

Pillon, B., Dubois, B., Lhermitte, F., and Agid, Y. (1986). Heterogeneity of cognitive impairment in progressive supranuclear palsy, Parkinson's disease, and Alzheimer's disease. *Neurology* **36**, 1179–85.

—— *et al.* (1987). Evaluation of cognitive slowing in Parkinson's disease: the 'fifteen objects test'. (Submitted for publication.)

Price, J.L. and Stern, J. (1985). Individual cells in the nucleus basalis-diagonal band complex have restricted axonal projections to the cerebral cortex in the rat. *Brain Research* **269** 352–6.

Raisman, R., Cash, R., and Agid, Y. (1986). Parkinson's disease: decreased density of ^{3}H-imipramine and ^{3}H-paroxetine binding sites in the putamen. *Neurology* **36**, 556–60.

Riklan, M., Whelihan, W., and Cullinan, T. (1976). Levodopa and psychometric test performance in Parkinsonism—five years later. *Neurology* **26**, 173–9.

Rogers, D. (1986). Bradyphrenia in Parkinsonism: a historical review. *Psychological Medicine* **16**, 257–65.

Ruberg, M., Ploska, A., Javoy-Agid, F., and Agid, Y. (1982). Muscarinic binding and choline acetyltransferase in Parkinsonian subjects with reference to dementia. *Brain Research* **232**, 129–39.

—— *et al.* (1985). Dopaminergic and cholinergic lesions in progressive supranuclear palsy. *Annals of Neurology* **18**, 523–9.

—— and Agid, Y. (1987). Dementia in Parkinson's disease. In *Handbook of Psychopharmacology, vol. 20: Psychopharmacology of the Aging Nervous System* (ed. L. Iversen, S. Iversen, and S. Snyder). Plenum Publishing Co., New York (in press).

Sadeh, M., Brahim, J., and Modan, M. (1982) Effects of anticholinergic drugs on memory in Parkinson's disease, *Archives of Neurology* **39**, 666–7.

Scatton, B., Javoy-Agid, F., Rouqier, L., Dubois, B., and Agid, Y. (1983). Reduction of cortical dopamine, noradrenaline, serotonin and their metabolites in Parkinson's disease. *Brain Research* **275**, 321–8.

Simon, H., Scatton, B., and Le Moal, M. (1980). Dopaminergic A10 neurones are involved in cognitive functions. *Nature* **286**, 150–1.

Steele, J.C., Richardson, J.C., and Olszewski, J. (1964). Progressive supranuclear palsy. *Archives of Neurology* **10**, 333–59.

Stern, Y. and Langston, J.W. (1985). Intellectual changes in patients with MPTP-induced Parkinsonism. *Neurology* **35**, 1506–9.

——, Mayeux, R., and Cote, L. (1984). Reaction time and vigilance in Parkinson's disease. *Archives of Neurology* **41**, 1086–9.

Stinus, L., Gaffori, O., Simon, H., and Le Moal, M. (1978). Disappearance of hoarding and disorganization of eating behavior after ventral mesencephalic tegmentum lesion in rats. *Journal of Comparative Physiology and Psychology* **92**, 288–96.

Strang, R.R. (1976) Imipramine in treatment of Parkinsonism: a double blind placebo study. *British Medical Journal* **2**, 33–4.

Sweet, R.D., McDowell, H., and Feigesen J.S. (1976). Mental symptoms in Parkinson's disease during treatment with levodopa. *Neurology (Minneapolis)* **26**, 305–10.

Taghzouti, K. *et al.* (1986). Lesions of noradrenergic pathways supress the behavioral deficits induced by electrolytic lesions within the ventral tegmental area *Abstracts, Meeting of the American Society for Neuroscience*, 1986.

Tagliavini, F., Pillieri, G., Gemignani, F., and Lechi, A. (1983). Neuronal loss in the basal nucleus of Meynert in progressive supranuclear palsy. *Acta Neuropathology (Berlin)* **61**, 157–60.

Taylor, A.E. Saint-Cyr, J.A., and Lang A.E. (1986). Frontal lobe dysfunction in Parkinson's disease. *Brain* **109**, 845–83.

van Praag, H.M. (1982). Depression. *Lancet* **ii**, 1259–64.

Whitehouse, P.J. (1986). The concept of subcortical and cortical dementia: another look. *Annals of Neurology* **19**, 1–6.

——, Hedreen, J.C., White, C.L., and Price, D.L. (1983). Basal forebrain neurons in the dementia of Parkinson Disease. *Annals of Neurology* **13**, 243–8.

Williams, M. and Pennybaker, J. (1954). Memory disturbances in third ventricle tumors. *Journal of Neurology, Neurosurgery and Psychiatry* **17**, 115–23.

Wilson, R.S., Kaszyniak, A.W., Klawans, H.L., and Garron, D.C. (1980). High speed memory scanning in Parkinsonism. *Cortex* **16**, 67–72.

Wishaw, I.Q., O'Connor, W.T., and Dunnett, S.B. (1985). Disruption of central cholinergic mechanisms in the rat by basal forebrain lesions or atropine: effects on feeding, sensorimotor behaviour, locomotor activity and spatial navigation. *Behavioural Brain Research* **17**, 103–15.

Zweig, R.M., Whitehouse, P.J., Casanova, M.F., Walker, L.C., Jankel, W.R., and Price, D.L. (1985). Loss of putative cholinergic neurons in progressive supranuclear palsy (abst.). *Annals of Neurology* **18**, 144.

15

Problems in the cognitive neurochemistry of Alzheimer's disease

DANIEL COLLERTON

INTRODUCTION

Cognitive neurochemistry is a comparatively new discipline. Although its roots can be traced back to Hippocrates and Galen, it is only in the past decade that it has become a recognizable subsection of the neurosciences. A major impetus for this development has been fundamental discoveries about neurochemical deficiencies in certain forms of neuropsychiatric diseases and their treatment with drugs. The recognition in the early sixties that the motor symptoms of Parkinson's disease are associated with a lack of brain dopamine (Hornykiewicz 1978), the development of effective psychotropic drugs which influence brain neurotransmitter function (Lipton, 1978), and more recent discoveries of marked neurochemical abnormalities in the brains of patients with illnesses such as Huntington's chorea (Bird and Spokes 1982) and Alzheimer's disease (Hardy *et al*. 1985; Bowen and Davison 1986), have led to a general belief that higher mental functions such as language, praxis, problem solving ability, and in particular memory can profitably be studied in terms of changes in the neurochemical activity of the brain (Zornetzer 1978).

In this chapter, I shall examine the neurochemical approach to Alzheimer's disease (AD). Post-mortem studies in the mid 1970s discovered abnormalities in cholinergic transmission in AD (Bowen *et al*. 1976; Davis and Maloney 1976; Perry *et al*. 1977); a finding which led to the so-called cholinergic hypothesis of AD—the suggestion that a substantial contributing cause of the dementia öf AD is loss of cholinergic function (Smith and Swash 1978; Rossor 1981; Bartus *et al*. 1982; Coyle *et al*. 1983; E.K. Perry 1986; Pomara and Stanley 1986). Over time, this hypothesis has developed into what is undoubtedly the major focus of human cognitive neurochemistry: the relationship between the neurochemical and behavioural changes of AD.

This interest is fuelled by the increasing pressure which society is coming under as the number of demented elderly rises (Brody 1985; Rocca *et al*. 1986). AD is the commonest cause of this dementia. By itself, it accounts for approximately half of all cases of dementia and it contributes to perhaps another quarter (Kay *et al*. 1964, 1970; Blessed 1980; Hendersen 1986). Clinically, it is characterized by a relentlessly progressive decline in intellectual abilities in the absence of major changes in mood, alertness, or mobility (see, for example, Cummings and Benson 1983; Huppert and Tym 1986). Histopathologically, it is marked by granulovacuolar degeneration, perivascular amyloidosis, and senile plaques and neurofibrillary tangles (see, for example, Alzheimer 1907; Simchowicz 1911; Glenner 1983; Tomlinson and Corsellis 1984). In addition, there are severe and widespread, though selective, changes in neurochemicals (see, for example, Hardy *et al*. 1985). Its cause is unknown and there is no effective treatment (Heyman *et al*. 1984; Amaducci *et al*. 1986; Hendersen 1986, but see also Summers *et al*. 1986; Barnes 1987 for recent advances).

PROBLEMS IN THE DIAGNOSIS OF ALZHEIMER'S DISEASE AND THE SELECTION OF EXPERIMENTAL GROUPS

The difficulties in investigating the cognitive neurochemistry of AD begin with the problems of firmly diagnosing the disease itself, especially in life (McKhann *et al*. 1984; Khachaturian 1985; Huppert and Tym 1986; Katzman 1986). A definitive diagnosis of AD can only be made by examining the brain for specific neuropathological features, and not always then (Alafuzoff 1985). Often in behavioural studies post-mortem or biopsy confirmation of diagnosis is not available in which case the experimenter must depend upon previously verified diagnostic criteria to select subjects. These can identify cases of AD with a high specificity as determined by post-mortem (Blessed 1980) or biopsy (Fox *et al*. 1985) follow-up. The counterpart to this high specificity, however, is a loss of sensitivity since the available criteria depend heavily upon excluding other possible causes of dementia. To take one example, AD patients may present with a parietal lobe syndrome (Crystal *et al*. 1982). These patients may have a similarly localized pathology in the cortex and thus be valuable cases for fractionating the dementia of AD. The use of diagnostic criteria which stress the necessity for a global impairment of cognition, or the invariable presence of amnesia (McKhann *et al*. 1984; Huppert and Tym 1986), may miss such cases. The converse of this, though, is that if diagnostic criteria are relaxed so that sensitivity is increased, the specificity of the diagnosis appears to fall (Table 15.1).

Variations in the efficiency of diagnostic criteria may account for some of the controversy over the effects of age of onset on the symptoms and pathology of AD (Jorm 1985). It has been suggested that there are two forms of AD: a young onset form which has a more severe and rapid course and is accompanied by severe, widespread pathology, and an older onset form with a slower progression and milder symptoms and pathology (Bondareff 1983;

TABLE 15.1. *Sensitivity and specificity of the diagnosis of Alzheimer's disease as determined at post-mortem*

Authors	Clinical diagnosis	Age range	Sensitivity	Specificity
Blessed (1980)	Alzheimer type dementia	over 65	0.28	0.99
	Alzheimer type dementia plus senile dementia	over 65	0.64	0.92
Molsa *et al.* (1985)	Alzheimer disease	59–65	0.71	0.73
Constantinidis (1978)	Presenile Alzheimer disease	under 65		0.78
	Senile Alzheimer disease	over 65		0.60
	Presenile and senile Alzheimer disease		0.69	0.66
Todorov *et al.* (1975)	Alzheimer's presenile dementia	under 65	0.88	0.78
	Alzheimerized senile dementia	over 65	0.28	0.43
	Alzheimerized senile dementia plus senile dementia	over 65	0.44	0.77
	Alzheimer's presenile dementia plus Alzheimerized senile dementia and senile dementia		0.69	0.78

Notes: sensitivity and specificity were taken directly from papers or calculated from figures given. Diagnostic criteria are not exactly specified in these papers and the ground for distinguishing between 'Alzheimer-type dementia', 'Alzheimerized senile dementia', and 'senile dementia' are not clear. All would now be considered as later onset Alzheimer's disease. For this reason, sensitivity and specificity are given for the groups separately and combined. It seems safe, however, to conclude that Alzheimer's presenile dementia and Alzheimerized senile dementia as defined by Todorov *et al.* were diagnosed by the same criterion.

Byrne and Arie 1985). Although it is unlikely that these variations reflect qualitatively different forms of the disease (Collerton 1985; Jorm 1985), the possibility remains that there are systematic variations in the neurochemistry (Bird, *et al.* 1983; Rossor *et al.* 1984) and psychology of AD (Sulkava and Amberla 1982; Filley *et al.* 1986) depending on the age of onset. The effects of different neurochemical losses on behaviour within AD might therefore be open to investigation by comparing different age groups. However, the sensitivity and specificity of diagnosis appear to vary across the age range. Criteria which select earlier onset AD patients with a relatively good sensitivity and selectivity are much less successful at selecting later onset patients (see, for example, Todorov *et al*. 1975 in Table 15.1). This may be due to a tendency for elderly people to have multiple illnesses which increase their likelihood of being excluded from the experimental group. If this process were random, it would not be of importance. If, however, it were biased in that AD was exacerbated by other illnesses and those patients were excluded, then, as a whole, younger patients would appear to have a more severe disease.

Even if representative groups of the whole population were compared, it would not necessarily be correct to assume that any differences between groups were due to the direct effects of differing pathologies. It is well recognized that the apparent deficits of demented patients vary markedly with environment (Mace and Rabins 1982; Woods and Britton 1985). Cohort effects also need to be considered. Cross-sectional studies, such as post-mortem examination, biopsies, and most psychological assessments, confound age with date of birth. For example, brain weight at a particular age has tended to increase during this century (Miller and Corsellis 1977) and most, but not all, neurochemicals decline over the age range (Kubanis and Zornetzer 1981; Rossor *et al*. 1984; Rogers and Bloom 1985). Thus, comparisons of AD patients with age-matched controls may tend to underestimate the pathological changes in elderly AD patients (Collerton 1985). A final complication is the age-related increase of mild Alzheimer pathology in the 'normal' elderly (Tomlinson *et al*. 1968; Dayan 1970; Ball 1977). Since the pathological diagnosis of AD is based upon the density of plaques or tangles exceeding a certain threshold, different thresholds have been suggested for different ages (McKhann *et al*. 1984). Again, old and young groups are not directly comparable.

These confounding factors, though small and tending to cancel each other, do suggest that age-related divisions may be less valuable than ones based solely on symptoms. As noted above, there is considerable behavioural heterogeneity within AD (see also Chui *et al*. 1985; Neary *et al*. 1986*a*, *b*); variation which must reflect differences in pathology.

PROBLEMS IN THE BEHAVIOURAL ASSESSMENT OF ALZHEIMER'S DISEASE PATIENTS

The major practical problem with symptom-based divisions is the extreme rarity or non-existence of AD patients with restricted abnormalities. In most cases, there are multiple, though selective, neurochemical abnormalities and multiple, though again selective, behavioural changes. (For examples of mnemonic deficits see Butters 1984; Miller 1971; Weingartner *et al*. 1981, 1983; Wilson *et al*. 1981, 1982, 1983; Butters *et al*. 1983, *et al* 1984; Davis and Mumford 1984; Muramoto 1984; Nebes *et al*. 1984; for examples of language problems see Miller and Hague 1975; Appell *et al.* 1982; Bales 1982; Gewirth *et al.* 1984; Kirshner *et al.* 1984; Cummings *et al.* 1985.) It is almost certainly too much to hope that these will map simply onto each other. For example, the overall dementia of AD is probably not a simple summation of the various component disorders in language, praxis, memory and other skills (Huppert and Tym 1986). Even within specific skill areas, the deficits of AD patients may be more than the sum of the deficits expected from the restricted lesions of classical neuropsychology—the drawing problems of AD patients are not simply related to the drawing deficits of patients with restricted left or right hemisphere lesions (Moore and Wyke 1984). When these cognitive complexities are added to the emotional, motivational, sensory, and motor problems common in AD patients, the cross-sectional studies commonly employed are unlikely to untangle the inter-relationships of inter-correlated behavioural changes.

Longitudinal studies may be more useful. For instance, the fact that the amnesia of AD usually predates language problems (Cummings and Benson 1983; Huppert and Tym 1986) is a strong argument that the amnesia is not caused by the aphasia, a conclusion reinforced by the rare cases in which aphasia predates amnesia (Mesulam 1982).

PROBLEMS IN THE PATHOLOGY OF ALZHEIMER'S DISEASE

The inter-correlation of deficits is even more of a problem in pathological studies. The pathological abnormalities of AD tend to correlate not only with each other but also with estimates of the dementia of the subject in life (Table 15.2). How is one to determine the functional significance of these changes?

TABLE 15.2. *Correlations of selected neuropathological and behavioural indices in Alzheimer's disease*

	Cellular changes	Plaques	Neurofibrillary tangles	Cholinergic changes
Dementia	0.5–0.7[1] 0.5–0.8[5]	0.5–0.8[2] 0.4[7]	0.5–0.6[3]	0.5–0.8[4] 0.6, 0.4[6]
Cellular abnormalities		0.5–0.7[8] 0.4–0.5[11] Non-significant[14]	0.4–0.6[9] 0.4–0.7[12] 0.6[15]	0.3–0.4[10] 0.5–0.6[13]
Plaques			0.8[16]	0.4[17] 0.4–0.6[18] 0.7[19] Non-significant[20]
Neurofibrillary tangles				0.5[21] 0.5[22] 0.4[23]

Notes: Size, but not direction, of correlation is noted to 1 significant figure.

[1]Neary *et al*. (1986*b*). Temporal lobe biopsy. Significant correlations between clinical ratings, WAIS performance, token test, reaction time and pyramidal cell loss, nuclear and nucleolar volume, and cytoplasmic RNA.

[2]Blessed *et al*. (1968). Whole brain post-mortem. Significant correlations between plaques and global rating and memory-information-concentration task in combined demented and non-demented groups and demented group alone.

[3]Neary *et al*. (1986*b*). Significant correlations between tangles and WAIS verbal IQ and token test but not with WAIS performance IQ, clinical ratings, or visual reaction time.

[4]Perry *et al*. (1978). Neocortex post-mortem. Significant correlations between global rating and choline acetyltransferase (ChAT) activity.

[5]Sumpter *et al*. (1986). Temporal lobe biopsy. Significant correlations between areal proportion and surface area of rough endoplasmic reticulum and choice reaction time, WAIS, clinical rating, and token test. No significant correlations with cell body area, cytoplasmic area, nuclear area, nucleolar area, mitochondrial area, and lipofuscin area.

[6]Neary *et al*. (1986*b*). Significant correlations between acetylcholine (ACh) synthesis and clinical rating and visual reaction time. No significant correlations between ACh synthesis and WAIS or token test. No significant correlations between ChAT activity and any behavioural measures.

[7]Neary *et al*. (1986*b*). Significant correlations between plaques and WAIS verbal IQ but not with any other behavioural measures.

[8]Mann *et al*. (1985). Temporal cortex post-mortem. Significant correlations between plaques and cell loss and nucleolar volume.

[9]Neary *et al*. (1986*b*). Significant correlations between tangles and cell loss, nuclear and nucleolar volume and cytoplasmic RNA.

[10]Mountjoy *et al*. (1984). Neocortex post-mortem. Significant correlations between cell loss and ChAT activity in frontal and temporal cortex across demented and control groups, but not in cingulate, parietal, or occipital cortex, or in any regions in the demented group alone.

[11]Neary *et al*. (1986*b*). Significant correlations between plaques and nuclear and nucleolar volume, and cytoplasmic RNA, but not with cell loss.

[12]Mann *et al*. (1985). Significant correlations between tangles and cell loss and nucleolar volume.

[13]Neary *et al*. (1986*b*). Significant correlations between ChAT activity and cell loss, but not nuclear and nucleolar volume or cytoplasmic RNA. Significant correlations between ACh synthesis and cell loss and nuclear and nucleolar volume, but not cytoplasmic RNA.

[14]Terry *et al*. (1981). Whole brain post-mortem.

[15]Ball (1977). Hippocampus post-mortem. Significant correlation between tangles and cell loss.

[16]Neary *et al*. (1986*b*).

[17]Perry *et al*. (1978).

[18]Mountjoy *et al*. (1984). Significant correlations between plaques and ChAT activity in frontal and temporal, but not cingulate, parietal, or occipital cortex.

[19]Neary *et al*. (1986*b*). Significant correlation between plaques and ChAT activity, but not ACh synthesis.

[20]Wilcock *et al*. (1982). Frontal and temporal cortex post-mortem. No significant correlations between ChAT activity and plaques.

[21]Neary *et al*. (1986*b*). Significant correlations between tangles and ChAT activity, but not ACh synthesis.

[22]Mountjoy *et al*. (1984). Significant correlations between tangles and ChAT activity in frontal and temporal, but not cingulate, parietal, or occipital cortex.

[23]Wilcock *et al*. (1982). Significant correlations between tangles and ChAT activity in temporal, but not frontal lobes.

It would be wrong to assume that they are all equally important, or even that some are important at all (Mountjoy 1986).

Various methods are available. The simplest, and least useful, is to assume that the greater the neurochemical abnormality, the greater its importance. This is unlikely to be true. There is a strong possibility that different neurochemical systems would react in different ways to similar numerical changes. To illustrate this it is commonly suggested that a loss of over 80 per cent of striatal dopamine is necessary before clinical symptoms of Parkinsonism become apparent (Bernheimer *et al.* 1973). If a similar threshold occurs in the cholinergic projections to the cortex, it appears to be around 30 per cent (T. Arendt, personal communication). Furthermore, in experimental models in animals marked changes in the activity of neurones (Aston-Jones *et al.* 1985*a*, *b*) can occur in the absence of measurable changes in neurotransmitter markers (Rogers and Bloom 1985) and in AD compensatory growth may reduce the functional effects of lesions (Geddes *et al.* 1985).

A more sophisticated approach is to correlate pre-mortem psychological results and post-mortem pathological findings in selected groups of patients. However, the repeated tests needed to minimize the delay between psychological assessments and death, and the wide range of ability which must be encompassed, in practice precludes extensive psychological testing. Thus, these studies have produced valuable evidence for the quantitative relationship between global measures of dementia and pathological abnormalities in the cortex and cholinergic systems (Blessed *et al.* 1968; Perry *et al.* 1978), without, however, demonstrating any qualitative relationship between specific behavioural and pathological abnormalities. The problem of limited psychological assessments may be overcome by biopsy studies, but these in their turn suffer from the limited pathological assessment which can be performed on the small areas of cortex which are sampled (Francis *et al.* 1985; Neary *et al.* 1986*b*).

One could also try to compensate statistically for intercorrelated abnormalities, though threshold effects, and truncated and non-linear distributions of pathological markers make unambiguous answers difficult.

In support of hypotheses drawn from pathological data, evidence has therefore been drawn from other areas.

DEFINING THE QUALITATIVE RELATIONSHIP BETWEEN NEUROCHEMISTRY AND BEHAVIOUR IN ALZHEIMER'S DISEASE

The role of drug studies

Drug studies have been a major focus of research, with particular attention being paid to cholinergic pharmacology. This is a vast field with over 80

TABLE 15.3 *Effects of cholinergic drugs on performance on cognitive tasks in man*

Study	Cholinergic drug	Task	Effects
Safer and Allen (1971)	Scopolamine	Variable delay recall of digits: 0–20 s delay	– (TD, DD)
		Auditory, visual vigilance	0 (DD)
		Addition	0 (DD)
		Motor co-ordination	–
Dundee and Pandit (1972)	Scopolamine	Picture recognition	– (DD)
Ghoneim and Mewaldt (1975)	Scopolamine	Immediate, delayed (10, 20, 35, 55 min) recall of semantically categorized and non-categorized word lists	0 (TAE)
		Delayed (20, 40, 55 min) recognition of semantically categorized and non-categorized word lists	0 (TAE)
		Delayed picture recognition: 2 min delay	0
		Immediate recall of 8 digits	0
Ghoneim and Mewaldt (1977)	Scopolamine	Immediate, delayed (15, 25, 45 min) recall and recognition of word lists	0 (TAE)
Petersen (1977)	Scopolamine	Immediate, delayed (30, 60 min) recall of word lists	0 (TAE, DD)
Drachman (1978)	Scopolamine	Digit span	0
		Supraspan digit learning	–
		Immediate free recall of word list	–
		Generation of category exemplars	–
		Wechsler Adult Intelligence Scale: VIQ	0
		PIQ	–
		FSIQ	–
	Physostigmine	As above for scopolamine	0
Sitaram *et al.* (1978*b*)	Scopolamine	Serial word list learning	–
	Arecholine		+
	Choline		+

TABLE 15.3 *continued*

Study	Cholinergic drug	Task	Effects
Crow (1979)	Scopolamine Atropine	Scanning task	0(Sc, At)
		Number-colour paired associate learning	– (Sc) 0(At)
		Free recall of word lists:	0 (Sc) 0(At)
		immediate	0 (Sc) 0(At)
		delayed (60s)	– (Sc) 0(At)
Jones *et al.* (1979)	Scopolamine	Immediate recall of 9 digits	0
		Immediate recall of semantically categorized word lists	0
Liljequist and Mattila (1979)	Scopolamine	Memory for chess problems	–
	Physostigmine		±
Mewaldt and Ghoneim (1979)	Scopolamine	Immediate, delayed (25, 35 min) recall of word lists	0 (TAE)
		Delayed (45, 55 min) recognition of word lists	0
		Immediate recall of 9 digits	–
Rasmusson and Dudar (1979)	Scopolamine	Supraspan digit learning	0
		Maze learning	0
Drachman and Sahakian (1980)	Physostigmine	As Drachman (1978)	0
Caine *et al.* (1981)	Scopolamine	Brown–Peterson short-term forgetting: 0–18 s delay	0 (TD)
		Serial word list learning	–
		Selective reminding	–
		Immediate recall of semantically or phonologically categorized word lists	0
		Delayed (10 min) recall of categorized word lists	–
		Cued, delayed (10 min) recall of categorized word lists	0
		Verbal generation of category exemplars	–
		Auditory vigilance for:	

TABLE 15.3 *continued*

Study	Cholinergic drug	Task	Effects
		gaps in noise	0
		order of tones	–
Nuotto (1983)	Scopolamine	Immediate recall of 8 digits: forward	0
		back	–
		Paired associate word learning	–
		Hand-eye coordination	–
		Simple reaction time	–
		Multiple choice reaction time	–
		Tapping rate	–
		Flicker fusion frequency	–
Wesnes and Warburton (1983)	Scopolamine	Identification of pauses in rotation of clock hand:	
		sensitivity	– (DD)
		response bias	– (DD)
Callaway (1984)	Scopolamine	Reaction time	0̲ (DD)
		Latency of P300 component of EEG	0̲ (DD)
Frith *et al.* (1984)	Scopolamine	Immediate recall of categorized word lists: semantic, phomemic, abstract, concrete categories	0̲
Wesnes and Revell (1984)	Scopolamine	Stroop test	– (Sc) 0(Ni)
	Nicotine	Digit sequence identification	– (Sc) 0(Ni)
Wesnes and Warburton (1984)	Scopolamine	Digit sequence identification	0̲ (DD)
	Nicotine		$\overset{+}{0}$ (DD)
Dunne and Hartley (1985)	Scopolamine	Immediate, delayed (50s) recall:	$\overset{+}{\underline{0}}$
		Delayed recognition: of dichotically presented words	0
Loviaux *et al.* (1985)	Xanthinol Nicotinate	Immediate free recall of semantically, phonologically and neutrally categorized word lists	$\overset{+}{0}$

TABLE 15.3 *continued*

Study	Cholinergic drug	Task	Effects
		Paired associate word learning	0
		Verbal generation of category exemplars	0
		Recognition of category exemplars	0
		Free recall of word lists	0
		Simple reaction time	0
		Groniger intelligence task	0
Beatty *et al.* (1986)	Scopolamine	Repeated immediate recall of high and low imagery word lists	–
		Delayed (30 min) recall of word lists	–
		Delayed (30 min) recognition of word lists	0
		Symbol—digit paired associate learning	0
		Brown–Peterson short term forgetting	–
		Verbal generation of category exemplars	0

Notes: 0, no effect; – , impairment; + , facilitation; combinations indicate mixed effects. DD (non-DD), dose-dependent (non-dose-dependent) effects—results are (are not) monotonically related to dose of drug; TD (non-TD), time-dependent (non-time-dependent) effects—results are (are not) related to the size of delay in a delayed task; TAE, time of administration effect— results vary with the stage of the task at which the drug is administered. Sc, scopolamine; At, atropine; Ni, nicotine. Further details of these studies are in Collerton (1986) from which this table is condensed.

published studies (see Weiss and Heller 1969; Zornetzer 1978; Collerton 1986 for reviews; Baratti *et al.* 1984; Levy *et al.* 1984; Prado-Alcala *et al.* 1984; Aigner and Mishkin 1986; Beatty *et al.* 1986; Levin and Bowman 1986; Mattingly 1986 for uncited studies). Though the evidence that they have produced has been influential in the development of the cholinergic hypothesis, their present relevance to AD is unclear (Collerton 1986).

Much of the behavioural pharmacology of the cholinergic system has suffered from the side-effects of the available cholinergic drugs. Although their peripheral effects [mainly nausea, inco-ordination, and loss of visual acuity (Safer and Allen 1971; Martindale 1982; Nuotto 1983)] seem, on the whole, not to account for their behavioural effects (see, for example,

Bohdanecky *et al.* 1967; Wagman and Maxey 1969; Ksir 1974; Evans 1975; Kokkinidis and Anisman 1976*a*, *b*; Drachman 1978; Sitaram *et al.* 1978*a*, *b*; Liljequist and Matilla 1979; Eckerman *et al.* 1980; Caine *et al.* 1981; Wesnes and Warburton 1983, 1984; Levy *et al.* 1984; Dunnett 1985), their effects on arousal would seem to be an important factor which is generally ignored (Safer and Allen 1971; Mewaldt and Ghoneim 1979; Nuotto 1983; Dunne and Hartley 1985). When this is added to the failure to carry out proper dose-response studies and the selection of insufficiently powerful tests and experimental designs (Zornetzer 1978; Collerton 1986), it is not surprising that there is a certain amount of controversy over the exact effects of cholinergic drugs. Restricting ourselves to human studies, while it cannot be denied that cholinergic drugs affect performance upon tasks of memory (Table 15.3), neither can it be denied that they affect performance upon many tasks that cannot easily be classified as tests of memory (Table 15.3), unless, that is, one adopts a definition of memory so wide as to be effectively meaningless. The situation in animal experiments is, if possible, even more confused (Collerton 1986). Nor is it yet possible to do more than speculate as to whether their wide range of effects are because they affect a single psychological function which underlies all these tasks; whether the systemically administered drugs simultaneously affect several different pathways each of which has a separate psychological function; or whether their results are due to a combination of general and specific effects. It is therefore not unexpected that there should be a large number of discrepant interpretations of the behavioural effects of cholinergic drugs (Table 15.4).

This uncertainty renders parallels between the effects of drugs and AD potentially misleading, especially when, as is usual, drug and disease are compared in terms of their effects on hypothetical psychological systems (memory, arousal, attention, etc.) rather than compared directly on the same tests. When such comparisons have been performed, the results are equivocal. Thus, the effects of the anticholinergic drug, scopolamine, on the performance of young normals on the Wechsler Adult Intelligence Scale (WAIS) (Drachman 1978; Drachman and Leavitt 1974) was similar to the performance of only 50 per cent of groups of AD patients studied subsequently on the same test (Brinkman and Braun 1984; Fuld 1984). Since the pattern of results on the WAIS after scopolamine treatment (mean of scores on information and vocabulary subtests > mean of similarities and digit span > mean of digit symbol and block design less than object assembly) is also extremely difficult to account for by a pure effect on memory and since scopolamine and AD produce different patterns of errors on memory tests (Beatty *et al.* 1986), the cholinergic hypothesis in its specifically mnemonic form may be incorrect.

In a similar manner to cholinergic drugs, though less extensively, the effects of noradrenergic drugs have been explored as a model of Alzheimer's

TABLE 15.4. *Interpretations of the behavioural effects of cholinergic drugs*

Area	Specific deficits	Authors
Memory	Encoding/consolidation	Caine *et al.* (1981)
		Crow (1979)
		Liljequist and Mattila (1979)
		Mewaldt and Ghoneim (1979)
		Petersen (1977)
		Ridley *et al.* (1984*a*)*
		Ridley *et al.* (1984*b*)*
	Recent memory	Pazzagli and Pepeu (1964)*
	Storage	Haroutunian *et al.* (1985)*
		Weiner and Deutsch (1968)*
	Retrieval	Buresova *et al.* (1964)*
		Caine *et al.* (1981)
		Moss *et al.* (1981)*
		Petersen (1977)
		Ridley *et al.* (1984*a, b*)*
	Learning	Buresova *et al.* (1964)*
		Daly (1968)*
		Glick *et al.* (1973)*
	Working memory	Stevens (1981)*
		Wirsching *et al.* (1984)*
	Short-term memory	Alpern and Marriott (1973)*
	State-dependent memory	Berger and Stein (1969)*
		Gardner *et al.* (1972)*
		Overton (1966)*
Non-memory	Stimulus sensitivity	Brown and Warburton (1971)
		Wesnes and Warburton (1984)
	Stimulus selection	Rick *et al.* (1981)*
	Sensory discrimination	Heise *et al.* (1976)*
	Vision	Evans (1975)*
	Stimulus control of	Eckerman *et al.* (1980)*
	behaviour	Heise and Lilie (1970)*
		Heise and Milar (1984)*
		Ksir (1974)*
		Russel *et al.* (1961)*
	Set perseveration	Giardini *et al.* (1983)*
	Disruption of habituation	Kokkinidis and Anisman (1976*a*)*
		Kokkinidis and Anisman (1976*b*)*
	Disruption of strategies	Sutherland *et al.* (1982)*
		Wishaw (1985)*
	Attention	Callaway (1984)
		Callaway *et al.* (1985)
		Dunne and Hartley (1985)

Notes: Interpretations marked * are drawn from animal experiments. Further details of these studies are given in Collerton (1986).

disease since a substantial noradrenergic deficit exists in some patients with AD (Perry *et al.* 1981; Bondareff *et al.* 1982; Mann *et al.* 1984; Rossor *et al.* 1984); patients suffering from Korsakoff's psychosis, another disease associated with amnesia, have a noradrenergic deficit (McEntee and Mair 1980; McEntee *et al.* 1984), although this is controversial (Martin *et al.* 1984) and some evidence suggests that manipulations of noradrenergic function in man (see, for example, Frith *et al.* 1985) and animals (see, for example, Everitt *et al.* 1983) can alter performance upon memory tasks.

One could multiply these examples almost indefinitely. The recent reports of a substantial glutaminergic deficit in AD (Greenamyre *et al.* 1985; Hardy *et al.* 1987), and a role for the N-methyl-D-aspartate receptor in memory (Morris *et al.* 1986) are a case in point.

That these various neurochemical abnormalities coexist in AD is an added complication. It is potentially extremely important to discover whether such deficiencies are additive or subtractive or whether one potentiates the other. Drug studies provide an extremely valuable means of investigating such questions.

The crucial problem with all such studies lies in their interpretation. What is the significance of similarities or dissimilarities in the effects of drugs and disease? Would, for example, the discovery that both AD and an anticholinergic drug affected performance upon a task in an identical way mean that the poor performance of AD patients was due to their cholinergic deficiency, or might it merely be a reflection of the fact that the cells which normally receive a cholinergic input were themselves disrupted? If AD and the drugs had different effects, would that mean that the patients' cholinergic abnormalities were unimportant, or could it be that the acute administration of anticholinergic drugs to young normal volunteers does not sufficiently mimic the cholinergic disruption produced by a chronic loss of cholinergic cells in the aged brains of AD patients? Without careful systematic studies which assess the effects of numbers of drugs on numbers of tasks and which carefully match the effects of drugs and disease on one or more reference tasks, we can do little but speculate (Collerton 1986).

The role of lesion studies

Although lesion studies share some of the difficulties inherent in the interpretation of drug experiments, they have the advantage that specific neurotransmitter pathways can be manipulated, an impossible procedure with systemically administered drugs (see for example, Dunnett 1985). However, in the case of cholinergic systems, this is offset by the lack of a specific cholinergic neurotoxin (Jarrard *et al.* 1984) which, combined with the diffuse distribution of the cholinergic neurones projecting to the neocortex and medial temporal lobes, means that selective cholinergic lesions are difficult to make (Flicker *et al.* 1983; Dunnett *et al.* 1985; Fine *et al.* 1985; Wishaw *et al.* 1985).

Furthermore, the lesion produced is unstable, resolving over several months to produce a complex pattern of behavioural deficits (Bartus *et al.* 1986). The inconclusive effects of these lesions in animals have recently been reviewed by Collerton (1986) and Salamone (1986). In man, lesions which include the basal forebrain cholinergic systems cause amnesia and dementia (Damasio *et al.* 1985*a*, ,*b*).

Despite this, lesion experiments in combination with drug investigations are extremely useful in attempting to separate the effects of different neurochemical manipulations. An excellent example is afforded by the demonstration that selective neurotoxin lesions of dopamine terminals in sulcus principalis of the monkey frontal cortex, have the same effect as surgical removal of that area of cortex, while selective noradrenaline depletions in the same area do not share this effect (Brozoski *et al.* 1979). It is probably overoptimistic to hope that drug or lesion experiments, at least in animals would be directly applicable to AD. Their use may be more in demonstrating that there are psychological functions that are neurochemically dissoluble, rather than in delineating exactly what those functions are (Bond 1984).

The role of animal models

Of even greater use would be animal models which mimicked the developing histopathological and neurochemical changes of AD. Such models are not yet available. Although in susceptible animals, the administration of aluminium salts can result in neurofibrillary tangles and administration of scrapie agent causes plaque formation, these pathological features are not similar to those seen in AD (Overstreet and Russell 1984; Tomlinson and Corsellis 1984). At the moment, which neurochemicals are deemed worthy of investigation depends mainly upon the results of neurochemical investigations after death. The danger in this is that the neurotransmitter abnormalities found post-mortem may occur at a late stage of the disease, and the functionally important neurochemical changes take place much earlier. By the time that there are the major structural changes in the brain needed to make a diagnosis of AD, neurochemical changes may be relatively unimportant. An animal model which allowed the progression of the disease to be followed would therefore be extremely valuable.

At present, it is not altogether clear whether the progression of AD results from the gradual intensification of a single pathological process within a single region or pathway; whether it is caused by one process spreading throughout a number of regions; whether it results from a number of different pathological processes operating within the same region; or whether there are a number of different processes operating simultaneously within different regions.

It seems highly unlikely that the dementia of AD results from one or more processes operating within a single brain area. Although it has been suggested that, for example, hippocampal damage (Ball *et al*. 1985) or cholinergic abnormalities (Rossor 1981) by themselves could account for all the dementia of AD, evidence from classical neuropsychology on the effects of localised lesions of particular brain areas shows that no areas of the brain have been identified which when damaged produce an AD-like dementia (see, for example, Collerton and Fairbairn 1985).

It is more likely that the progression results from one or more processes spreading throughout the brain. Recent evidence has shown that the distribution of tangles in the AD brain follows anatomical pathways and is understandable as a trans-synaptic transmission of some degenerative process (Pearson *et al*. 1985). The multiple neurochemical changes of AD may therefore be secondary to whatever is the primary pathological process: more a reflection of which neurotransmitters are found in a particular brain area, than neurotransmitter-specific pathology. An illustration of this is the discussion of the primary or secondary importance of the cholinergic abnormalities in AD (Arendt and Bigl 1986; Collerton 1986; Mesulam 1986; Neary *et al*. 1986*b*).

That is not to say that neurotransmitter changes are therefore unimportant. Though they may be the proximal rather than the ultimate cause of the dementia, they may offer a focus for therapeutic intervention. While there is as yet no means of replacing missing neurones (though current experiments in neuronal transplants suggest that such a possibility lies in the foreseeable future (Gash *et al*. 1985; Fine 1986) the technology of manipulating synaptic transmission is well advanced.

The role of neuroimaging

New developments in neuroimaging also hold out considerable promise for longitudinal studies of the pathology of AD. Sequential measurement of brain metabolism by Positron Emission Tomography (PET) has allowed the regional spread of AD to be followed and the impact of changes in brain areas to be assessed (see, for example, Foster *et al*. 1983, 1984, 1986; Cutler *et al*. 1985; Duara *et al*. 1986; Haxby *et al*. 1986), while, in the near future, advances in PET scanning using radiolabelled neurochemicals and receptor ligands (Leenders *et al*. 1984) and, in high resolution, *in vivo* Nuclear Magnetic Resonance spectroscopy (Radda 1986), mean that *in vivo* measurement of neurochemicals may become possible. The potential use of this has been shown by studies of metabolism or blood flow in the so-called 'subcortical dementias'. These have shown that while the pathological lesion may lie subcortically, the functional lesion may be in the cortex (D'Antona *et al*. 1985; Skyhoj *et al*. 1986).

Comparing AD with other diseases

Although the problem of multiple end-stage pathological deficits in AD can be minimized by selecting mildly affected cases at post-mortem on the assumption that these are earlier stages of the disease, or by performing biopsy studies on early cases (Francis *et al.* 1985), there is always a nucleus of inter-correlated abnormalities (Table 15.2): a nucleus which, as more neuro-chemicals are measured in AD, is likely to grow.

One valuable technique available to circumvent this problem is to compare the dementia of AD with the intellectual deficits associated with other diseases which manifest similar though more restricted pathologies. In the case of the cholinergic abnormalities of AD for example, comparison with selected cases of Parkinson's disease with cortical cholinergic loss has shown that the dementia of AD is more severe than might be expected from the loss of cholinergic function alone (Rossor *et al.* 1982; Perry *et al.* 1985). This means at the least that another process is contributing to the dementia of AD (although whether that is a quantitative or qualitative contribution is unclear) and it is consistent with the suggestion that the other process alone is respon-sible. Similarly, the Parkinsonian-dementia complex of Guam (Hirano *et al.* 1961) and dementia pugilistica (Corsellis *et al.* 1973) produce the fibrillary tangles, but not the plaques of AD. Careful quantification of the dementia and pathology in these diseases and direct comparisons with AD will show which, if any, of these abnormalities *could* cause the dementia of AD.

More useful still would be the demonstration that a particular pathological abnormality could occur in the absence of behavioral changes. This would be a strong argument that the particular pathological feature was unimportant in AD. For example, the normal elderly may have plaque counts in the AD range, without tangles and without dementia (Tomlinson 1980; Terry *et al.* 1982) suggesting that even though plaque counts correlate with dementia they do not necessarily contribute to it. When dealing with neurochemical changes, however, one must be cautious. Normal levels of a neurotransmitter do not necessarily mean that there has not been a loss of that neurochemical. By chance alone, some cases will have abnormally high initial levels of a transmitter and so can suffer an appreciable loss, perhaps sufficient to cause functional abnormality, but still end up within the normal range.

Reversing the deficits of Alzheimer's disease

The difficulties with all the preceding approaches is that they are necessarily based upon arguments by analogy or data from correlations; both of which are inconclusive. The most direct technique, and perhaps the only one which could demonstrate the actual as opposed to potential relevance of a particular abnormality, is to assess the effects of reversing the neurochemical deficiencies of AD. For example, there have been many attempts to reverse the cholinergic abnormality of AD (Bartus *et al*. 1982; Davis *et al*. 1982; Kay *et al*. 1982; Muramoto *et al*. 1984; Beller *et al*. 1985; Little *et al*. 1985; Davidson *et al*. 1986; Hollander *et al*. 1986; Summers *et al*. 1986), with, on the whole, disappointing results. Even when it is effective, as it appears to be in a small minority of patients [who may not actually have AD, Rossor *et al*. (1982) and Perry *et al*. (1985) having reported patients who presented with an AD-like dementia, but who turned out at post-mortem to have cholinergic deficit in the absence of other AD pathology], the results are seldom clinically relevent (Davidson *et al*. 1986) and may be similar to the effects of other, non-cholinergic, drugs (Hope 1982).

However, while success of a therapy would be strong evidence for the importance of a particular change, failure is only weak evidence against. The lack of effect of cholinergic therapy, for instance, might be due to suboptimal drugs, dosages, or route of administration, or the incorrect selection or assessment of subjects, rather than any intrinsic unimportance of the cholinergic abnormalities (Davis *et al*. 1982; Bowen and Davison 1986; Harbaugh 1986; Hollander *et al*. 1986). It might also be a reflection of the widespread pathology of AD. Unlike Parkinson's disease, the paradigmatic exemplar of neurological replacement therapy (Hornykiewicz 1978), in which the primary lesion is in the input neurones (nigrostriatal dopamine system) to a target area (striatum), in AD both the targets (neocortex and medial temporal lobes) and their input neurones (forebrain cholinergic systems) are disrupted (E.K. Perry 1986; R.H. Perry 1986). These cholinergic projections may additionally have a more specific role than do the mainly regulatory dopaminergic systems (Gray 1982; Aston-Jones *et al*. 1985*a*, *b*). Replacing a highly patterned input with a constant level of stimulation might therefore be ineffective. Even if cholinergic function could be sucessfully restored, it might still be unimportant in reversing impairments. If the cholinergic function in cortex were normal but all the other pathological features of AD were still present, then the AD patient might still be demented. Equally if all the other features of AD were corrected, a cholinergic abnormality in itself might be sufficient to cause severe disablement (Perry *et al*. 1985).

CONCLUSIONS

Cognitive neurochemistry is the study of the neurochemical (often, though not strictly correctly, taken to mean neurotransmitter) basis of higher mental processes. The thesis presented should show that, from this viewpoint, AD is not a particularly useful disease to study. The abnormalities are simply too numerous, and too closely related for uncomplicated answers to emerge readily. From a practical perspective though, AD is far too important a health problem to neglect.

Classical neuropsychology has advanced by studying groups of patients with specific psychological deficits or known localized damage to the brain. This approach has been successfully used to delineate the areas of brain particularly concerned with memory (medial temporal lobes and dorsomedial thalamus), language (temporal/parietal lobes), praxis (parietal/occipital lobes), and response selection (frontal lobes) (see, for example, Lezak 1982). One of the great hopes of cognitive neurochemistry is that it may be possible to replace anatomical areas with neurochemical pathways. Thus, for example, memory has been suggested to be particularly dependent upon intact cholinergic (Smith and Swash 1978) or noradrenergic (Zornetzer 1978) function. By analogy with neuropsychology, advances in cognitive neurochemistry will come about through the study of restricted neurochemical syndromes.

ACKNOWLEDGEMENTS

I would like to thank friends and collegues, in particular John Hardy, for useful discussions. This work was supported by the Research Funds of the Royal Bethlem and Maudsley Hospitals and the Northern Regional Health Authority.

REFERENCES

Aigner, T.G. and Mishkin, M. (1986). The effects of physostigmine and scopolamine on recognition memory in monkeys. *Behavioural and Neural Biology* 45, 81-7.

Alafuzoff, I. (1985). *Histopathological and immunocytochemical studies in age-associated dementias. The importance of rigorous histopathological criteria for the classification of progressive dementing disorders.* Umea University Medical Dissertations, Umea.

Alpern, H.P. and Marriott, J.G. (1973). Short-term memory: facilitation and disruption with cholinergic agents. *Physiology and Behaviour* 11, 571-5.

Alzheimer, A. (1907). Uber ein eigenartige erkrankung der hirnrinde. *Allgemeine Zeitschrift fur Psychiatrie* 64, 146-8.

Amaducci, L.A. *et al.* (1986). Risk factors for clinically diagnosed Alzheimer's disease: a case control study of an Italian population. *Neurology* 36, 922-31.

Appell, J., Kertesz, A., and Fisman, M. (1982). A study of language functioning in Alzheimer Patients. *Brain and Language* 17, 73-91.

Arendt, T. and Bigl V. (1986). Alzheimer plaques and cortical cholinergic innervation. *Neuroscience* 17, 275-6.

Aston-Jones, G., Rogers, J., Shaver, R.D., Dinan, T.G., and Moss, D.E. (1985a). Age-impaired impulse flow from nucleus basalis to cortex. *Nature* 318, 462-4.

——, Shaver, R., and Dinan, T.G. (1985*b*). Nucleus basalis neurones exhibit axonal branching with decreased impulse conduction velocity in rat cerebrocortex. *Brain Research* **325**, 271–85.

Ball, M.J. (1977). Neuronal loss, neurofibrillary tangles and granulovacuolar degeneration in the hippocampus with aging and dementia. A quantitative study. *Acta Neuropathologica* **37**, 111–18.

——, *et al.* (1985). A new definition of Alzheimer's disease: a hippocampal dementia. *Lancet* **i**, 16–26.

Baratti, C.M., In roini, I.B., and Huygens, P. (1984). Possible interaction between central cholinergic muscarinic and opioid peptide systems during memory consolidation in mice. *Behavioural and Neural Biology* **40**, 155–69.

Barnes, D.M. (1987). Defect in Alzheimer's is on chromosome 21. *Science* **235**, 846–7.

Bartus, R.T., Dean, R.L., Beer, B., and Lippa, A.S. (1982). The cholinergic hypothesis of geriatric memory dysfunction. *Science* **217**, 408–17.

——., Pontecorvo, M.J., Flicker, C., Dean, R.L., and Figueiredo, J.C. (1986). Behavioural recovery following bilateral lesions of the nucleus basalis does not occur spontaneously. *Pharmacology, Biochemistry, and Behaviour* **24**, 1287–92.

Bayles, K.A. (1982). Language function in senile dementia. *Brain and Language* **16**, 265–80.

Beatty, W.W., Butters, N., and Janowsky, D.S. (1986). Patterns of memory failure after scopolamine treatment: implications for cholinergic hypotheses of dementia. *Behavioural and Neural Biology* **45**, 196–211.

Beller, S.A., Overall, J.E., and Swann, A.C. (1985). Efficacy of oral physostigmine in primary degenerative dementia. A double-blind study of response to different dose level. *Psychopharmacology* **87**, 147–51.

Berger, B.D. and Stein, L. (1969). An analysis of the learning deficits produced by scopolamine. *Psychopharmacologia* **14**, 271–83.

Bernheimer, H., Birkmayer, W., Hornykeiwicz, O., Jellinger, K., and Seitelberger, F. (1973). Brain dopamine and the syndromes of Parkinson and Huntington. *Journal of Neurological Science* **20**, 415–55.

Bird, E.D., and Spokes, E.G.S. (1982). Huntington's chorea. In *Disorders of Neurohumoural Transmission* (ed. T.J. Crow), pp. 145–82. Academic Press, New York.

——, Stranahan, S., Sumi, S.M., and Raskind, M. (1983). Alzheimer's disease: choline acetyltransferase activity in brain tissue from clinical and pathological subgroups. *Annals of Neurology* **14**, 284–93.

Blessed, G. (1980). Clinical aspects of the senile dementias. In *Biochemistry of Dementia* (ed. P.J. Roberts), pp. 1–14. John Wiley & Sons, London.

——, Tomlinson, B.E., and Roth, M. (1968). The association between quantitative measures of dementia and senile changes in the cerebral gray matter of elderly subjects. *British Journal of Psychiatry* **114**, 797–811.

Bohdanecky, Z., Jarvik, M.E., and Carley, J.L. (1967). Differential impairment of delayed matching in monkeys by scopolamine and scopolamine methylbromide. *Psychopharmacologia* **11**, 293–9.

Bond, N.W. (1984). Animal models in psychopathology: an introduction. In *Animal Models of Psychopathology* (ed. N.W. Bond), pp. 1–21. Academic Press, London.

Bondareff, W. (1983). Age and Alzheimer's disease. *Lancet* **i**, 1447.

——, Mountjoy, C.Q., and Roth, M. (1982). Loss of neurons of origin of the adrenergic projection to cerebral cortex (nucleus locus ceruleus) in senile dementia. *Neurology* **32**, 164–8.

Bowen, D.M. and Davison, A.N. (1986). Biochemical studies of nerve cells and energy metabolism in Alzheimer's disease. *British Medical Bulletin* **42**, 75–80.

——, Smith, C.B., White, P., and Davison, A.N. (1976). Neurotransmitter-related enzymes and indicies of hypoxia in senile dementia and other abiotrophies. *Brain* **99**, 459–96.

Brinkman, S.D. and Braun, P. (1984). Classification of dementia patients by a WAIS profile related to central cholinergic deficiencies. *Journal of Clinical Neuropsychology* **6**, 393–400.

Brody, J.A. (1985). Prospects for an ageing population. *Nature* **315**, 463–6.

Brown, K. and Warburton, D.M. (1971). Attenuation of stimulus sensitivity by scopolamine. *Psychonomic Science* **22**, 297–8.

Brozoski, T.J., Brown, R.M., Rosvold, H.E., and Goldman, P.S. (1979). Cognitive deficit caused by regional depletion of dopamine in the prefrontal cortex of rhesus monkeys. *Science* **206**, 929–32.

Buresova, O., Bures, J., Bohdanecky, Z., and Weiss, T. (1964). Effect of atropine upon learning, extinction, retention and retrieval in rats. *Psychopharmacologia* **5**, 255–63.

Butters, N. (1984). The clinical aspects of memory disorders: contributions from experimental studies of amnesia and dementia. *Journal of Clinical Neuropsychology* **6**, 17–36.

——, Albert, M.S., Sax, D.S., Miliotis, P., Nagode, J., and Sterste, A. (1983). The effect of verbal mediators on the pictorial memory of brain-damaged patients. *Neuropsychologia* **21**, 307–23.

——, Miliotis, P., Albert, M.S., and Sax, D.S. (1984). Memory assessment: evidence of the heterogeneity of amnesic symptoms. In *Advances in Clinical Neuropsychology* (ed. G. Goldstein), pp. 127–59. Plenum Press, New York.

Byrne, J. and Arie, T. (1985). Rational drug treatment of dementia? *British Medical Journal* **291**, 1845–6.

Caine, E.D., Weingartner, H., Ludlow, C.L., Cudahy, E.A., and Wehry, S. (1981). Qualitative analysis of scopolamine-induced amnesia. *Psychopharmacology* **74**, 74–80.

Callaway, E. (1984). Human information-processing: some effects of methylphenidate, age, and scopolamine. *Biological Psychiatry* **19**, 649–62.

——, Halliday, R., Naylor, H., and Schechter, G. (1985). Effects of oral scopolamine on human stimulus processing. *Psychopharmacology* **85**, 133–8.

Chui, H.C., Teng, E.L., Hendersen, V.W., and Moy, A.C. (1985). Clinical subtypes of dementia of the Alzheimer type. *Neurology* **35**, 1544–50.

Collerton, D. (1985). Rational drug treatment of dementia? *British Medical Journal* **291**, 347–8.

—— (1986). Cholinergic function and intellectual decline in Alzheimer's disease. *Neuroscience* **19**, 1–28.

—— and Fairbairn. A. (1985). Alzheimer's disease and the hippocampus. *Lancet* **i**, 278–9.

Constandinidis, J. (1978). Is Alzheimer's disease a major form of senile dementia? Clinical, anatomical, and genetic data. In *Alzheimer's Disease, Senile Dementia, and Related Disorders* (ed. R. Katzman, R.D. Terry, and K.L. Bick), pp. 15–25. Raven Press, New York.

Corsellis, J.A.N., Bruton, C.J., and Freeman-Browne, D. (1973). The aftermath of boxing. *Psychological Medicine* , 270–303.

Coyle, J.T., Price, D.L., and DeLong, M.R. (1983). Alzheimer's disease: a disorder of cortical cholinergic innervation. *Science* **219**, 1184–90.

Crow T.J. (1979). Action of hyoscine on verbal learning in man: evidence for a cholinergic link in the transition from primary to secondary memory. In *Brain Mechanisms in Memory and Learning: from the Single Neuron to Man* (ed M.A.B. Brazier), pp. 269–75. Raven Press, New York.

Crystal, H.A., Horoupian, D.S., Katzman, R., and Jotkowitz, S. (1982). Biopsy-proved Alzheimer disease presenting as a right parietal lobe syndrome. *Annals of Neurology 12*, 186–8.

Cummings, J.L., and Benson, D.F. (1983). *Dementia: a Clinical Approach.* Butterworths, London.

——, Hill, M.A., and Read, S. (1985). Aphasia in dementia of the Alzheimer type. *Neurology* **35**, 394–7.

Cutler, N.R. *et al.* (1985). Brain metabolism as measured with positron emission tomography: serial assessment in a patient with familial Alzheimer's disease. *Neurology* **35**, 1556–61.

Daly, H.B. (1968). Disruptive effects of scopolamine on fear conditioning and on instrumental escape learning. *Journal of Comparative and Physiological Psychology* **66**, 579–83.

Damasio, A.R., Eslinger, P.J., Damasio, H., Van Hoesen, G.W., and Cornell, S. (1985*a*). Multimodal amnesic syndrome following bilateral temporal and basal forebrain damage. *Archives of Neurology 42*, 263–71.

——, Graff-Radford, N.R., Eslinger, P.J., Damasio, H., and Kassell, N. (1985*b*. Amnesia following basal forebrain lesions. *Archives of Neurology 42*, 263–71.

D'Antona R. *et al.* (1985). Subcortical dementia. Frontal cortex hypometabolism detected by positron tomography in patients with progressive supranuclear palsy. *Brain* **108**, 785–99.

Davidson, M. *et al.* (1976). Physostigmine in patients with Alzheimer's disease. *Psychopharmacology Bulletin* **22**, 101–5.

Davies, P. and Maloney, A.J. (1976). Selective loss of central cholinergic neurones in Alzheimer's disease. *Lancet* **ii**, 1403.

Davis, K.L. (1982). Cholinergic treatment in Alzheimer's disease: implications for future research. In *Alzheimer's Disease—a Report of Progress in Research* (ed. S. Corkin, K. Davis, J.H. Crowdon, E. Usdin, and R.J. Wurtman), pp. 124–31. Raven Press, New York.

Davis, P.E., and Mumford, S.J. (1984). Cued recall and the nature of the memory disorder in dementia. *British Journal of Psychiatry* **144**, 383–6.

Dayan, A.D. (1970). Quantitative histological studies on the aged human brain. I. Senile plaques and neurofibrillary tangles in 'normal' patients. *Acta Neuropathologica* **16**, 85–94.

Drachman, D.A. (1978). Central cholinergic systems and memory. In *Psychopharmacology: a Generation of Progress* (ed. M.A. Lipton, A. DiMascio, and K.F. Killam), pp. 651–2. Raven Press, New York.

—— and Leavitt J. (1974). Human memory and the cholinergic system. A relationship to aging? *Archives of Neurology* **30**, 113–21.

—— and Sahakian B.J. (1980). Memory and cognitive function in the elderly. *Archives of Neurology* **37**, 674–5.

Duara, R. *et al.* (1986). Positron emission tomography in Alzheimer's disease. *Neurology* **36**, 879–87.

Dundee, J.W. and Pandit, S.K. (1972). Anterograde amnesic effects of pethidine, hyoscine, and diazepam in adults. *British Journal of Pharmacology* 44, 140–4.

Dunne, M.P. and Hartley, L.R. (1985). The effects of scopolamine upon verbal memory: evidence for an attentional hypothesis. *Acta Psychologica* 58, 205–17.

Dunnett, S.B. (1985). Comparative effects of cholinergic drugs and lesions of nucleus basalis or fimbria-fornix on delayed matching in rats. *Psychopharmacology* 87, 357–63.

——, Toniolo, G., Fine, A., Ryan, C.N., Bjorklund, A., and Iversen, S.D. (1985). Transplantation of embryonic ventral forebrain neurones to the neocortex of rats with lesions of the nucleus basalis magnocellularis—II. Sensorimotor and learning impairments. *Neuroscience* 16, 787–97.

Eckerman, D.A., Gordon, W.A., Edwards, J.D., MacPhail, R.C., and Gage, M.I. (1980). Effects of scopolamine, pentobarbital, and amphetamine on radial arm maze performance in the rat. *Pharmacology, Biochemistry and Behaviour* 12, 595–602.

Evans, H.L. (1975). Scopolamine effects upon visual discrimination: modifications related to stimulus control. *Journal of Pharmacology and Experimental Therapeutics* 195, 105–13.

Everitt, B.J., Robbins, T.W., Gaskin, M., and Fray, P.J. (1983). The effects of lesions to ascending noradrenergic neurons on discrimination learning and performance in the rat. *Neuroscience* 10, 397–410.

Filley, C.M., Kelly, J., and Heaton, R.K. (1986). Neuropsychological features of early- and late-onset Alzheimer's disease. *Archives of Neurology* 43, 574–6.

Fine, A. (1986). Transplantation in the central nervous system. *Scientific American* August, 42–50.

Fine, A., Dunnett, S.B., Bjorklund, A., and Iversen, S.D. (1985). Cholinergic ventral forebrain grafts improve passive avoidance memory in a rat model of Alzheimer disease. *Proceedings of the National Academy of Sciences USA* 82, 5227–30.

Flicker, C., Dean, R.L., Watkins, D.L., Fisher, S.K., and Bartus, R.T. (1983). Behavioural and neurochemical effects following neurotoxic lesions of a major cholinergic input to the neocortex in the rat. *Pharmacology, Biochemistry and Behaviour* 18, 973–81.

Foster, N.L., Chase, T.N., Fedio, P., Patronas, N.J., Brooks, R.A., and Di Chiro, G. (1983). Alzheimer's disease: focal cortical changes shown by positron emission tomography. *Annals of Neurology* 15, 145–50.

—— et al. (1984) Cortical abnormalities in Alzheimer's disease. *Annals of Neurology* 16, 649–54.

——, Chase, T.N., Patronas, N.J., Gillespie, M.M., and Fedio, P. (1986). Cerebral mapping of apraxia in Alzheimer's disease by positron emission tomography. *Annals of Neurology* 19, 139–43.

Fox, J.H., Penn, R., Clasen, R., Martin, E., Wilson, R., and Savoy S. (1985). Pathological diagnosis in clinically typical Alzheimer's disease. *New England Journal of Medicine* 313, 1419–20.

Francis, P.T. et al. (1985). Neurochemical studies of early onset Alzheimer's disease: possible influence on treatment. *New England Journal of Medicine* 313, 7–11.

Frith, C.D., Richardson, J.T.E., Samuel, M., Crow, T.J., and Mckenna, P.J. (1984). The effects of intravenous diazepam and hyoscine upon human memory. *Quarterly Journal of Experimental Psychology* 36A, 133–44.

——, Dowdy, J., Ferrier, I.N., and Crow T.J. (1985). Selective impairment of paired associate learning after administration of a centrally-acting adrenergic agonist (clonidine). *Psychopharmacology* 87, 490–493.

Fuld, P.A. (1984). Test profile of cholinergic dysfunction and of Alzheimer-type dementia. *Journal of Clinical Neuropsychology* 6, 380–92.

Gardner, E.L., Glick, S.D., and Jarvik, M.E. (1972). ECS dissociation of learning and one-way cross-dissociation with physostigmine and scopolamine. *Physiology and Behaviour* 8, 11–5.

Gash, D.M., Collier, T.J., and Sladek J.R. (1985). Neuronal transplantation: a review of recent developments and potential applications to the aged brain. *Neurobiology of Aging* 6, 1–30.

Geddes, J.W., Monaghan, D.T., Cotman, C.W., Lott, I.T., Kim, R.C., and Chui H.C. (1985). Plasticity of hippocampal circuitry in Alzheimer's disease. *Science* 230, 1179–81.

Gewirth, L.R., Shindler, A.G., and Hier, D.B. (1984). Altered patterns of word associations in dementia and aphasia. *Brain and Language* 21, 307–17.

Ghoneim, M.M. and Mewaldt, S.P. (1975). Effects of diazepam and scopolamine on storage, retrieval and organisational processes in memory. *Psychopharmacology* 44, 257–62.

—— (1977). Studies on human memory: the interactions of diazepam, scopolamine, and physostigmine. *Psychopharmacology* 52, 1–6.

Giardini, V., Amorico, L., De Acetis, L., and Bignami, G. (1983). Scopolamine and acquisition of go—no go avoidance: a further analysis of the perseverative antimuscarinic deficit. *Psychopharmacology* 80, 131–7.

Glenner, G.G. (1983). Alzheimer's disease. The commonest form of amyliodosis. *Archives of Pathology and Laboratory Medicine* 107, 281–2.

Glick, S.D., Mittag, T.W., and Green J.P. (1973). Central cholinergic correlates of impaired learning. *Neuropharmacology* 12, 291–6.

Gray, J.A. (1982). Multiple book review of *The Neuropsychology of Anxiety: An Enquiry into the Functions of the Septo-hippocampal System. Behaviour and Brain Sciences* 5, 469–534.

Greenamyre, J.T., Penney, J.B., Young, A.B., D'Amato, C.J., Hicks, S.P., and Shoulson I. (1985). Alterations in L-glutamate binding in Alzheimer and Huntington diseases. *Science* 227, 1496–9.

Harbaugh, R.E. (1986). Intracranial drug administration in Alzheimer's disease. *Psychopharmacology Bulletin* 22, 106–9.

Hardy, J. *et al.* (1985). Transmitter deficits in Alzheimer's disease. *Neurochemistry International* 7, 545–63.

—— *et al.* (1987). Region specific loss of glutamate innervation in Alzheimer's disease. *Neuroscience Letters* 73, 77–80.

Haroutunian, V., Barnes, E., and Davis, K.L. (1985). Cholinergic modulation of memory in rats. *Psychopharmacology* 87, 266–71.

Haxby, J.V., Grady, C.L., Duara, R., Schlageter, N., Berg, G., and Rapoport S.I. (1986). Neocortical metabolic abnormalities preceed non-memory deficits in early Alzheimer-type dementia. *Archives of Neurology* 43, 882–5.

Heise, G.A. and Lilie, N.L. (1970). Effects of scopolamine, atropine, and d-amphetamine on internal and external control of responding on non-reinforced trials. *Psychopharmacologia* 18, 38–49.

—— and Millar, K.S. (1984). Drugs and stimulus control In *Handbook of Psychopharmacology, Volume 18: Drugs, Neurotransmitters and Behaviour* (ed.

L.L. Iversen, S.D. Iversen, and S.H. Snyder), pp. 129–83. Plenum Press, New York.

——, Conner, R., and Martin, R.A. (1976). Effects of scopolamine on variable intertrial interval spatial alternation and memory in the rat. *Psychopharmacology* **49**, 131–7.

Henderson, A.S. (1986). The epidemiology of Alzheimer's disease. *British Medical Bulletin* **42**, 3–10.

Heyman, A,, Wilkinson, W.E., Stafford, J.A., Helms, M.J., Sigmon, A.H., and Weinberg, T. (1984). Alzheimer's disease: a study of epidemiological aspects. *Annals of Neurology* **15**, 335–41.

Hirano, A., Malamud, N., and Kurland L.T. (1961). Parkinson-dementia complex, an endemic disease on the island of Guam. II. Pathologic features. *Brain* **84**, 662–79.

Hollander, E., Mohs, R.C., and Davis, K.L. (1986). Cholinergic approaches to the treatment of Alzheimer's disease. *British Medical Bulletin* **42**, 97–100.

Hope, K. (1982). The changing profile of hydergine. *British Journal of Clinical Practice*, Symposium Supplement 16.

Hornykiewicz, O. (1978). Dopamine in the basal ganglia. Its role and therapeutic implications (including the clinical use of L-dopa). *British Medical Bulletin* **29**, 172–8.

Huppert, F.A. and Tym, E. (1986). Clinical and neuropsychological assessment of dementia. *British Medical Bulletin* **42**, 11–18.

Jarrard, L.E., Kant, G.J., Meyerhoff, J.L., and Levy, A. (1984). Behavioural and neurochemical effects of intraventricular AF64A administration in rats. *Pharmacology, Biochemistry, and Behaviour* **21**, 273–80.

Jones, D.M., Jones, M.E.L., Lewis, M.J., and Spriggs, T.L.B. (1979). Drugs and human memory: effects of low doses of nitrazepam and hyoscine upon retention. *British Journal of Clinical Pharmacology* **7**, 479–83.

Jorm, A.F. (1985). Subtypes of Alzheimer dementia: a conceptual analysis and critical review. *Psychological Medicine* **15**, 543–53.

Katzman, R. (1986). Alzheimer's disease. *New England Journal of Medicine* **314**, 964–73.

Kay, D.W.K., Beamish, P., and Roth, M. (1964). Old age mental disorder in Newcastle upon Tyne 1. A study of prevalence. *British Journal of Psychiatry* **110**, 146–58.

——, Bergmann, K., Forster, E.M., McKechnie, A.A., and Roth, M. (1970). Mental illness and hospital usage in the elderly: a random sample followed up. *Comprehensive Psychiatry* **11**, 26–35.

Kay, W.H., Sitaram, N., Weingartner, H., Ebert, M.H., Smallberg, S., and Gillin, J.C. (1982). Modest facilitation of memory in dementia with combined lecithin and anticholinesterase treatment. *Biological Psychiatry* **17**, 275–80.

Khachaturian, Z.S. (1985). Diagnosis of Alzheimer's disease. *Archives of Neurology* **42**, 1097–105.

Kirshner, H.S., Webb, W.G., and Kelly, M.P. (1984). The naming disorder of dementia. *Neuropsychologia* **22**, 23–30.

Kokkinidis, L. and Anisman, H. (1976*a*). Dissociation of the effects of scopolamine and d-amphetamine on a spontaneous alternation task. *Pharmacology, Biochemistry, and Behaviour* **5**, 293–7.

—— and Anisman, H. (1976*b*). Interaction between cholinergic and catecholaminergic agents in a spontaneous alternation task. *Psychopharmacology* **48**, 261–70.

Ksir, C.J. (1974). Scopolamine effects on two-trial delayed-response performance in the rat. *Psychopharmacology* **34**, 127–34.

Kubanis, P. and Zornetzer, S.F. (1981). Age-related behavioural and neurobiological changes: a review with emphasis on memory. *Behavioural and Neural Biology* **31**, 115–72.

Leenders, K.L., Gibbs, J.M., Frackowiak, R.S.J., Lammertsma, A.A., and Jones, T. (1984). Positron emission tomography of brain: new possibilities for the investigation of human cerebral pathophysiology. *Progress in Neurobiology* **23**, 1–38.

Levin, E.D. and Bowman, R.E. (1986). Scopolamine effects on Hamilton search task performance in monkeys. *Pharmacology, Biochemistry, and Behaviour* **24**, 819–21.

Levy, A., Elsmore, T.F., and Hursh, S.R. (1984). Central vs. peripheral anticholinergic effects on repeated acquisition of behavioural chains. *Behavioural and Neural Biology* **40**, 1–4.

Lezak, M.D. (1982). *Neuropsychological Assessment*, 2nd edn. Oxford University Press, New York.

Liljequist, R. and Mattila, M.J. (1979). Effect of physostigmine and scopolamine on the memory functions of chess players. *Medical Biology* **57**, 402–5.

Lipton, M.A., DiMascio, A., and Killam, K.F. (1978). *Psychopharmacology, a Generation of Progress*. Raven Press, New York.

Little, A., Levy, R., Chaaqui-Kidd, P., and Hand, D. (1985). A double-blind placebo controlled trial of high dose lecithin in Alzheimer's disease. *Journal of Neurology, Neurosurgery and Psychiatry* **48**, 736–42.

Mace, N.L. and Rabins, P.V. (1982). *The 36 Hour Day: a Family Guide to Caring for Persons with Alzheimer's Disease*. Johns Hopkins University Press, Baltimore.

Mann, D.M.A., Yates, P.O., and Marcyniuk, B. (1984). A comparison of changes in the nucleus basalis and locus caeruleus in Alzheimer's disease. *Journal of Neurology, Neurosurgery and Psychiatry* **47**, 201–3.

——, ——, and —— (1985). Correlations between senile plaque and neurofibrillary tangle counts in cerebral cortex and neuronal counts in cortex and subcortical structures in Alzheimer's disease. *Neuroscience Letters* **56**, 51–5.

Martin, P.R. (1984). Central nervous system catecholamine metabolism in Korsakoff's psychosis. *Annals of Neurology* **15**, 184–7.

Martindale, W. (1982). *Extra Pharmacopoeia*. Pharmaceutical Press, London.

Mattingly, B.A. (1986). Scopolamine disrupts lever-press shock escape learning in rats. *Pharmacology, Biochemistry, and Behaviour* **24**, 1635–8.

McEntee, W.J. and Mair, R.G. (1980). Memory enhancement in Korsakoff's psychosis by clonidine: further evidence for a noradrenergic deficit. *Annals of Neurology* **7**, 466–70.

——, ——, and Langlais, P.J. (1984). Neurochemical pathology in Korsakoff's psychosis: implications for other cognitive disorders. *Neurology* **34**, 648–52.

McKhann, G., Drachman, D., Fclstein M., Katzman, R., Price, D., and Stadlon, E.M. (1984). Clinical diagnosis of Alzheimer's disease: report of the NINCDS-ADRDA work group under the auspices of the Department of Health and Human Services task force on Alzheimer's disease. *Neurology* **34**, 939–44.

Mesulam, M.-M. (1986). Alzheimer plaques and cortical cholinergic innervation. *Neuroscience* **17**, 275–6.

—— (1982). Slowly progressive aphasia without generalised dementia. *Annals of Neurology* 11, 592–8.

Mewaldt, S.P. and Ghoneim, M.M. (1979). The effects and interactions of scopolamine, physostigmine, and methamphetamine on human memory. *Pharmacology, Biochemistry, and Behaviour* 10, 205–10.

Miller, A.K.H. and Corsellis, J.A.N. (1977). Evidence for a secular increase in human brain weight during the past century. *Annals of Human Biology* 4, 253–7.

Miller, E. (1971). On the nature of the memory disorder in presenile dementia. *Neuropsychologia* 9, 75–81.

——, and Hague, F. (1975). Some characteristics of verbal behavior in presenile dementia. *Psychological Medicine* 5, 255–9.

Molsa, P., Paljarvi, L., Rinne, J.O., Rinne, U.K.; and Sako, E. (1985). Validity of clinical diagnosis in dementia: a prospective clinicopathological study. *Journal of Neurology, Neurosurgery and Psychiatry* 48, 1085–90.

Moore, V. and Wyke, M.A. (1984). Drawing disability in patients with senile dementia. *Psychological Medicine* 14, 97–105.

Morris, R.G.E., Anderson, E., Lynch, G.S., and Baudry, M. (1986). Selective impairment of learning and blockage of long-term potentiation by a N-methyl-D-aspartate receptor antagonist AP 5. *Nature* 319, 774–6.

Moss, D.E., Rogers, J.B., Deutsch, J.A., and Salome, R.R. (1981). Time dependent changes in anterograde scopolamine-induced amnesia in rats. *Pharmacology, Biochemistry, and Behaviour* 14, 321–3.

Mountjoy, C.Q. (1986). Correlations between neuropathological and neurochemical changes. *British Medical Bulletin* 42, 81–5.

——, Rossor, M., Iversen, L.L., and Roth, M. (1984). Correlation of cortical cholinergic and GABA deficits with quantitative neuropathological findings in senile dementia. *Brain* 107, 507–18.

Muramoto, O. (1984). Selective reminding in normal and demented old people: auditory verbal versus visual spatial task. *Cortex* 20, 461–78.

——, Sugishita, M., and Ando, K. (1984). Cholinergic system and constructional praxis: a further study of physostigmine in Alzheimer's disease. *Journal of Neurology, Neurosurgery and Psychiatry* 47, 485–91.

Neary, D. *et al.* (1986*a*). Neuropsychological syndromes in presenile dementia due to cerebral atrophy. *Journal of Neurology, Neurosurgery and Psychiatry* 49, 163–74.

—— *et al.* (1986*b*). Alzheimer's disease: a correlative study. *Journal of Neurology, Neurosurgery and Psychiatry* 49, 129–237.

Nebes, R.D., Martin, D.C., and Horn, L.C. (1984). Sparing of semantic memory in Alzheimer's disease. *Journal of Abnormal Psychology* 93, 321–30.

Nuotto, E. (1983). Psychomotor, physiological and cognitive effects of scopolamine and ephenidrine in healthy man. *European Journal of Clinical Pharmacology* 24, 603–9.

O'Connor, W.T. and Dunnett, S.B. (1985). Disruption of central cholinergic mechanisms in the rat by basal forebrain lesions or atropine: effects on feeding, sensorimotor behaviour, locomotor activity and spatial navigation. *Behavioural Brain Research* 17, 103–15.

Overstreet, D.H. and Russell, R.W. (1984). Animal models of memory disorders. In *Animal Models in Psychopathology* (ed N.W. Bond), pp. 257–78. Academic Press, London.

Overton, D.A. (1966). State-dependent learning produced by depressant and atropine-like drugs. *Psychopharmacologia* **10**, 6–31.

Pazzagli, A. and Pepeu, G. (1984). Amnesic properties of scopolamine and brain acetylcholine in the rat. *International Journal of Neuropharmacology* **4**, 291–9.

Pearson, R.C.A., Esiri, M.M., Hiorns, R.W., Wilcock, G.K., and Powell, T.P.S. (1985). Anatomical correlates of the distribution of the pathological changes in the neocortex in Alzheimer's disease. *Proceeedings of the National Academy of Sciences USA* **82**, 4531–4.

Perry, E.K. (1986). The cholinergic hypothesis—10 years on. *British Medical bulletin* **42**, 63–9.

——, Perry, R.H., Blessed, G., and Tomlinson, B.E. (1977). Necropsy evidence of central cholinergic deficits in senile dementia. *Lancet* **i**, 189.

——, Tomlinson, B.E., Blessed, G., Bergmann, K., Gibson, P.H., and Perry, R.H. (1978). Correlation of cholinergic abnormalities with senile plaques and mental test scores in senile dementia. *British Medical Journal* **2**, 1457–9.

——, Tomlinson, B.E., Blessed, G., Perry, R.H., Cross, A.J., and Crow, T.J. (1981). Neuropathological and biochemical observations on the noradrenergic system in Alzheimer's disease. *Journal of the Neurological Sciences* **51**, 279–87.

—— *et al.* (1985). Cholinergic correlates of cognitive impairment in Parkinson's disease: comparison's with Alzheimer's disease. *Journal of Neurology, Neurosurgery and Psychiatry* **48**, 413–21.

Perry, R.H. (1986). Recent advances in neuropathology. *British Medical Bulletin* **42**, 34–41.

Petersen, R.C. (1977). Scopolamine induced learning failures in man. *Psychopharmacology* **52**, 283–9.

Pomara, N. and Stanley M. (1986). The cholinergic hypothesis of memory dysfunction in Alzheimer's disease—revisited. *Psychopharmacology Bulletin* **22**, 110–8.

Prado-Alcala, R.A., Signoret-Edward, L., Figueroa, M., Giordano, M., and Barrientos, M.A. (1984). Post-trial injection of atropine into the caudate nucleus interferes with long-term but not short-term retention of passive avoidance. *Behavioural and Neural Biology* **42**, 81–4.

Radda G.K. (1986). The use of N.M.R. spectroscopy for the understanding of disease. *Science* **233**, 640–5.

Rasmusson, D.D. and Dudar, J.D. (1979). Effect of scopolamine on maze learning performance in humans. *Experientia* **35**, 1069–70.

Rick, J.T., Whittle, K.L., and Cross, S.H. (1981). Disruption and facilitation of cue discrimination in the rat by cholinergic agents. *Neuropharmacology* **20**, 747–52.

Ridley, R.M., Barratt, N.G., and Baker, H.F. (1984*a*). Cholinergic learning deficits produced by scopolamine and ICV hemicholinium. *Psychopharmacology* **83**, 340–5.

——, Bowes, P.M., Baker, H.F., and Crow, T.J. (1984*b*). An involvement of acetylcholine in object discrimination learning and memory in the marmoset. *Neuropsychologia* **22**, 253–63.

Rocca, W.A., Amaducci, L.A., and Schoenberg, B.S. (1986). Epidemiology of clinically diagnosed Alzheimer's disease. *Annals of Neurology* **19**, 415–24.

Rogers, J. and Bloom, F.E. (1985). Neurotransmitter metabolism and function in the aging central nervous system. In *Handbook of the Biology of Aging* (ed. C.E. Finch and E. Schneider), pp. 645–91. Van Nostrand Reinhold, New York.

Rossor, M.N. (1981). Parkinson's disease and Alzheimer's disease as disorders of the isodendritic core. *British Medical Journal* **283**, 1588–90.

——, Garrett, N.J., Johnson, A.L., Mountjoy, C.Q., Roth, M., and Iversen, L.L. (1982). A post-mortem study of the cholinergic and GABA systems in senile dementia. *Brain* **105**, 313–30.

——, Iversen, L.L., Reynolds, G.P., Mountjoy, C.Q., and Roth, M. (1984). Neurochemical characteristics of early onset and late onset types of Alzheimer's disease. *British Medical Journal* **288**, 961–4.

Russell R.W., Watson, R.H.J., and Frankenhaeuser, M. (1961). Effects of chronic reductions of brain cholinesterase activity on acquisition and extinction of a conditioned avoidance response. *Scandinavian Journal of Psychology* **2**, 21–9.

Safer, D.J. and Allen, R.P. (1971). The central effects of scopolamine in man. *Biological Psychiatry* **3**, 347–55.

Salamone, J.D. (1986). Behavioural functions of the nucleus basalis magnocellularis and its relationship to dementia. *Trends in Neuroscience* **9**, 256–8.

——, Beart, P.M., Alpert, J.E., and Iversen, S.D. (1984). Impairment of T-maze reinforced alternation performance following nucleus basalis magnocellularis lesions in rats. *Behavioural Brain Research* **13**, 68–70.

Simchowicz, T. (1911). Histologische studien uber die senile demenz. *Histologische und histopathologische Arbeiten über die Grosshirnerinde* **4**, 267–444.

Sitaram, N., Weingartner, H., Caine, E.D., and Gillin, J.C. (1978*a*). Choline: selective enchancement of serial learning and encoding of low imagery words in man. *Life Sciences* **22**, 1555–60.

——, —— , and Gillin, J.C. (1978*b*). Human serial learning: enhancement with arecholine and choline and impairment with scopolamine. *Science* **201**, 274–6.

Skyhoj Olsen, T., Bruhn, P., and Oberg, R.G.E. (1986). Cortical hypoperfusion as a possible cause of subcortical aphasia. *Brain* **109**, 393–410.

Smith, C.M. and Swash, M. (1978). Possible biochemical basis of memory disorder in Alzheimer's disease. *Annals of Neurology* **3**, 471–3.

Stevens, R. (1981). Scopolamine impairs spatial maze performance in rats. *Physiology and Behaviour* **27**, 285–6.

Sulkava, R. and Amberla, K. (1982). Alzheimer's disease and senile dementia of the Alzheimer's type. A neuropsychological study. *Acta Neurologica Scandinavica* **65**, 651–60.

Summers, W.K., Majovski, L.V., Marsh, G.M., Tachiki, K., and Kling, A. (1986). Oral tetrahydroaminoacridine in long term treatment of senile dementia, Alzheimer type. *New England Journal of Medicine* **315**, 1241–5.

Sumpter, P.Q., Mann, D.M.A., Davies, C.A., Neary, D., Snowdon, J.S., and Yates, P.O. (1986). A quantitative study of the ultrastructure of pyramidal neurones of the cerebral cortex in Alzheimer's disease in relation to the degree of dementia. *Neuropathology and Applied Neurobiology* **12**, 321–9.

Sutherland, R.J., Wishaw, I.Q., and Regehr, J.C. (1982). Cholinergic receptor blockade impairs spatial localisation by use of distal cues in the rat. *Journal of Comparative and Physiological Psychology* **96**, 563–73.

Terry, R.D., Deck, A, Deteresa, R., and Schechter, R. (1981). Some morphometric aspects of the brain in senile dementia of the Alzheimer type. *Annals of Neurology* **10**, 184–92.

——, Davies, P., DeTeresa, R., and Katzman, R. (1982). Are both plaques and tangles required to make it Alzheimer's disease? *Journal of Neuropathology and Experimental Neurology* **41**, 364.

Todorov, A.B., Go, R.C.P., Constantinidis, J., and Elston, R.C. (1975). Specificity of the clinical diagnosis of dementia. *Journal of the Neurological Sciences* **26**, 81–98.

Tomlinson, B.E. (1980). The structural and quantitative aspects of the dementias. In *Biochemistry of Dementia* (ed. P.J. Roberts), pp. 15–51. John Wiley and Sons, London.

—— and Corsellis, J.A.N. (1984). Ageing and the dementias In *Greenfield's Neuropathology* (ed. J.H. Adams, J.A.N. Corsellis, and L.W. Duchen), pp. 951–1025. Edward Arnold, London.

——, Blessed, G., and Roth, M. (1968). Observations on the brains of non-demented old people. *Journal of the Neurological Sciences* **7**, 331–56.

Wagman, W.D. and Maxey, G.C. (1969). The effects of scopolamine methylbromide and methyl scopolamine hydrobromide upon the discrimination of interoceptive and exteroceptive stimuli. *Psychopharmacologia* **15**, 280–8.

Weiner, N. and Deutsch, J.A. (1968). Temporal aspects of anticholinergic and anticholinesterase-induced amnesia for an appetitive habit. *Journal of Comparative Physiology and Psychology* **66**, 613–17.

Weingartner, H., Kaye, W., Smallberg, S.A., Ebert, M.H., Gillin, J.C., and Sitaram, N. (1981). Memory failures in progressive idiopathic dementia. *Journal of Abnormal Psychology* **90**, 187–96.

——, Grafman, J., Boutelle, W., Kaye, W., and Martin, P.R. (1983). Forms of memory failure. *Science* **221**, 380–2.

Weiss, B. and Heller, A. (1969). Methodological problems in evaluating the role of cholinergic mechanisms in behaviour. *Federation Proceedings* **28**, 135–46.

Wesnes, K. and Revell, A. (1984). The separate and combined effects of scopolamine and nicotine on human information processing. *Psychopharmacology* **84**, 5–11.

—— and Warburton, D.M. (1983). Effects of scopolamine on stimulus sensitivity and response bias in a visual vigilance task. *Neuropsychobiology* **9**, 154–7.

—— and ——. (1984). Effects of scopolamine and nicotine on human rapid information processing performance. *Psychopharmacology* **82**, 147–50.

Wilcock, G.K., Esiri, M.M., Bowen, D.M., and Smith C.C.T. (1982). Alzheimer's disease. Correlation of cortical choline acetyltransferase activity with the severity of dementia and histological abnormalities. *Journal of the Neurological Sciences* **57**, 407–17.

Wilson, R.S., Kasniak, A.W., and Fox, J.H. (1981). Remote memory in senile dementia. *Cortex* **17**, 41–8.

——, Bacon, L.D., Fox, J.H., and Kelly, M.P. (1982). Facial recognition memory in dementia. *Cortex* **18**, 329–36.

——, —— L.D., Kramer, R.L., Fox, J.H., and Kaszniak, A.W. (1983). Word frequency effect and recognition memory in dementia of the Alzheimer type. *Journal of Clinical Neuropsychology* **5**, 97–104.

Wirsching, B.A., Beninger, R.J., Jhamandas, K., Boegman, R.J., and El-Defrawy, S.R. (1984). Differential effects of scopolamine on working and reference memory of rats in the radial maze. *Pharmacology, Biochemistry and Behaviour* **20**, 659–62.

Wishaw, I.Q. (1985). Cholinergic receptor blockade in the rat impairs locale but not taxon strategies for place navigation in a swimming pool. *Behavioural Neuroscience* **99**, 979–1005.

Woods, R.T. and Britton, P.B. (1985). *Clinical Psychology with the Elderly*. Croom Helm, London.

Zornetzer, S.F. (1978). Neurotransmitter modulation and memory: a new neuro-pharmacological phrenology. In *Psychopharmacology, a Generation of Progress* (ed. M.A. Lipton, A. DiMascio, and K.F. Killam), pp. 637–49. Raven Press, New York.

16

How far could cholinergic depletion account for the memory deficits of Alzheimer-type dementia or the alcoholic Korsakoff syndrome?

MICHAEL D. KOPELMAN

INTRODUCTION

Progress is limited [in memory research] since the strictly 'psychological' approach to the analysis of human memory functions often avoids any interpretations in terms of neural function; while studies at a neuronal, synaptic, or molecular level relate only indirectly (if at all) to observable human behaviour. (Drachman and Leavitt 1974)

As this quotation suggests, it is indeed remarkable how often the animal, human, clinical, pharmacological, and neuropathological literatures have discussed similar topics, whilst making only minimal cross-reference to one another. Fortunately, this stubborn isolationism by the differing disciplines is now breaking down; and the current, intense interest in the role of the cholinergic neurotransmitter system in human memory, and in its depletion in dementia, has been an important catalyst for this change. Although there is a risk that issues will be over-simplified in making reference across disciplines, and there is a danger of a crude reductionism, such cross-fertilization can only be regarded as healthy.

The present paper will review human studies which have attempted to relate memory impairments to cholinergic depletion. Neuropathological studies will not be discussed, as these have been reviewed elsewhere (Kopelman 1986a; Rossor, this volume); but the paper will mainly consider Alzheimer-type dementia, as this is the disorder in which neuropathological evidence of cholinergic depletion is most clearly established. I propose to give a (brief) account of current concepts of memory for those unfamiliar with the topic; to summarize the changes seen in the early stages of Alzheimer-type dementia in terms of those concepts; and to compare these impairments with those obtained on administering a cholinergic antagonist 'blocker' to healthy subjects (as a 'model' for cholinergic depletion). A fourth section will consider very briefly the possible role of the cholinergic neurotransmitter system in some related disorders, including the Korsakoff syndrome, in which there is also some evidence that cholinergic depletion may occur. Much of this research is at a relatively early stage in its development, and the present paper will give an introduction to the main themes and difficulties. Particular aspects of this topic have been reviewed in fuller detail elsewhere (Kopelman 1986a; Morris and Kopelman 1986).

CURRENT CONCEPTS IN MEMORY RESEARCH

Much memory research still postulates three main levels in memory processes. These have been labelled sensory memory, primary (or 'short-term') memory, and secondary ('long-term') memory. Traditionally, these were assumed to represent separate memory stores operating in series but, nowadays, greater emphasis is placed upon the processes they entail and their interaction, and there is evidence that parallel as well as serial processing occurs.

Sensory memory refers to a peripheral system operating within each modality. This represents the first stage in information-processing, having a relatively large storage capacity, but involving the formation of a memory trace of a very brief duration (less than 1 second). In the visual modality, sensory memory is known as 'iconic' memory, and in the auditory modality 'echoic' memory (Sperling 1963; Neisser 1967).

The concept of *primary memory*, holding information for a few seconds only, is usually attributed to William James (1890), although James himself drew heavily from other writers. James argued that our most recent perceptions are held in consciousness for 'several seconds' with a special intensity and vividness as a 'primary memory image', which appears to belong to the 'rearward portion of the present space of time'. He quoted Exner who wrote that this 'primary memory-image vanishes if not caught by attention in the course of a few seconds. Even when the initial impression is attended to, the liveliness of its image in memory fades fast'. By contrast, secondary memory referred to 'the knowledge of a former state of mind once dropped from consciousness . . . and now revived anew . . . an object [in secondary memory] is brought up, recalled, fished up, so to speak, from a reservoir in which, with countless other objects, it lay buried and lost from view'.

The modern conception of primary memory involves tasks or processes in which small quantities of information are recalled over a short retention period, not longer than approximately 30 seconds. James' essentially passive view of primary memory is reflected in traditional tests of primary memory such as 'span' tests and the recency component of free recall. In span tests, the subject is required to recall in correct sequence a series of digits, consonants or words, or to touch a series of blocks, immediately after presentation. The recency effect in free recall refers to the fact that the last few items of a (supraspan) series of words or numbers are remembered particularly well relative to earlier items (the precise method of determining the 'recency component' varies from study to study). However, modern conceptions of primary memory differ from the nineteenth-century view in that they also emphasize its active, information-processing role. Primary memory in this sense is often known as 'working memory', emphasising that material can be held temporarily in tasks such as problem-solving and speech comprehension, and that it may be manipulated and transformed (Baddeley 1983; Craik 1984; Hitch 1984). The so-called Brown–Peterson test measures 'short-term' forgetting during a period of up to 30 seconds whilst the subject performs a 'distractor' task (whose difficulty can be varied), and it is a more stringent measure of primary or working memory as an active system than are span or recency tests.

In Baddeley's conception of working memory, there are three subcomponents: an 'articulatory loop', containing a passive, auditory-verbal input store and an articulatory rehearsal mechanism; an equivalent system for visuo-spatial information; and a central processor which operates as a co-ordinating attentional system of limited capacity (Baddeley 1983; Hitch 1984). By contrast, Craik (1984) conceives of working memory as a unitary system of limited information-processing capacity, and he proposes that tasks vary in the degree of information-processing resources which they require. Tests which Baddeley would see as measuring the functioning of the articulatory loop, Craik would describe as utilizing fewer processing resources.

Secondary (or long-term) memory encompasses all material recalled beyond a period of a few seconds, and items within it may have occurred within the remote or recent past. It is characterized by an immensely greater capacity and a much slower forgetting rate than are evident in primary memory. Within secondary memory, recent research has postulated a number of (somewhat overlapping) distinctions: for example, between explicit and implicit memory, conscious recollection and procedural memory, episodic and semantic memory, remote and recent memory.

Explicit memory refers to the conscious recollection of episodes, which may include earlier learning on a psychological test (such as the learning of paired associates) as well as memory for autobiographical events. This kind of memory is, of course, severely impaired in organic amnesia, and there can be both a retrograde and an anterograde component to that impairment. There is a considerable literature examining whether the anterograde deficit in organic amnesia lies in the *acquisition* process occurring during and shortly after the initial 'registration' of information, or in its subsequent *retention and retrieval* (Meudell and Mayes 1982; Kopelman, 1987). Acquisition processes are usually assumed to involve some kind of (psychological) *encoding*, and/or a brief period of (physiological) 'consolidation'.

Impicit memory encompasses both procedural memory and the so-called priming phenomenon. Claparede (1911) demonstrated that 'the feeling that (a memory) belongs to the person's experience can be absent in situations in which retention is evident'. Modern experiments have confirmed that severely amnesic patients (e.g. Korsakoff cases) can acquire and show retention of certain skills, and this has become known as *procedural memory*. For example, these patients can learn to perform jigsaw puzzles or read back-to-front writing, in the absence of any recall or recognition of having performed the tasks before (e.g. Brooks and Baddeley 1976; Cohen and Squire 1980). In a particularly striking example, Starr and Philips (1970) described a pianist who was taught a new piece of music, but the next day had no recall of it. On being hummed the first bars, he was able to continue playing the piece, although still unable to recall the name of the piece or that he had been taught it before. Procedural learning does not appear to depend on well-established pre-morbid learning, and most researchers postulate that it depends on a subsystem of memory independent of that which subserves 'conscious recollection' (Moscovitch 1982*a*; Squire *et al.* 1984). *Priming* is a much more transient phenomenon, in which a learning episode has a facilitative effect upon the performance of a subsequent task. For example, reading a series of words during a study session makes it especially likely that subjects will give those words in response to some subsequent test, and amnesic subjects (including Korsakoff patients) show this effect even though their performance at a recognition test of those same words may be severely impaired (Shimamura 1986). Shimamura and Squire (1984) demonstrated that the effect disappears completely after 2 hours, and they proposed that priming involves the activation of the same 'processing structures' which are required for skill learning to develop. By contrast, Schachter (1985) has argued that these two types of implicit memory may in fact reflect separate subsystems.

Semantic memory has been characterized in terms of a subject's 'knowledge of the world' (Baddeley 1984), as it entails a person's memory for facts, concepts, and language (as opposed to memory for personal experiences or events). The distinction between semantic and episodic memory was first proposed by Tulving in 1972; and, since that time, there have been claims that Korsakoff patients are selectively impaired in episodic memory (e.g. Weingartner *et al.* 1983) whereas other patients are selectively impaired in semantic memory (e.g. Warrington 1975). However, there are problems in postulating semantic memory as a distinct memory subsystem. Huppert and Piercy (1982) pointed out that many tests of so-called semantic memory require the recall of overlearned, well-established material, whereas tests of episodic memory characteristically involve the learning of new materials. Moreover, the concept of 'semantic memory' is itself something of a conglomerate, in that the memory system which subserves our use of language (commonly spared in amnesia) may well differ from that which enables us to name the present or past prime minister (characteristically affected in amnesia). It seems that episodic and semantic memory are useful concepts in pointing to different types of knowledge along a continuum, but it does not necessarily follow that recall of these different types of knowledge requires the functioning of distinct memory subsystems.

Remote memory and *recent memory* tend to be rather loosely defined. They generally refer to the relative *age* of particular memories, and remote memory impairment is often used synonymously with retrograde amnesia. By contrast, retrograde and anterograde amnesia are precisely defined in terms of the *onset* of an injury or illness. Recent amnesia research has examined the nature of different types of retrograde loss, and whether there is a relative sparing of the most distant memories, known as a 'temporal gradient' (reviewed in Butters, this volume; Kopelman 1987).

MEMORY DEFICITS IN ALZHEIMER-TYPE DEMENTIA

The impairments produced by Alzheimer-type dementia upon these different components of memory will be considered briefly, as they have previously been reviewed in detail by Morris and Kopelman (1986). The memory impairments in Huntington's disease have been reviewed by Butters (this volume), and those produced by other forms of dementia have yet to be studied in detail.

Sensory memory

Sensory memory in dementia has been investigated only in respect of the visual (iconic) modality. In a preliminary study, Miller (1977a) demonstrated impaired recognition by Alzheimer patients in a test involving the tachistoscopic presentation of an array of letters; the patients were completely unable to report letters exposed for less than 100 msec. Schlotterer (1977) also examined the recognition of tachistoscopically presented letters, and, in addition, he manipulated the complexity of a 'mask' presented immediately after brief exposure of the letters. By so doing, he demonstrated that Alzheimer patients are impaired only when a complex mask is used (fragments of letters), rather than a simple mask (such as a light flash). Citing Turvey (1973), Schlotterer inferred from this that the more central components of iconic memory are impaired in Alzheimer patients, whereas the more peripheral components are spared.

Primary memory

There is considerable evidence that Alzheimer patients are impaired at 'span' tests such as digit, word, or block span (Miller 1973; Kaszniak *et al.* 1979;

Corkin 1982; Morris 1984; Kopelman 1985*a*). The only studies which fail to find this are those in which the clinical description of the patients suggests either that minimally impaired cases (of uncertain diagnosis) or non-Alzheimer cases have been included (Weingartner *et al.* 1981; Whitehead 1973). Moreover, Corkin (1982) demonstrated that the impairment on such tests is proportional to a clinical rating of the severity of dementia.

Most studies also show impairment in the recall of the last few items of a list of words presented to the subject (the recency component in free recall). Such a result has been reported by Miller (1971), Wilson *et al.* (1983), and Martin *et al.* (1985). Wilson *et al.* (1983) reported that the impairment in the recency component was relatively small compared with the more substantial loss of earlier items, whereas Martin *et al.* (1985) reported a uniform impairment at all points along the so-called serial position curve.

Severe impairment on the Brown–Peterson test (of 'short-term' forgetting) has consistently been reported (Corkin 1982; Kopelman 1985*a*; Morris 1986). Corkin found that this impairment was correlated with a clinical rating of the severity of dementia; and Kopelman (1985*a*) demonstrated a reversal of the normal age correlation such that younger Alzheimer patients (as well as those with the lowest IQ scores) performed worst at this test. Alzheimer patients also showed this reversal of the normal age correlation at a test involving the immediate recall of short sentences (Kopelman 1986*b*). These latter findings are consistent with neuropathological and neurochemical observations, which indicate that abnormalities are more severe and widespread in patients who have had an earlier age of onset (e.g. Rossor *et al.* 1984), suggesting that performance at primary memory tasks may be sensitive to the degree of underlying, neuropathological change.

This pattern of results differs from that seen in Korsakoff patients, in whom digit span and the recency component of free recall are characteristically unimpaired (Baddeley and Warrington 1970). Performance on the Brown–Peterson test has proved controversial in Korsakoff patients (Baddeley and Warrington 1970; Butters and Cermak 1974), but, in a direct comparison of Korsakoff and Alzheimer patients, Kopelman (1985*a*) found only very minimal evidence of impairment in the former group compared with a substantial impairment in the latter. In general, it is felt that Korsakoff patients' performance on primary memory tests shows only mild impairment (Moscovitch 1982*a*). Similarly, Morris (1984, 1986) has contrasted the performance of Alzheimer patients with patients described as having a selective impairment of auditory verbal short-term memory (Shallice and Warrington 1970). Morris argued that the latter group have an impairment of the so-called phonological input store of the articulatory loop in working memory, whereas Alzheimer patients have a deficit in the 'central processing' component.

Secondary memory

Undoubtedly, Alzheimer patients have a severe impairment of secondary memory, as manifested on such tests as free recall of a word list, recall of a short story or paragraph (logical memory), paired-associate learning, and picture and face recognition memory (e.g. Miller 1977a; Corkin 1982; Wilson *et al.* 1983; Martin *et al.* 1985; Kopelman 1986c). In part, this secondary memory deficit may be a consequence of the primary or working memory deficit (cf. Miller 1971; Wilson *et al.* 1983), although some theorists oppose the view that material must pass through primary memory before being stored in secondary memory (e.g. Shallice and Warrington 1970).

Recent studies have investigated whether Alzheimer patients' anterograde impairment lies in the initial *acquisition* of information or in its subsequent *retention*. As indicated above, Kopelman (1985a) showed faster forgetting of verbal material from 0 to 20 seconds after presentation by Alzheimer patients, relative to Korsakoff patients or healthy controls, on the Brown–Peterson test. Recently, Moss *et al.* (1986) have shown that Alzheimer patients continue to forget faster than Korsakoff patients or controls on a word recall test between a 15-second and a 2-minute delay interval. On the other hand, Kopelman (1985a) exposed a series of pictures from magazines to Alzheimer and Korsakoff patients for prolonged exposure times, and thereby matched their performance on a recognition test at 10 minutes as closely as possible to that of controls (cf. Huppert and Piercy 1978). Following this 'titration' procedure, the rate of forgetting over the course of a week was virtually identical in all three groups, and this result held good whatever the scoring procedure (correct recognition scores, percentage retention, or d prime). Freed *et al.* (cited in Corkin *et al.* 1984) have obtained a similar result in Alzheimer patients and several groups have obtained this result in Korsakoff patients (Huppert and Piercy 1978; Squire 1981; Martone *et al.* 1986)*. Taken together, these results suggest that the impairment in Alzheimer (and other amnesic) patients occurs in the earliest (or 'acquisition') stages of memory processing.

Other studies have examined *encoding processes* in order to try to delineate the nature of this 'acquisition' deficit more closely. For example, Corkin (1982) examined encoding processes in a recognition memory test by asking subjects questions about the words to be learned in a list: either 'Does a man/woman say the word?' (sensory orientation), or 'Does the word rhyme with ____?' (phonemic orientation), or 'Is the word a type of ____?' (semantic orientation). She found that recognition memory in healthy controls was most enhanced by semantic orientation, but that Alzheimer patients did not

---

*Mayes (1986) has criticised the technique employed in these studies, but there are several reasons, both theoretical and empirical, for believing that this criticism will prove to be of negligible importance (Huppert and Kopelman 1987).

show any benefit from encouraging semantic encoding. On the other hand, Martin *et al.* (1985) used a somewhat similar procedure in a cued recall test, and these authors reported normal benefits from encouraging semantic encoding in Alzheimer patients. A difficulty in interpreting these experiments arises from the fact that a 'floor' effect may have been operating in Corkin's Alzheimer group, and a 'ceiling' effect in Martin *et al.*'s control group. However, both studies were consistent in that the difference between the performance of the Alzheimer and control groups was sufficiently wide that it could not be accounted for by subtle discrepancies in semantic encoding alone.

Implicit memory has been studied relatively little to date in Alzheimer patients. Moscovitch (1982*b*) reported a normal priming effect (and severely impaired recognition memory) in three early Alzheimer patients. Likewise, Nebes *et al.* (1984) found that seeing a word had a normal effect in reducing the time it subsequently took Alzheimer patients to read a semantically related word (e.g. 'doctor' followed by 'nurse'). As discussed by Morris and Kopelman (1986), there are other studies (Miller 1975; Morris *et al.* 1983) whose results can be interpreted in terms of preserved priming by Alzheimer patients. On the other hand, Huppert and Goddard (reported by Huppert and Tym 1986), and Shimamura *et al.* (1986) have demonstrated impaired priming in Alzheimer patients. Similarly, Kopelman (1986*c*) found that Alzheimer patients, unlike Korsakoff patients, were impaired in the recall of even very easy paired-associates (e.g. hand-foot), whose normal recall in Korsakoff patients has been attributed to preserved priming (Shimamura and Squire 1984; Shimamura 1986). The apparent discrepancy in these findings may result either from the differing tests employed or, perhaps more likely, from Alzheimer patients being tested at different stages in their illness.

Semantic memory in Alzheimer patients has mainly been studied in connection with their use of language. Various studies have demonstrated impairments in naming and comprehension, in the organization of semantic knowledge, and in the ability of Alzheimer patients to make use of semantic context, e.g. to interpret homophones such as week/weak, nun/none (Schwartz *et al.* 1979; Appell *et al.* 1982; Martin and Fedio 1983; Gewirth *et al.* 1984; Kirshner *et al.* 1984). Recently, Ober *et al.* (1986) have demonstrated that, as the illness advances, Alzheimer patients show qualitative as well as quantitative alterations in their retrieval from semantic memory; and Huff *et al.* (1986) have attributed the naming disorder of Alzheimer patients to a loss of specific, semantic information about items within a certain category or class. On the other hand, in the early stages of the disorder, the general comprehension and expression of language remain relatively intact; and several studies have drawn attention to the residual abilities of Alzheimer patients on some tests of semantic memory (Miller 1977*b*; Nebes. *et al.* 1984; Martin *et al.* 1985; Kopelman, 1986*b*). For example, Kopelman (1986*b*) found that Alzheimer patients were disproportionately impaired in the immediate recall of sentences from which semantic cues had been removed (e.g. 'colourless green ideas sleep furiously'), and that some of their errors consisted of 'normalizing' (correcting) these sentences.

A few studies have examined the *remote memory* (retrograde) impairment in Alzheimer-type dementia, in particular to see if there is any sparing of the most distant memories as clinical 'lore' suggests there should be. Using tests which require subjects to answer questions or recognize pictures about famous people or news events from the past, Wilson *et al.* (1981) and Sagar *et al.* (1985) failed to find any sparing of the most remote memories, whereas Moscovitch (1982*b*) reported very marked sparing of earlier memories. Again, this discrepancy may result from differences in the patients tested. In a preliminary analysis of his own results, the present author found a gentle 'temporal gradient', i.e. a mild degree of sparing of the most distant memories (M.D. Kopelman, in preparation).

MEMORY IMPAIRMENTS IN CHOLINERGIC BLOCKADE

Beyond the laboratory, the effect of cholinergic blockers upon human memory has been noticed in the context of anaesthetics (e.g. Hardy and Wakely 1962; Dundee and Pandit 1972), the treatment of extrapyramidal symptoms in the elderly and schizophrenics (Potamianos and Kellett 1982; Frith 1984), and in young subjects abusing the anticholinergic agent benzhexol (benztropine; Cogentin) in high doses for prolonged periods (Crashaw and Mullen 1984). Within the laboratory, studies have investigated the nature of the memory impairment which cholinergic blockade imposes. The earliest of these studies preceded the (relatively recent) discovery of cholinergic depletion in Alzheimer-type dementia; but, nowadays, they allow us to attempt to delineate how far cholinergic depletion could account for the deficits in dementia, although we are still at a very early stage in doing this.

Studies of cholinergic blockade have generally involved the use of hyoscine (scopolamine) because of its easy penetration across the blood-brain barrier. However, the studies have varied in terms of the route by which the drug has been administered; the dosage employed; the time after administration that testing was commenced; the tests employed; and the type of placebo used. A saline placebo has been employed in many experiments, but this gives the risk that any hyoscine effects obtained may be spurious (because the peripheral effects of hyoscine are easily detectable). Methscopolamine, a peripheral cholinergic blocker, has been used in some experiments, but it is very poorly absorbed when administered orally, and it has not been employed in studies involving intravenous administration because of its possible adverse effects. The peripheral cholinergic blocker glycopyrrolate can be administered intravenously, but there is some evidence that it may cross the blood-brain barrier in small quantities providing the risk that negative hyoscine effects may be spurious. In short, the ideal design would employ both a saline placebo and a peripheral cholinergic blocker, but this requires a large sample or repeated testing and this strategy has seldom been adopted.

It should also be noted that the brain contains (at least) two types of cholinergic receptor: muscarinic receptors (which are much more abundant and have further subtypes) and nicotinic receptors. These two types of receptor have different functional effects in the autonomic and peripheral nervous systems, and it is very plausible that this may also be the case in the central nervous system. Hyoscine preferentially blocks muscarinic receptors, but at high doses it also blocks nicotinic receptors (Weiner 1980).

The results of most of these studies have been summarized and tabulated previously (Kopelman 1985*b*, 1986*a*). The present paper will discuss these results in terms of the aspects of memory delineated in the previous two sections. Clinical trials of cholinergic agonists will not be considered, as these have been discussed in detail elsewhere, and they tend to shed less light on the specific role of the cholinergic system in memory (Kopelman 1986*a*; Hollander *et al*. 1986). Laboratory studies of cholinergic agonists have produced a pattern of results which is generally consistent with that obtained in studies of cholinergic blockade (e.g. Warburton *et al*. 1986).

Sensory memory

This has been very little investigated in cholinergic blockade studies. Callaway *et al*. (1985) flashed an array of one to sixteen numbers on a display for 50 msec, and their subjects were required to identify whether the numbers contained a five or a six. Performance on this test is known to decline with increasing age, but oral doses of hyoscine (0.6 to 1.2 mg) did not produce any impairment, although reaction times were slowed on other tests.

Primary memory and attention

The more passive aspects of primary memory are typically unaffected following hyoscine administration. Various studies have shown that digit and word span are unaffected (Ostfeld and Aruguete 1962; Drachman and Leavitt 1974; Ghonheim and Mewaldt 1975; Mohs and Davis 1985). My colleague, Dr Tim Corn, and I have confirmed that both digit span and block span are unimpaired (M.D. Kopelman and T. Corn, in preparation). By contrast, the learning of a supraspan series of nine digits or more is impaired following hyoscine administration (Drachman and Leavitt 1974; Mewaldt and Ghonheim 1979). Other studies have shown that hyoscine does not affect performance within the recency component in the free recall of a list, but that it does affect the secondary memory component, i.e. items coming in the early or middle portion of a (supraspan) list (Crow and Grove-White 1973;

Crow 1979; Jones *et al.* 1979; Mewaldt and Ghonheim 1979; Frith *et al.* 1984). These results contrast with the studies in Alzheimer-type dementia, cited in the previous section, which typically reveal impairment on span tests and the recency component of free recall.

Caine *et al.* (1981) reported impairment on the Brown-Peterson test, using a small sample and a different version (and scoring procedure) from that used in Alzheimer studies. In an earlier review, Kopelman (1986a) suggested that, if confirmed, this impairment might reflect the greater information-processing load which this test imposes. Preliminary analysis of my results with Dr T. Corn, using a procedure identical to that employed in an Alzheimer study, reveals a mild, dose-related impairment (not statistically significant); and a further study has examined the effect of manipulating the information load in this test (M.D. Kopelman and T. Corn, in preparation). Taking this result together with those of span tests and studies of the recency effect, it seems unlikely that cholinergic depletion alone will account for the deficits in primary memory which are so prominent in Alzheimer-type dementia. In this connection, it may be pertinent that younger Alzheimer patients are those who perform worst at primary memory tests (see above) and are also those in whom an array of several depleted neurotransmitters is characteristically found at autopsy (Rossor *et al.* 1984). These depleted trans-mitters include the catecholamines, which have frequently been implicated in attention processes (Robbins 1984; Clark *et al.* 1986).

This raises the related question of whether hyoscine itself exerts its principal effect upon attention processes. This issue has been investigated in a series of elegant pharmacological studies, which compared the effect of administering various stimulants and sedatives with that of manipulating the cholinergic system (Crow *et al.* 1975; Ghonheim and Mewaldt 1977; Drachman 1977; Crow 1979). The conclusion of these studies was that the effect of hyoscine upon memory could not be accounted for by a general effect upon attention or arousal. On the other hand, higher doses of hyoscine certainly give rise to a subjective feeling of drowsiness, and this has been documented on self-rating scales (Ostfeld and Aruguete 1962; Ghonheim and Mewaldt 1975, 1977; Nuotto 1983; Dunne and Hartley 1985). This drowsiness is seen during the later, but not usually the earlier, stages of dementia, and it is a potential hazard in making inferences concerning the effect of cholinergic depletion in that disorder. Whilst some studies have failed to find impairment at tests of attention and vigilance at doses at which memory impairment has been demonstrated (Crow 1979; Caine *et al.* 1981), other studies have reported impairments (Safer and Allen 1971; Nuotto 1983; Wesnes and Warburton 1983, 1984; Dunne and Hartley 1985; Parrott 1986). The details of most of these studies are tabulated in Kopelman (1986a), where it was argued that impairment at tests of attention or vigilance is a dose-dependent phenomenon, occurring at higher doses of hyoscine or where the information-processing load of the task is relatively great (e.g. dichotic listening). A preliminary analysis of findings by Dr T. Corn and myself are consistent with this conclusion; and it seems likely that the pattern of results on primary memory tests may turn out to be very similar.

Secondary memory

There is considerable evidence that hyoscine produces impairment at tests of *'explicit'* or *'episodic' memory*. As mentioned above, impairment is seen at supraspan tasks and at performance within the secondary memory component of the serial position curve. Other studies have demonstrated deficits in the recall of word lists presented aurally or visually or in the two modalities simultaneously, in the recall of (meaningless) pattern configurations or (meaningful) chess configurations, at learning number-colour associations or other paired-associates, and in the recall of short paragraphs (logical memory) or pictures of objects (Ostfeld and Aruguete 1962; Safer and Allen 1971; Crow and Grove-White 1973; Crow *et al*. 1975; Drachman and Leavitt 1974; Ghonheim and Mewaldt 1975, 1977; Sitaram *et al*. 1978; Liljequist and Mattila 1979; Caine *et al*. 1981; Frith *et al*. 1984; Richardson *et al*. 1984). The details of many of these studies are tabulated elsewhere (Kopelman 1986*a*), and it suffices to say that an impairment of explicit memory is the most characteristic early feature in Alzheimer-type dementia.

As in the Alzheimer studies, researchers have tried to examine whether this impairment lies in the initial *acquisition* of information (input into storage) or in its subsequent *retention* and *retrieval*. These studies have generally compared the effect of administering hyoscine before or immediately after learning, and the conclusion without exception has been that cholinergic blockade affects the initial learning of information rather than its subsequent retention (Safer and Allen 1971; Ghonheim and Mewaldt 1975, 1977; Peterson 1977; Mewaldt and Ghonheim 1979). This conclusion is consistent with that obtained in Alzheimer patients (Kopelman 1985*a*), although the experimental procedures and types of material employed have been very different. This issue has recently been investigated using a test previously employed in an Alzheimer study (M.D. Kopelman and T. Corn, in preparation).

Encoding processes have been examined in three published studies, as well as the M.D. Kopelman and T. Corn study (in preparation). Caine *et al*. (1981) failed to find any interaction between the type of encoding (acoustic v. semantic) with which words were learned and the hyoscine-induced impairment in their recall. Richardson *et al*. (1984) found that hyoscine impaired subjects' ability to reject acoustically or semantically confusable 'distractors' in a word recognition test; but in a recall test, these authors found that the drug effect could be overcome by attending to the acoustic or semantic properties of the words when the structure of the word lists encouraged this (Frith *et al.* 1984). Taken together, these studies suggest that cholinergic blockade does not have a specific effect on either phonemic or semantic encoding processes, a conclusion consistent with that drawn by Warburton *et al.* (1986) in a recent cholinergic agonist (nicotine) study.

The effect of hyoscine administration on *implicit memory* has been largely neglected in the literature to date. Caine *et al.* (1981) reported that hyoscine-treated subjects showed a normal response to cueing in a recall test, and this result may reflect a preserved priming phenomenon, as mentioned elsewhere (Kopelman 1986*a*). A very preliminary analysis of the results in a study nearing completion suggests that the response to priming and cued recall is normal following hyoscine administration, despite the impaired performance at free recall, and that procedural learning may also be normal (M.D. Kopelman and T. Corn, in preparation). This result is consistent with what appears to be the finding in the earlier stages of Alzheimer-type dementia (see above).

Semantic memory has also been little explored. Drachman and Leavitt (1974) found that verbal IQ was preserved following hyoscine administration, indicating that long-established semantic knowledge remains unimpaired. In this connection, Brinkman and Braun (1984) and Fuld (1984) have argued that the overall pattern of IQ scores which Drachman and Leavitt obtained is broadly consistent with the pattern of scores seen in early Alzheimer patients. On the other hand, Drachman and Leavitt reported impairment in naming items from a given category, and Caine *et al.* (1981) also reported impairment at a verbal fluency (word retrieval) test to a letter or category stimulus. This might appear to implicate a deficit in retrieval from semantic memory: however, the impairment in the Caine study became apparent only after the subjects had attempted retrieval for prolonged intervals and therefore it may have reflected increased fatigue.

Remote memory has not been formally investigated in published hyoscine studies to date, but the clinical impression is that cholinergic blockade does not produce retrograde impairment (Crashaw and Mullen 1984). Consistent with this are studies showing that administering hyoscine immediately after learning does not affect subsequent retention or retrieval (see above), and a cholinergic agonist study which failed to find improvement on a test of remote memory (Davis and Mohs 1982). Preliminary analysis of results by M.D. Kopelman and T. Corn appears to confirm that hyoscine does not affect performance on a remote memory task, whether tested by recall or recognition.

Problems of method

A number of difficulties with these studies have recently been raised. It is certainly desirable that the tests employed should be as close as possible to those used in Alzheimer studies, and that a wide range of functions (beyond memory and attention) should also be investigated. The type of placebo employed has been discussed above, and ceiling and floor effects are, of course, a potential hazard in any psychological investigation. Whilst it is desirable that measures of central cholinergic blockade should be made, it is not necessary to possess a PET scanner to do so, and it is a great pity that simple physiological measurements have not been reported in many studies. Although it might appear sensible, at first sight, to perform these studies on older subjects, there are sound clinical and ethical reasons for wishing to use relatively young, healthy subjects in these experiments. Moreover, the evidence suggests that the loss of cholinergic neurons and enzymes in normal ageing is minimal, relative to that which occurs in dementia (Chui *et al*. 1984; Rossor *et al*. 1984). Although there is conflict between the results of animal and human studies, a reasonably consistent pattern is beginning to emerge within the human studies; and only by a selective reporting of the evidence could it be suggested that there is unacceptable conflict with respect, for example, to digit span or the recency component in free recall.

CHOLINERGIC DEPLETION IN THE KORSAKOFF SYNDROME AND OTHER DISORDERS

The author's interest in this topic arose, in part, from the belief that there might be a case for investigating cholinergic function in memory disorders other than Alzheimer-type dementia.

If (these) hypotheses are upheld, there is a case for concentrating on looking for a neurochemical remedy for the condition with a single psychological deficit (Korsakoff's syndrome) rather than that with multiple deficits (Alzheimer-type dementia). Victor *et al.* (1971) found that early administration of thiamine results in a partial alleviation of the memory disorder in Korsakoff's. There is not any neurochemical evidence of cholinergic deficiency in Korsakoff's; but, if acetylcholine mediates the transfer (of information) into secondary memory (Crow 1979) and if this is the deficit in Korsakoff's, a trial of combined thiamine and cholinergic therapy might be worthwhile.' (Kopelman 1981)

Evidence supporting the argument that the anterograde amnesia of Korsakoff's syndrome may involve a single memory deficit, whereas Alzheimer-type dementia involves multiple memory deficits, has subsequently been obtained (Kopelman 1985*b*). However, the author was wrong to assume in 1981 that neuropathological evidence implicating the cholinergic system in Korsakoff's syndrome did not already exist. Witt (1985) (see also E. Joyce, this volume) has recently reviewed this evidence, and she has drawn attention to the hypothesis that depletion of the active form of thiamine

(thiamine pyrophosphate, TPP) might produce diminished levels of acetyl-CoA or high-energy phosphates (ATP, ADP, AMP), thereby resulting in depleted acetylcholine synthesis. Although the data concerning acetylcholine levels following thiamine deficiency remain controversial, a consistent finding is that acetylcholine turnover is indeed reduced following dietary thiamine deficiency, and there is some evidence that inhibition of (TPP-dependent) pyruvate decarboxylase causes a measurable impairment of acetylcholine synthesis (Gibson *et al.* 1975). An alternative possible link between Korsakoff's syndrome and the cholinergic system has been mooted by Arendt *et al.* (1983), who reported a substantial reduction of the neuron count in the nucleus basalis of three Korsakoff brains at autopsy, thereby implicating an involvement of the ascending cholinergic projections. This led Butters (1985) to propose that Korsakoff's syndrome primarily involves a cholinergic, rather than a diencephalic, pathology. On the other hand, Lishman (1986) has hypothesized that these cholinergic pathways may be implicated in the Wernicke–Korsakoff pathology only in those cases involving more generalized cognitive impairment, better labelled as 'alcoholic dementia'.

The studies of Gibson *et al.* (1975) and Arendt *et al.* (1983) require replication in larger samples and the precise role, if any, of the cholinergic system in the neuropathology of Korsakoff's syndrome remains to be resolved. However, Table 16.1 provides a summary view of the pattern of deficits reported in studies of Alzheimer-type dementia, the Korsakoff syndrome, and cholinergic blockade. The pattern of impairment evident in the anterograde amnesia of the Korsakoff syndrome may prove to be more consistent with that obtained in cholinergic blockade than the more widespread memory deficits of Alzheimer-type dementia, supporting the view that a trial of a cholinergic agent in the Korsakoff syndrome might be beneficial. Furthermore, Victor *et al.* (1971) reported that the locus coeruleus is sometimes implicated in the Wernicke–Korsakoff pathology. Variable involvement of the ascending noradrenergic pathways in Korsakoff patients (as an additional, but not an essential part of the pathology) might explain some of the conflicts in the literature, e.g. differing levels of adrenergic metabolites in the CSF (McEntee *et al.* 1984; Martin *et al.* 1984) and variable performance on those tests of primary memory, such as the Brown–Peterson, which make particularly heavy attentional demands (cf. Kopelman 1985*b*).

There is also, of course, neuropathological evidence that the cholinergic system is depleted in other disorders such as Parkinson's disease and Down's syndrome. Perry (1986) has argued that this cholinergic depletion may account for the mild dementia seen in some Parkinson patients, even if it turns out to make only a partial contribution to the profound dementia of Alzheimer's disease. In this connection, it is also of interest that studies of mental handicap, including Down's syndrome, implicate an impairment in

TABLE 16.1 *Summary of the memory impairments in studies of Alzheimer-type dementia, Korsakoff's syndrome, and cholinergic blockade*

Memory component	Alzheimer-type dementia	Korsakoff syndrome*	Cholinergic blockade
Sensory memory Visual/iconic	The more peripheral aspects intact; the more central components impaired.	?Unimpaired.	?Unimpaired (one study only).
Primary ('working') memory Span, recency Brown–Peterson	Impaired. Severely impaired.	Unimpaired. Typically, only very minimal impairment, although there are variable results between studies.	Unimpaired. Probably, a mild, dose-related impairment (see text).
Secondary memory Explicit memory: acquisition *v*. retention	Severe 'acquisition' deficit ± secondary retrieval deficit.	Severe 'acquisition' deficit ± secondary retrieval deficit.	'Acquisition' deficit ± secondary retrieval deficit.
Explicit memory: encoding	Probably, impaired semantic encoding, but does not account alone for the anterograde amnesia.	Probably, impaired semantic encoding but does not account alone for the anterograde amnesia.	Non-specific impairment.
Implicit memory	Early stages: ?preserved. later stages: ?impaired.	Preserved.	?Preserved (see text).
Semantic memory (language-based tests)	Well-rehearsed aspects intact in the early stages, other aspects impaired.	Typically, unaffected.	Probably unimpaired, but ?impaired word fluency.
Remote memory	Severe impairment.	Severe impairment with variable sparing of the most distant memories.	Very probably, unaffected (see text).

*Space does not permit full discussion in this paper of all the evidence on which these summary statements about the Korsakoff syndrome are based, but further details will be found in Kopelman (1987) and Butters (this volume).

the initial acquisition of information rather than its retention, once learning has been accomplished (Haywood and Heal 1968; Taylor 1979).

CONCLUSION

The present paper has reviewed some of the concepts employed in memory and amnesia research, and it has considered the impairments seen in studies of Alzheimer-type dementia and those induced by cholinergic blockade in the light of these concepts. Brief reference has also been made to other disorders, particularly the Korsakoff syndrome. This research is at a relatively early stage in its development, and many of the studies to date have employed relatively crude methods of investigation. Nevertheless, a fairly consistent pattern of results is beginning to emerge, which is summarized in Table 16.1 and which was discussed in more detail in previous publications.

In general, sensory memory and the more passive aspects of primary memory are unaffected by cholinergic blockade. They do appear to be impaired (sometimes quite severely) in Alzheimer-type dementia, but not in the Korsakoff syndrome. On the other hand, performance at a test which makes greater information-processing demands, such as the Brown–Peterson, does show a very mild, dose-related impairment after cholinergic blockade, minimal impairment in the Korsakoff syndrome, and severe impairment in Alzheimer dementia. Taken together, these results would seem to implicate the depletion of other neurotransmitter systems in mediating performance at primary memory tests and, as indicated above, the noradrenergic system is a principal candidate.

The deficits in secondary memory seen in cholinergic blockade resemble those seen in the earlier stages of Alzheimer-type dementia and in the Korsakoff syndrome (with a normal forgetting rate, once learning has been accomplished, and relatively preserved implicit and semantic memory). However, the pharmacological studies suggest that cholinergic depletion cannot account for the severe impairment of remote memory seen in dementia and the Korsakoff syndrome, and may not explain the impairments of implicit and semantic memory which occur as a dementing illness progresses.

Elsewhere, it has been argued that the structural abnormalities in dementia may place a severe limitation on the potential benefits of any neurotransmitter replacement therapy; and therefore, it is of interest that structural changes may correlate more closely with psychological function than does cholinergic function (Neary *et al.* 1986). Neuropathologists are now paying increasing attention to failures in protein synthesis (Perry 1986; Mann *et al.* 1986), and, simultaneously, there has been a revival of interest in the relationship of protein synthesis to memory (see Kopelman and Lishman 1986). As everybody clambers aboard the bandwagon to Nucleus Basalis (perhaps anxious that the horse, called Alzheimer, may already have bolted), we might ask ourselves where correlational studies should next be done.

POSTCRIPT ADDED IN PRESS

O'Donnell, Pitts, and Fann (1986) have recently reported a trial of methylphenidate and cholinergic agents (physostigmine, choline) in Korsakoff patients. Their findings appear to contradict the predictions made in this chapter: they report improvement following methylphenidate but not following cholinergic administration. However, the authors tested a small sample of elderly patients (N = 6, aet 55–67), in whom psychometric data regarding (pre-treatment) memory and general cognitive status were not described; they did not employ a dose-determining phase in the administration of physostigmine, and there was a single measure of secondary memory. In short, the interpretation of this study must remain somewhat equivocal until a more fully assessed trial has been conducted.

ACKNOWLEDGEMENTS

The author's work was conducted within the Section of Neuropsychiatry, in the Department of Psychiatry Institute of Psychiatry, London, and he is very grateful to Professor W.A. Lishman for encouragement and advice throughout. The author would also like to acknowledge collaboration with Dr T. Corn in his recent empirical work, and with Dr R. Morris in writing an earlier paper from which some material in the second section of this paper was drawn. The author would also like to thank Mrs M. Fisher and Mrs P. Mott for patiently typing the manuscript. He is supported by a Wellcome Trust Lectureship.

REFERENCES

Appell, J., Kertesz, A., and Fisman, M.A. (1982). A study of language functioning in Alzheimer's disease. *Brain and Language* 17, 73–91.

Arendt, T., Bigl, V., Arendt, A., and Tennstedt, A. (1983). Loss of neurons in the nucleus basalis of Meynert in Alzheimer's disease, paralysis agitans and Korsakoff's disease. *Acta Neuropathologica (Berlin)* 61, 101–8.

Baddeley, A.D. (1983). Working memory. *Philosophical Transactions of the Royal Society of London,* B302, 311–23.

—— (1984). The fractionation of human memory. *Psychological Medicine* 14, 259–64.

—— and Warrington, E.K. (1970). Amnesia and the distinction between long- and short-term memory. *Journal of Verbal Learning and Verbal Behaviour* 9, 176–89.

Brinkman, S.D. and Braun, P. (1984). Classification of dementia patients by a WAIS profile related to central cholinergic deficiencies. *Journal of Clinical Neuropsychology* 6, 393–400.

Brooks, D.N. and Braddeley, A.D. (1976). What can amnesic patients learn? *Neuropsychologia* **14**, 111–22.

Butters, N. (1985). Alcoholic Korsakoff's syndrome: some unresolved issues concerning etiology, neuropathology, and cognitive deficits. *Journal of Clinical and Experimental Neuropsychology* **7**, 181–210.

—— and Cermak, L.S. (1974). Some comments on Warrington's and Baddeley's report of normal short-term memory in amnesic patients. *Neuropsychologia* **12**, 283–5.

Caine, E.D., Weingartner, H., Ludlow, C.L., Cudhay, E.A., and Wehry, S. (1981). Qualitative analysis of scopolamine-induced amnesia. *Psychopharmacology* **74**, 74–80.

Callaway, E., Halliday, R., Naylor, H., and Schechter, G. (1985). Effects of oral scopolamine on human stimulus evaluation. *Psychopharmacology* **85**, 133–8.

Chui, H.C., Bondareff, W., Zarow, C., and Slager, U. (1984). Stability of neuronal number in the human nucleus basalis of Meynert with age. *Neurobiology of Aging* **5**, 83–8.

Claparede, E. (1911). Recognition and 'me-ness'. (Recognition et moiite). *Archives Psychologique, Geneve* **11**, 79–90.

Clark, C.R., Geffen, G.M., and Geffen, L.B. (1986). Role of monoamine pathways in the control of attention: effects of droperidol and methylphenidate in normal adult humans. *Psychopharmacology* **90**, 28–34.

Cohen, N.J. and Squire, L.R. (1980). Preserved learning and retention of a pattern-analysing skill in amnesia: dissociation of knowing how and knowing that. *Science* **210**, 207–10.

Corkin, S. (1982). Some relationships between global amnesias and the memory impairments in Alzheimer's disease. In *Alzheimer's Disease: a Report of Research in Progress* (ed. S. Corkin, K.L. Daivs, J.H. Growdon, E. Usdin, and R.J. Wurtman). Raven Press, New York.

——, Growdon, J.H., Nissen, M.J., Huff, F.J., Freed, D.M., and Sagar, H.J. (1984). Recent advances in the neuropsychological study of Alzheimer's disease. In *Alzheimer's disease: Advances in Basic Research and Therapies. Proceedings of the Third Meeting of the International Study Group: On the Treatment of Memory Disorders Associated with Ageing* (ed. R.J. Wuroman, S Corkin, and J.H. Growdon).

Craik, F.I.M. (1984). Age differences in remembering. In *The Neuropsychology of Memory* (ed. L.R. Squire and N. Butters), pp 3–12. Guilford Press, New York & London.

Crashaw, J.A. and P.E. (1984). A study of benzhexol abuse. *British Journal of Psychiatry* **145**, 300–3.

Crow, T.J. (1979). Action of hyoscine on verbal learning in man: evidence for a cholinergic link in the transition from primary to secondary memory? In *Brain Mechanism in Memory and Learning* (ed. M.A.B. Brazier). Raven Press: New York.

—— and Grove-White, I.G. (1973). An analysis of the learning deficit following hyoscine administration to man. *British Journal of Pharmacology* **49**, 322–7.

——, —— and Ross, D.G. (1975). The specificity of the action of hyoscine on human learning. *British Journal of Clinical Pharmacology* **2**, 367–8.

Davis, K.L. and Mohs, R.C. (1982). Enhancement of memory processes in Alzheimer's disease with multiple-dose intravenous physostigmine. *American Journal of Psychiatry* **139**, 1421–4.

Drachman, D.A. (1977). Memory and cognitive function in man: does the cholinergic system have a specific role? *Neurology* **27**, 783–90.

—— and Leavitt, J. (1974). Human memory and the cholinergic system. *Archives of Neurology* **30**, 113–21.

Dundee, J.W. and Pandit, S.K. (1972). Anterograde amnesic effects of pethidine, hyoscine, and diazepam in adults. *British Journal of Pharmacology* **44**, 140–4.

Dunne, M.P. and Hartley, L.R. (1985). The effects of scopolamine upon verbal memory: evidence for an attentional hypothesis. *Acta Pychologica* **58**, 205–17.

Frith, C.D. (1984). Schizophrenia, memory, and anticholinergic drugs. *Journal of Abnormal Psychology* **9**, 339–41.

——, Richardson, J.T.E., Samuel, M., Crow, T.J., and McKenna, P.J. (1984). The effects of intravenous diazepam and hyoscine upon human memory. *Quarterly Journal of Experimental Psychology* **36A**, 133–44.

Fuld, P.A. (1984). Test profile of cholinergic dysfunction and of Alzheimer-type dementia. *Journal of Clinical Neuropsychology* **6**, 380–92.

Gewirth, L.R., Shindler, A.G., and Hier, D.B. (1984). Altered patterns of word associations in dementia and aphasia. *Brain and Language* **21**, 307–17.

Ghonheim, M.M. and Mewaldt, S.P. (1975). Effects of diazepam and scopolamine on storage, retrieval, and organisational processes in memory. *Psychopharmacologia (Berlin)* **44**, 257–62.

—— and —— (1977). Studies on human memory: the interactions of diazepam, scopolamine, and physostigmine *Psychopharmacology* **52**, 1–6.

Gibson, G.E., Jope, R., and Blass, J.P. (1975). Decreased synthesis of acetylcholine accompanying impaired oxidation of pyruvic acid in rat brain minces. *Biochemical Journal* **148**, 17–23.

Hardy, T.K. and Wakely, D. (1962). The amnesic properties of hyoscine and atropine in pre-anaesthetic medication. *Anasthesia* **17**, 331–6.

Haywood, H.C. and Heal, L.W. (1968). Retention of learned visual associations as a function of IQ and learning levels. *American Journal of Mental Deficiency* **72**, 828–38.

Hitch, G.J. (1984). Working Memory. *Psychological Medicine* **14**, 265–71.

Hollander, E., Mohs, R.C., and Davis, K.L. (1986). Cholinergic approaches to the treatment of Alzheimer's disease. *British Medical Bulletin* **42**, 97–100.

Huff, F.J., Corkin, S., and Growdon, J.H. (1986). Semantic impairment and anomia in Alzheimer's disease. *Brain and Language* **28**, 235–49.

Huppert, F.A. and Kopelman, M.D. (1987). Rate of forgetting in normal ageing: a comparison with dementia. (Submitted for publication).

—— and Piercey, M. (1978) Dissociation between learning and remembering in organic amnesia. *Nature* **275**, 317–18.

—— and —— (1982). In search of the functional locus of amnesic syndromes. In *Human Memory and Amnesia* (ed. L.S. Cermak) pp. 123–38. Lawrence Erlbaum Associates, Hillsdale, New Jersey.

—— and Tym, E. (1986) Clinical and neuropsychological assessment of dementia. *British Medical Bulletin* **42**, 1–8.

James, W. (1890). *Principles of Psychology, Vol. 1.* Holt, New York.

Jones, D.M.M., Jones, M.E.L., Lewis, M.J., and Springs, T.L.B. (1979). Drugs and human memory: effects of low doses of nitrazepam and hyoscine on retention. *British Journal of Clinical Pharmacology* **7**, 479–83.

Kaszniak, A., Garron, D., and Fox, J. (1979). Differential effects of age and cerebral atrophy upon span of immediate recall and paired-associate learning in older patients suspected of dementia. *Cortex* 15, 285–95.

Kirshner, H.S., Webb, W.G., and Kelly, M.P. (1984). The naming disorder in dementia. *Neuropsychologia* 22, 23–30.

Kopelman, M.D. (1981). A psychological investigation of the amnesic deficits in Korsakoff's syndrome and Alzheimer's disease. Part 1 M. Phil. Thesis Institute of Psychiatry, University of London (unpubl.).

—— (1985a). Rates of forgetting in Alzheimer-type dementia and Korsakoff's syndrome. *Neuropsychologia* 23, 623–38.

—— (1985b). Multiple memory deficits in Alzheimer-type dementia: implications for pharmacotherapy. *Psychological Medicine* 15, 527–41.

—— (1986a). The cholinergic neurotransmitter system in human memory and dementia: a review. *Quarterly Journal of Experimental Psychology* 38a, 535–73.

—— (1986b). Recall of anomalous sentences in dementia and amnesia. *Brain and Language* 29, 154–70.

—— (1986c). Clinical tests of memory. *British Journal of Psychiatry* 148, 517–25.

—— (1987). Amnesia: organic and psychogenic. *British Journal of Psychiatry* 150, 428–42.

—— and Lishman, W.A. (1986). Pharmacological treatments of dementia (non-cholinergic). *British Medical Bulletin* 42, 101–5.

Liljequist, R. and Mattila, M.J. (1979). Effect of physostigmine and scopolamine on the memory functioning of chess players. *Medical Biology* 51, 402–5.

Lishman, W.A. (1986). Alcoholic dementia: a hypothesis. *Lancet,* i, 1184–6.

Mann, D.M.A., Yates, P.O., and Marcniuk, B. (1986). A comparison of nerve cell loss in cortical and subcortical structures in Alzheimer's disease. *Journal of Neurology, Neurosurgery, and Psychiatry* 49, 310–12.

Martin, A. and Fedio, P. (1983). Word production and comprehension in Alzheimer's disease: a breakdown of semantic knowledge. *Brain and Language* 19, 124–41.

Martin, P.R., *et al.* (1984). Central nervous system catecholamine metabolism in Korsakoff's psychosis. *Annals of Neurology* 15, 184–7.

Martin, A., Brouwers, P., Cox, C., and Fedio, P. (1985). On the nature of the verbal memory deficit in Alzheimer's disease. *Brain and Language* 25, 323–41.

Martone, M., Butters, N., and Trauner, D. (1986). Some analyses of forgetting of pictorial material in amnesic patients. *Journal of Clinical and Experimental Neuropsychology* 8, 161–78.

Mayes, A.R. (1986). Learning and memory disorders and their assessment. *Neuropsychologia* 24, 25–40.

McEntee, W.J., Mair, R.G., and Langlais, P.J. (1984). Neurochemical pathology in Korsakoff's psychosis: implications for other cognitive disorders. *Neurology* 34, 648–52.

Meudell, P. and Mayes, A.R. (1982). Normal and abnormal forgetting: some comments on the human amnesic syndrome. In *Normality and Pathology in Cognitive Functions* (ed. A.W. Ellis), pp. 203–38. Academic Press, London.

Mewaldt, S.P. and Ghonheim, M.M. (1979). The effects and interactions of scopolamine, physostigmine, and methamphetamine on human memory. *Pharmacology, Biochemistry and Behaviour* 10, 205–10.

Miller, E. (1971). On the nature of the memory disorder in presenile dementia. *Neuropsychologia* 9, 75–8.

—— (1973). Short- and long-term memory in presenile dementia (Alzheimer's disease). *Psychological Medicine* **3**, 221-4.

—— (1975). Impaired recall and the memory disturbance in presenile dementia. *British Journal of Social and Clinical Psychology* **14**, 73-9.

—— (1977a) A note on visual information processing in presenile dementia: a preliminary report. *British Journal of Social and Clinical Psychology* **16**, 99-100.

—— (1977b). *Abnormal Ageing: The Psychology of Senile and Presenile Dementia*. John Wiley, New York and London.

Mohs, R.C. and Davis, K.L. (1985). Interaction of choline and scopolamine in human memory. *Life Sciences* **37**, 193-7.

Morris, R.G. (1984). Dementia and the functioning of the Articulatory Loop System. *Cognitive Neuropsychology* **1**, 143-57.

—— (1986). Short-term forgetting in senile dementia of the Alzheimer's type *Cognitive Neuropsychology* **3**, 77-97.

—— and Kopelman, M.D. (1986). The memory deficits in Alzheimer-type dementia: a review. *Quarterly Journal of Experimental Psychology* **38** a, 575-602.

——, Wheatley, J., and Britton, P.G. (1983). Retrieval from long-term memory in senile dementia: cued recall revisited. *British Journal of Clinical Psychology* **22**, 141-2.

Moscovitch, M. (1982a). Multiple dissociations of function in amnesia. In *Human Memory and Amnesia*. (ed. L.S. Cermak). Lawrence Erlbaum, Hillsdale, New Jersey.

—— (1982b). A neuropsychological approach to perception and memory in normal and pathological aging. In *Aging and Cognitive Processes* (ed. F.I.M. Craik and S. Treub). Plenum Press, New York.

Moss, M.B., Albert, M.S., Butters, N., and Payne, M. (1986). Differential patterns of memory loss among patients with Alzheimer's disease, Huntington's disease, and alcoholic Korsakoff's syndrome. *Archives of Neurology* **43**, 239-46.

Neary, D. *et al.* (1986). Alzheimer's disease: a correlative study. *Journal of Neurology Neurosurgery and Psychiatry* **49**, 229-37.

Nebes, R.D., Martin, D.C., and Horn, L.C. (1984). Sparing of semantic memory in Alzheimer's disease. *Journal of Abnormal Psychology* **93**, 321-30.

Neisser, U. (1967). *Cognitive Psychology*. Appleton-Century-Crofts, New York.

Nuotto, E. (1983). Psychomotor, physiological and cognitive effects of scopolamine and ephedrine in healthy man. *European Journal of Clinical Pharmacology* **24**, 603-9.

Ober, B.A., Dronkers, N.F., Koss, E., Delis, D.C., and Friedland, R.P. (1986). Retrieval from semantic memory in Alzheimer-type dementia. *Journal of Clinical and Experimental Neuropsychology* **8**, 75-92.

O'Donnell, V.M., Pitts, W.M., and Fann, W.E. (1986). Noradrenergic and cholinergic agents in Korsakoff's syndrome. *Clinical Neuropharmacology* **9**, 67-70.

Osteld, A.M. and Aruguete, A. (1962). Central nervous system effects of hyoscine in man. *Journal of Pharmacology and Therapeutics* **137**, 133-9.

Parrott, A.C. (1986). The effects of transdermal scopolamine and 4 dose levels of oral scopolamine (0.15, 0.3, 0.6, and 1.2 mg) upon psychological performance. *Psychopharmacology* **89**, 347-54.

Perry, E.K. (1986). The cholinergic hypothesis—ten years on. *British Medical Bulletin* **42**, 63-9.

Peterson, R.C. (1977). Scopolamine-induced learning failures in man. *Psychopharmacologia* **52**, 283-9.

Potamianos, G. and Kellet, J.M. (1982) Anti-cholinergic drugs and memory: the effects of benzhexol on memory in a group of geriatric patients. *British Journal of Psychiatry* **140**, 470-2.

Richardson, J.T.E., Frith, C.D., Scott, E., Crow, T.J., and Cunningham-Owens, D. (1984). The effects of intravenous diazepam and hyoscine upon recognition memory. *Behavioural Brain Research* **14**, 193-9.

Robbins, T.W. (1984). Cortical noradrenaline, attention, and arousal. *Psychological Medicine* **14**, 13-21.

Rossor, M.N., Iverson, L.L., Reynolds, G.P., Mountjoy, C.Q., and Roth, M. (1984). Neurochemical characteristics of early and late onset types of Alzheimer's disease. *British Medical Journal* **288**, 961-4.

Safer, D.J. and Allen, R.P. (1971). The central effects of scopolamine in man. *Biological Psychiatry* **3**, 347-55.

Sagar, H.J., Cohen, N.J., Corkin, S., and Growdon, J.H. (1985). Dissociations among processes in remote memory. *Annals of the New York Academy of Sciences* **444**, 533-5.

Schachter, D. (1985). Multiple forms of memory in humans and animals. In *Memory Systems of the Brain* (ed. N.M. Weinberger, J.L. McGaugh, and G. Lynch), pp. 351-79. Guilford Press, New York and London.

Schlotterer, G. (1977). Changes in visual information processing with normal aging and progressive dementia of the Alzheimer type. PhD Thesis, University of Toronto (unpubl.).

Schwartz, M.F., Marin, O.S.M., and Saffran, E.M. (1979). Dissociation of language function in dementia: a case study. *Brain and Language* **7**, 277-306.

Shallice, T. and Warrington, E.K. (1970). Independent functioning of verbal memory stores: a neuropsychological study. *Quarterly Journal of Experimental Psychology* **22**, 261-73.

Shimamura, A.P. (1986). Priming effects in amnesia: evidence for a dissociable memory function. *Quarterly Journal of Experimental Psychology* **38A**, 619-44.

——, Salmon, D., Squire, L.R., and Butters, N. (1986). Memory dysfunction unique to Alzheimer's disease: impairment in word priming. Cited in Shimamura (1986) *op.cit.*

—— and Squire, L.R. (1984). Paired-associate learning and priming effects in amnesia: a neuropsychological study. *Journal of Experimental Psychology: General* **113**, 556-70.

Sitaram, N., Weingartner, H., and Gillin, J.C. (1978). Human serial learning. Enhancement with arecholine and impairment with scopolamine. *Science* **201**, 274-6.

Sterling, G. (1963). A model for visual memory tasks. *Human Factors* **5**, 19-31.

Squire, L.R. (1981). Two forms of human amnesia: an analysis of forgetting. *Journal of Neurosciences* **1**, 635-40.

——, Cohen, N.J., and Nadel, L. (1984). The medial temporal region and memory consolidation: a new hypothesis. In *Memory Consolidation* (ed. H. Weingartner and E. Parker). Lawrence Erlbaum, Hillsdale, New Jersey.

Starr, A. and Phillips, L. (1970). Verbal and motor memory in the amnesic syndrome. *Neuropsychologia* **8**, 75-88.

Taylor, E. (1979). Mental retardation, In *Essentials of Postgraduate Psychiatry* (ed. P. Hill., R. Murray, and A. Thorley), pp. 151-78. Academic Press, London.

Tulving, E. (1972). Episodic and semantic memory. In *Organisation of Memory* (ed. E. Tulving and W. Donaldson). Academic Press, New York and London.

Turvey, M. (1973). On peripheral and central processes in vision: inferences from an information-processing analysis with pattern masking stimuli. *Psychological Review* **80**, 1-50.

Victor, M., Adams, R.D., and Collins, G.H. (1971). The Wernicke-Korsakoff Syndrome. F.A. Davis Co., Philadelphia.

Warburton, D.M., Wesnes, K., Shergold, K., and James, M. (1986). Facilitation of learning and state dependency with nicotine. *Psychopharmacology* **89**, 55-9.

Warrington, E.K. (1975). The selective impairment of semantic memory. *Quarterly Journal of Experimental Psychology* **27**, 635-57.

Weiner, N. (1980). Atropine, scopolamine, and related antimuscarinic drugs. In *The Pharmacological Basis of Therapeutics* (ed. L.S. Goodman and A. Gilman), pp. 120-37. Macmillan, New York.

Weingartner, H., Kaye. W., Smallberg, S.A. Ebert, H., Gillin, J.C., and Sitaram, N. (1981). Memory failures in progressive idiopathic dementia. *Journal of Abnormal Psychology* **90**, 187-96.

——, Grafman, J., Boutelle, W., Kay, W., and Martin, P.R. (1983). Forms of memory failure. *Science* **221**, 380-2.

Wesnes, K. and Warburton, D.M. (1983). Effects of scopolamine on stimulus sensitivity and response bias in a visual vigilance task. *Neuropsychobiology* **9**, 154-7.

—— and Warburton, D.M. (1984). Effects of scopolamine and nicotine on human rapid information processing performance. *Psychopharmacology* **82**, 147-50.

Whitehead, A. (1973). Verbal learning and memory in elderly depressives. *British Journal of Psychiatry* **123**, 203-8.

Wilson, R.S., Kaszniak, A.W., and Fox, J.H. (1981). Remote memory in senile dementia. *Cortex* **17**, 41-8.

——, Bacon, L.D., Fox, J.H., and Kaszniak, A.W. (1983). Primary and secondary memory in dementia of the Alzheimer type. *Journal of Clinical Neuropsychology* **5**, 337-44.

Witt, E.D. (1985). Neuroanatomical consequences of thiamine deficiency: a comparative analysis. *Alcohol and Alcoholism* **20**, 201-21.

17

The neurochemistry of Korsakoff's syndrome

EILEEN M. JOYCE

INTRODUCTION

The discovery that the pathology of Alzheimer's dementia includes lesions of subcortical neurochemical systems has catalysed research into neurotransmitter function and cognition. Furthermore, several other diseases in which there is a recognized dementia are considered to have a significant neurochemical component, namely, Parkinson's disease and Huntington's chorea. Korsakoff's syndrome is considered to be a relatively pure amnestic disorder and can usefully be defined as an 'abnormal mental state in which learning and memory are altered out of all proportion to other cognitive functions in an otherwise alert and responsive patient' (Victor *et al*. 1971). Given the impetus for neurochemical evaluation of neuropsychological syndromes, this article reviews the available evidence for proposing a neurochemical basis to Korsakoff's syndrome and considers whether this approach is likely to be as worthwhile as it has been in understanding Alzheimer's dementia.

RE-EVALUATION OF TRADITIONAL CONCEPTS

Korsakoff's syndrome is considered to occur as a chronic sequel to Wernicke's encephalopathy. The most common clinical presentation of Wernicke's encephalopathy is thought to be a rapidly developing triad of ophthalmoplegia, gait ataxia and global confusion often accompanied by peripheral neuropathy (Wernicke 1881; Victor *et al*. 1971). It has become clear that there is also a subclinical variant of this disorder, where the diagnosis is made after death, there being little evidence of the above features documented in life (Harper, 1983). The pathological basis of Wernicke's encephalopathy is well described and characteristically consists of small, punctate lesions distributed symmetrically around the ventricular systems of the diencephalic region of the forebrain and brainstem; damage can also occur within the cerebral hemispheres, cerebellum and spinal cord but less commonly.

There also appears to be a set pattern to the distribution of the periventricular lesions. Within the diencephalon, the mammillary bodies and the mediodorsal nucleus of the thalamus are most consistently affected. In the brainstem, there are three segments that are characteristically involved with sparing in between:

(1) the level of the superior colliculus in the midbrain;
(2) the level of the upper pons extending into the lower midbrain; and
(3) the level of the upper medulla extending into the lower pons.

These lesions appear to lie within grey matter. However, this may reflect the fact that the majority of periventricular structures are nuclei, the important detail being proximity to the ventricles. Except for the immediate subependymal layer which is characteristically spared, there is certainly a gradient of pathology which decreases away from the ventricles. Within each structure the damage tends to be confined to the medial portion (Victor *et al.* 1971). Although Wernicke's original histological description was of small haemorrhages, it appears that this is not a common finding and is mainly a feature of some cases dying in the acute stages of the illness (Malamud and Skillicorn 1956; Victor *et al.* 1971; Harper 1979). More often, the nature of the lesion ranges from minimal neuronal damage to nerve cell death, accompanied by reactive gliosis and vascular changes, the characteristics of which depend upon the age of the lesion. The aetiology of Wernicke's encephalopathy is almost certainly thiamine deficiency; prompt administration of thiamine can reverse the majority of neurological signs within days or weeks (e.g. Joliffe *et al.* 1941; De Wardener and Lennox 1947; Cruickshank 1950; Boles and Boles 1951; Victor *et al.* 1971), and in animal studies the Wernicke lesions can be reproduced by decreasing brain thiamine content (see Witt 1985 for review).

It is well established that Wernicke's encephalopathy and Korsakoff's syndrome represent facets of one disease process in that they share the same underlying pathology and in the majority of cases Korsakoff's syndrome follows on from Wernicke's encephalopathy. However, the nature of Korsakoff's syndrome is not as clearly understood as that of Wernicke's encephalopathy. For example, with respect to the latter disorder, ophthalmoplegia can be readily ascribed to the lesions in brain stem nuclei controlling eye movements, the ataxia to cerebellar dysfunction and the global confusion to cortical dysfunction which often accompanies a metabolic encephalopathy. The contingent response to thiamine establishes the aetiology and suggests that the initial lesion is a reversible biochemical one. Delay of treatment presumably allows structural damage to ensue resulting in residual symptoms.

This is not the case with Korsakoff's syndrome. Firstly, the nature of the critical lesion required to produce the amnesic state is under dispute. Either the mammillary bodies, the mediodorsal nuclei of the thalamus or both have been implicated (see Victor *et al.* 1971; Mair *et al.* 1979). Secondly, it is unclear whether the administration of thiamine is potentially capable of reversing or preventing the features of Korsakoff's syndrome thus questioning the aetiological basis of Korsakoff's syndrome as a pure avitaminosis (Cutting 1978*b*). With respect to higher mental function, the natural history of the Wernicke–Korsakoff syndrome in the majority of cases is of an initial confusional state, the dense amnesia crystallising out of a more general cerebral derangement. This global confusional state is not a necessary prelude to the development of the amnesia. Korsakoff's syndrome itself can be a presenting feature either alone or with the ocular and ataxic signs of Wernicke's encephalopathy (Victor *et al.* 1971). The relationship between thiamine administration and change of symptoms is thus difficult to assess because in the former type of presentation, global confusion masks features of Korsakoff syndrome and in the latter, the date of onset of amnesia is hard to establish. It may even be that Korsakoff's syndrome can have an insidious onset, developing over many years of cumulative damage caused by chronic or subacute bouts of thiamine deficiency whereas Wernicke's encephalopathy is seen only when severe acute depletion ensues. The dramatic features of Wernicke's encephalopathy may bring the patient acutely to medical attention and therefore prove to be reversible with thiamine. In this respect it is of interest that Victor *et al.* (1971) found that only 21 per cent of patients showed complete recovery of Korsakoff's amnesia after thiamine replacement. Thirdly, the neuropsychological characteristics defining Korsakoff's syndrome are more complex than previously thought. The textbook description is of a relatively pure and global amnesia with a dense anterograde compononont and a retrograde amnesia which demonstrates a temporal gradient (for review, Butters 1985). Accompanying the amnesia is a mental state of apathy, but the sensorium is clear and other cognitive functions as demonstrated by I.Q. are normal; confabulation is often thought to be a key feature of Korsakoff's syndrome, but in reality is rare in chronic cases and tends to be, if present, a short-lived early feature (Victor *et al.* 1971). It is now clear that there are also deficits in other cognitive areas, especially those requiring discrimination of complex perceptual events, and this is true in several modalities, e.g. odour, auditory, and visual (for reviews see Mair and McEntee 1983; Butters, 1985). Furthermore, the amnesia is not as pervasive as once thought since Korsakoff's patients can master certain mnemonic tasks (Butters 1984).

Of relevance to the aetiology of Korsakoff's syndrome is the question of the role of alcohol. Although cases of Wernicke's encephalopathy secondary to pure nutritional deficiency have been well recognized (Wernicke 1881; for review see Victor *et al.* 1971), virtually all cases of Korsakoff's syndrome are seen in the context of chronic alcoholism. Freund (1973, 1982) has argued that alcohol consumption itself may contribute to the production of Korsakoff's syndrome as he was unable to find any reports of chronic Korsakoff's syndrome resulting from cases of Wernicke's encephalopathy which were induced by nutritional factors alone. It is indeed difficult to judge from the literature whether true Korsakoff's syndrome can develop in the non-alcoholic because of inadequate case descriptions. Joliffe *et al.* (1941) described three non-alcoholic cases of Wernicke's encephalopathy; only one survived and went on to develop Korsakoff's amnesia although the psychological features were not described. De Wardener and Lennox (1947) described 21 cases of prisoners of war surviving acute Wernicke's encephalopathy; 61 per cent had severe deficits of recent memory extending backwards for 2–3 weeks; all but one completely recovered with thiamine treatment and although some took up to 3 months, others improved within 2–7 days; the remaining patient developed a permanent memory disorder, but it is unclear whether this was typical Korsakoff's syndrome. Thus, in nutritionally deficient individuals where thiamine depletion has occurred over a shorter period of time than in alcoholics, it is possible that the memory disorder, like the neurological signs, being at this stage a reflection of a biochemical lesion rather than a structural one is completely reversible. Butters (1985) has also suggested that the memory deficit may be due to a combination of thiamine deficiency causing acute catastrophic damage superimposed upon chronic minor damage in the same area caused by long-term alcohol consumption. Indeed, alcohol *per se* can cause brain damage; CT scan studies have demonstrated both cortical atrophy and dilatation of the third ventricle in chronic alcoholics (e.g. Cala *et al.* 1980; Bergman *et al.* 1980; Ron 1983). However, it is not established whether alcoholic Korsakoff syndrome patients have essentially two lesions: the diencephalic lesion accounting for amnesia and cortical atrophy accounting for the other cognitive deficits; or whether all features can be explained by subcortical pathology (see below for a discussion).

The remaining discussion explores an alternative explanation of Korsakoff's syndrome by examining the effect of thiamine deficiency on neurotransmitter systems rather than discrete anatomical structures in the brain.

Neurotoxicity of thiamine deficiency

Theories concerning the neurotoxicity of thiamine deficiency stem from knowledge about the role of thiamine as an essential B vitamin. Ingested thiamine is phosphorylated to thiamine pyrophosphate (TPP) which serves as a coenzyme in several enzyme complexes. Firstly, TPP is a coenzyme for pyruvate dehydrogenase and for α-ketoglutarate dehydrogenase. Both of these enzymes catalyse steps in the tricarboxylic acid cycle which utilizes pyruvate as a substrate derived from glucose by non-oxidative means, and which produces the important product, energy, in the form of ATP. Secondly, TPP is a coenzyme for transketolase which is an enzyme involved in an extramitochondrial energy producing pathway, the so called pentose phosphate shunt, which is active mainly in tissues synthesizing fatty acids (and hence myelin) and steroids. This latter pathway is also a source of pentose sugars for the synthesis of nucleic acids. Finally, TPP is a coenzyme for the synthesis of the amino acids isoleucine, and valine from pyruvate. Thus, thiamine is an essential prerequisite for the production of energy from carbohydrate, and for the synthesis of fatty acids, nucleic acids, and some amino acids.

There are two possible mechanisms by which a lack of this vitamin could affect brain neurochemistry. First, by directly interfering with the metabolism of specific neurotransmitters, and secondly, by causing structural damage in the vicinity of the cell bodies or axonal paths of various neurotransmitter systems. Although there is some evidence to support the first mechanism (see Witt 1985) as a possible explanation for the initial biochemical lesion in Wernicke's encephalopathy, it is not clear that this biochemical lesion can proceed to permanent dysfunction or indeed to structural damage impervious to thiamine replacement therefore becoming the neuropathological basis of Korsakoff's syndrome.

The second mechanism seems more plausible and the various theories concerning this have been reviewed by Witt (1985). Perhaps the most appealing explanation for which there is some evidence is that within the brain, there are regional differences in glucose metabolism, with the brainstem and diencephalon having particularly high rates and thus creating a relatively high demand for thiamine in the form of TPP. Thiamine deficiency will therefore affect glucose metabolism in those areas first, and will create a local deficit of ATP and an accumulation of lactic acid caused by the shift from aerobic to anaerobic respiration. Since ATP is crucial for ion transport across membranes, deficiency will result in impaired neuronal function. If this persists or if sufficient lactic acid accumulates to produce direct toxicity, then neuronal damage and eventually cell death will ensue. This putative mechanism of course does not explain why the lesions develop medially in close proximity to the ventricles or why within this medial zone some areas are spared and others are not.

NEUROANATOMICAL CONSIDERATIONS

Probably the most detailed histopathological description of the topography of Wernicke–Korsakoff lesions is in the 1971 monograph of Victor, Adams, and Collins and it is noteworthy that many of the lesion sites correspond to areas known from animal studies to contain either the cell bodies or axonal paths of neurotransmitter systems.

Noradrenaline

Lindvall and Bjorklund (1978) have described a periventricular catecholaminergic system (predominantly noradrenaline) which is 'distributed along the periventricular and periaqueductal grey from the medulla oblongata up to the rostral diencephalon'. This can be further divided into dorsal and ventral periventricular systems. The cell bodies of the dorsal periventricular system are located in the nucleus of the solitary tract and the dorsal motor nucleus of the vagus within the medulla (A2 group of Dahlstrom and Fuxe 1964) and in the nucleus locus coeruleus of the pons (A6 group of Dahlstrom and Fuxe 1964). According to Victor *et al.* (1971), these structures are commonly included in the Wernicke–Korsakoff pathology (43.6, 47.4, and 67.9 per cent of cases, respectively). From its origin in the medulla, fibres of the dorsal periventricular noradrenergic system pass rostrally in proximity to the medial vestibular nucleus which is also often involved (71 per cent of cases). As this noradrenergic system passes forward in relation to the ventricular system, it is also vulnerable to transection by those lesions. Furthermore, the dorsal periventricular noradrenergic system terminates in areas also known to be commonly involved in Korsakoff's syndrome, e.g. inferior colliculus, midline thalamic nuclei, and dorsomedial nucleus of the hypothalamus. The *ventral* periventricular catecholaminergic pathway is partly derived from A2 and A6 cell groups, but in addition, receives fibres derived from midbrain cell groups. These periventricular cell systems are predominantly noradrenergic (Lindvall and Bjorklund 1978; A11 cell group of Dahlstrom and Fuxe 1964) but also include the dorsal extension of the mesencephalic dopamine cell system (A10 of Dahlstrom and Fuxe 1964). This ventral catecholaminergic periventricular system projects to the medial part of the mammillary bodies, the paraventricular and dorsomedial nucleus of the hypothalamus, the septum, and interstitial nucleus of the stria terminalis. Thus, the ventral catecholaminergic system can be seen as a potential target for the effects of thiamine depletion at several levels. There is yet a third noradrenergic system that is likely to be involved by way of its neuroanatomy: the dorsal tegmental bundle (Lindvall and Bjorklund 1978) also known as the dorsal bundle (Ungerstedt 1971). The dorsal tegmental bundle is exclusively derived from the locus coeruleus and provides a massive projection to forebrain structures, mainly cortex, hippocampus, and thalamus (anterior, ventrobasal, and midline nuclei) and to a lesser extent the hypothalamus (paraventricular nucleus). Although the dorsal tegmental bundle does not travel close to the ventricular system, its site of origin is often lesioned in Korsakoff's syndrome (67.9 per cent of cases, Victor *et al.* 1971). The locus coeruleus is a large nucleus sitting on the floor of the fourth ventricle and contains exclusively noradrenergic cell bodies; although this nucleus is relatively laterally placed, it is the medial portion that is regularly lesioned in Korsakoff's syndrome.

Serotonin

There are two midbrain raphe nuclei which give rise to ascending serotonin pathways: the medial and the dorsal. The dorsal raphe nucleus is more likely to be lesioned in Korsakoff's syndrome because it is situated just beneath the cerebral aqueduct (for a review see Azmitia 1978). The dorsal raphe cell group gives rise to four discrete fibre systems which collectively innervate mainly the cerebral cortex, hippocampus, septum, basal ganglia, and amygdala. One particularly vulnerable pathway, the dorsal raphe periventricular tract, travels rostrally below the aqueduct in the central grey matter, and surrounds the periventricular region of the thalamus and hypothalamus. Also of interest is the raphe medial tract, derived from both dorsal and medial raphe nuclei, as one of its projections is to the mammillary body, which is regularly lesioned in Korsakoff's syndrome.

Dopamine

The major mesencephalic dopamine cell bodies of the ventral tegmental area (A10) and substantia nigra (A9) which form the mesolimbic, mesocortical, and nigrostriatal pathways are unlikely to be involved in Korsakoff's syndrome because of their distance from the aqueduct and third ventricle (Ungerstedt 1971; Lindvall and Bjorklund 1978). However, there is a dorsal extension of the A10 cell group which probably innervates the hypothalamus by way of the ventral periventricular system described above, and which may be lesioned in Korsakoff's syndrome.

Acetylcholine

The relatively recent understanding of the forebrain cholinergic systems and their relationship to cognitive disorders makes cholinergic dysfunction in Korsakoff's syndrome an exciting possibility (for reviews see Sahakian 1987; Rossor, this volume; Kopelman, this volume). The main sources of long cholinergic tracts are several nuclei in the basal forebrain: the nucleus of the diagonal band projects to amygdala, the medial septum to hippocampus and the nucleus basalis of Meynert (nbM) to the entire neocortex in a topographic manner. Unfortunately, the state of the basal forebrain in Korsakoff's syndrome is rarely commented upon. Recently, however, Arendt *et al.* (1983) reported that the neuronal population of basal forebrain was decreased by 47 per cent in three post-mortem brains of Korsakoff syndrome patients; in two of these cases, it was more pronounced in the nucleus of the diagonal band and in the third, the nbM.

NEUROCHEMICAL EVIDENCE OF
NEUROTRANSMITTER DYSFUNCTION

The neurochemical evidence of neurotransmitter dysfunction in Korsakoff's syndrome which will be presented here is derived from two sources: first, from experimental work in which regional brain measures of neurotransmitter activity have been employed in animals rendered thiamine deficient either by dietary control alone or, to obtain a rapid effect, by diet plus the administration of the thiamine antagonist pyrithiamine; and secondly, from human studies in which neurotransmitter activity has been measured in lumbar cerebrospinal fluid (CSF). Both of these methods are somewhat unsatisfactory in establishing the nature of any neurochemical abnormality which is reflecting cognitive change: the former because the measures are made in most studies during the acute phase of thiamine depletion and thus may not represent the chronic lesion underlying Korsakoff's syndrome; and the latter, because lumber CSF measurements of neurotransmitter activity are indirect and cannot reflect accurately changes in brain regions or even in ventricular CSF because of contributions to the composition of this CSF from plasma and spinal cord as well as from brain.

Noradrenaline

The major brain metabolite of noradrenaline is 3-methoxy-4-hydroxy phenethylene glycol (MHPG) (for review see Leckman and Mass 1984). In 1978 McEntee and Mair demonstrated a 50 per cent reduction of CSF MHPG in nine Korsakoff's patients compared to a psychiatric control group. Martin *et al.* (1984), using the same method of gas chromatography, found no change in CSF MHPG in six Korsakoff's patients; in this latter study the controls were healthy volunteers. However, McEntee *et al.* (1984) subsequently analysed the CSF in a further sixteen Korsakoff's patients, this time using the more sensitive technique of high performance liquid chromatography (HPLC) and non-psychiatric controls. They again found a significant depletion of CSF MHPG (by 41 per cent). In order to control for the possibility that lumbar CSF may reflect plasma rather than brain MHPG, they estimated plasma MHPG in nine Korsakoff patients and found that this was not different from that of controls. With respect to animal studies, Iwata (1982) has reported a series of experiments using thiamine deficient rats. Noradrenaline levels were found to be elevated in cortex, probably secondary to decreased monoamine oxidase activity, but the synthesis of noradrenaline was diminished. These findings were in acutely thiamine deficient rats and were almost completely reversed by thiamine administration. In contrast Mair *et al.* (1985*b*) have reported noradrenaline levels in rats up to 15 weeks after the acute effects of thiamine depletion had been reversed with thiamine administration. Seven areas of brain were assayed (cortex and hippocampus, striatum, hypothalamus, olfactory bulb, midbrain and thalamus, pons and medulla, cerebellum) and of these a significant depletion was found in cortex (35 per cent) and olfactory bulb (27 per cent).

Serotonin

Although two studies of CSF in Korsakoff patients using gas chromatography reported no change in 5-hydroxy-indoleacetic acid, (5HIAA) the primary metabolite of serotonin (McEntee and Mair 1978; Martin *et al.* (1984), McEntee *et al.* (1984) reported a small (21 per cent), but significant depletion using the HPLC method. Botez *et al.* (1982) in a preliminary study of cases of human thiamine deficiency, although not necessarily with Korsakoff's syndrome, found decreased CSF thiamine and 5HIAA in three out of five cases and, although thiamine administration tended to reverse this deficit, in two of the cases 5HIAA remained at very low levels (27 and 25 per cent of control at 5 and 7 days, respectively) and these two cases were chronic alcoholics. In the rat, Mair *et al.* (1985*b*) were unable to find any significant differences in serotonin levels three months after acute thiamine depletion. However, there was a trend for serotonin to be increased in midbrain and thalamus, and in this area there was a significant increase in 5HIAA perhaps reflecting increased turnover. In the acute phase of thiamine depletion, Plaitakis *et al.* (1982) have also demonstrated increased serotonin turnover in the rat. They found increased 5HIAA levels and unaltered serotonin and tryptophan levels in all of the regions examined by Mair *et al.* (1985*b*). Using pargyline to inhibit monoamine oxidase, they also found an enhanced rate of serotonin accumulation, especially in cerebellum and hypothalamus. This enhanced serotonin turnover was also confirmed using the method by which radio-labelled serotonin is injected intracisternally followed by measurement of the regional accumulation of labelled 5HIAA.

On the other hand, these same authors found that serotonin uptake in cerebellar synaptosomes and to a lesser extent, in hypothalamic synaptosomes, was decreased. This effect was reversed by up to 92 per cent with thiamine replenishment. Plaitakis *et al.* (1982) argue that these changes in serotonin metabolism may reflect cellular membrane abnormalities in vesicular transport or storage mechanims during thiamine deficiency so that serotonin uptake is diminished and, once inside the cell, exposed to degradation by monoamine oxidase. The increased serotonin synthetic rate is argued to be a compensatory mechanism. However, whether this putative abnormality can give rise to cellular destruction is a matter of speculation. In this respect Chan-Palay (1977) demonstrated decreased autoradiographic labelling of serotonin neurones in the thiamine depleted rat. The brainstem was particularly affected with virtually no labelling of cell bodies in the raphe nuclei or of serotonin axons elsewhere. Within the diencephalon the mamillary bodies and periventricular region were most severely affected.

Dopamine

The neurochemical evidence implicating dopaminergic dysfunction is scarce. Again using the gas chromatography method, neither McEntee and Mair (1978) nor Martin *et al.* (1984) could demonstrate diminished levels of the dopamine metabolite homovanillic acid (HVA) in CSF. However, a small, but significant decrease (29 per cent) was observed using the HPLC method (McEntee *et al.* 1984). Although in their regional study of brain neurotransmitters in thiamine depleted rats Mair *et al.* (1985*b*) could find no alterations in dopamine levels, dihydroxyphenylacetic acid (DOPAC), another dopamine metabolite, was significantly reduced in cortex.

Acetylcholine

As yet there are no human CSF studies which have examined acetylcholine metabolism in Korsakoff's syndrome. However, Lal *et al.* (1985) have reported that the concentration of the acetylcholine precursor, choline, is elevated in red blood cells of patients with Korsakoff's syndrome. They argue that this finding may indicate changes in membrane transport of choline and might reflect neuronal membrane vulnerability within the central cholinergic systems of patients with this disorder. In animals, the effect of thiamine deficiency on brain acetylcholine levels is controversial. Gibson *et al.* (1982) have reviewed the data from whole brain, and regional studies in rats and pigeons, and show that half of the studies report no change and half a moderate (20–38 per cent) depletion of ACh. However, *in vivo* turnover studies in rats consistently demonstrate a decreased rate of ACh synthesis. Cheney *et al.* (1969) found decreased labelled acetylcholine in whole brain following injection of radio-labelled pyruvate and similarly Barclay *et al.* (1981) found diminished synthesis of acetylcholine from isotopes of both choline and glucose. Vorhees *et al.* (1977, 1978) have found decreased synthesis of acetylcholine after choline accumulation had been inhibited with hemicholinium in cortex, hippocampus, striatum, diencephalon, pons, and medulla. Again, it is not clear whether these changes reflect structural damage to cholinergic systems or represent reversible biochemical abnormalities. Indeed, it could be argued that the abnormal metabolism of acetylcholine is more likely to reflect a pure biochemical dysfunction than that of the monoamines, since it is synthesized from acetyl CoA which in turn is synthesized from pyruvate by pyruvate dehydrogenase, a thiamine dependent enzyme.

NEUROPSYCHOLOGICAL AND PSYCHOPHARMACOLOGICAL EVIDENCE OF NEUROTRANSMITTER DYSFUNCTION

There is very little behavioural evidence implicating serotonin or dopamine systems in either thiamine deficiency or Korsakoff's syndrome. With respect to acetylcholine, Barclay *et al.* (1981) reported that early in thiamine depletion a behavioural impairment in rats, on a task known as the string test, occurred and that it could be manipulated by cholinergic drugs. The string test involves placing a rat on a horizontal string hung between two vertical poles. The animal is then scored according to its behaviour, e.g. use of paws and tail, travelling and falling.

The administration of physostigmine, a cholinesterase inhibitor, improved scores early in thiamine deficiency, but not later when the animals were near to death. Administration of neostigmine, which is a peripheral cholinesterase inhibitor, had no effect, nor did nicotine. Similarly, the centrally acting muscarinic antagonist atropine blocked the effect of physostigmine, but methatropine, which does not enter the brain, did not, and nor did the centrally acting nicotinic receptor blocker mecamylamine. The authors conclude that impaired string test performance reflects a central muscarinic lesion in thiamine deficient rats. Administration of thiamine caused significant improvement of string test scores, but these remained below control levels. However, observations were made only 24 hours after thiamine replacement which is too soon to argue that a permanent irreversible deficit existed. Furthermore, it is difficult to know what the string test measures in behavioural terms. The authors suggest that it is a test of neurological competence. Certainly it seems very unlikely to reflect the equivalent of a higher, mental function.

O'Donnell *et al.* (1986) have examined the effect of cholinergic drugs on memory in six Korsakoff patients using a test of digit sequence recall and the more difficult list-learning test of Bushke and Fuld. They found no effect of a single dose of the anticholinesterase, physostigmine, compared to placebo control. However, treatment with choline chloride for seven days resulted in a non-significant improvement on the list-learning test thus hinting at an involvement of cholinergic systems in this syndrome.

The same authors have also looked at the effect of the indirect catecholamine agonist, methylphenidate. A single dose produced no effect, but after seven days treatment a significant improvement on the list-learning test was found.

The finding that methylphenidate can enhance memory in Korsakoff patients points to a possible catecholaminergic dysfunction in this disorder. Of the catecholamines, noradrenaline has been most strongly implicated by the finding that the specific adrenoreceptor agonist, clonidine, can improve performance on some tests of memory in Korsakoff's syndrome (McEntee and Mair 1980; Mair and McEntee 1986). In these studies, oral clonidine was administered daily for 2 weeks before administration of a battery of cognitive tests. Clonidine improved performance on two subtests of the Weschler Memory Scale, memory passages (with and without delay), visual reproduction, and also the consonant trigrams test. Although all of these are tests of memory, other tests which measured mnemonic processes were not improved. The authors concluded that improvement occurred mostly in easy perceptual tasks. The effects of other drugs have also been assessed in Korsakoff's syndrome: *d*-amphetamine, methysergide, ephedrine and L-dopa. Although several of these drugs are also catecholamine agonists, the improvement was only found with clonidine.

Clonidine is an α_2 adrenergic receptor agonist. α_2-Receptors exist post-synaptically and presynaptically, the latter mainly as heteroreceptors on serotonin terminals (Gothert *et al.* 1981; Raiteri *et al.* 1983), but autoreceptors on noradrenergic terminals may also exist (U' Pritchard *et al.* 1980; Titeler *et al.* 1978; Raiterri *et al.* 1983). In addition α_2-receptors exist on cell bodies of the locus coeruleus (Svensson *et al.* 1975) and the overall effect of clonidine on the intact brain is to decrease noradrenergic transmission by inhibiting firing of cells in the locus coeruleus (Svensson *et al.* 1975; Anden *et al.* 1970). In normal human volunteers, clonidine is sedative, produces a reduction of subjective arousal (Frith *et al.* 1985), and also impairs performance in a test of new learning, perhaps due to its causing interference from prior associations (Frith *et al.* 1985). In this respect its effect in normal volunteers resembles the performance of drug-free Korsakoff patients on paired associate learning (Winocur and Weiskrantz 1976; Cutting 1978*a*). Thus, the effect of clonidine in Korsakoff's syndrome appears paradoxical in two ways: first, Korsakoff patients are improved by clonidine, and secondly, because evidence suggests that noradrenergic transmission is reduced in Korsakoff's syndrome, this argues for a facilitating effect of clonidine on noradrenergic systems. The explanation may lie in the alteration of receptor balance that would occur after a lesion to the locus coeruleus in Korsakoff's syndrome. The inhibitory action of α_2-receptors on cell bodies would be decreased allowing clonidine to act mainly on post-synaptic receptors possibly made supersensitive by denervation. It is of interest to speculate which terminal areas are of importance in mediating the effects of clonidine in Korsakoff's syndrome. Although receptors are ubiquitous within noradrenergic systems (Young and Kuhar 1980) there is evidence to suggest that the frontal cortex may be of special importance. Firstly, U'Pritchard *et al.* (1980) demonstrated that neurotoxic 6-hydroxydopamine (6OHDA) lesions of the dorsal bundle in rats produced a massive proliferation of α_2 receptors in frontal cortex but not in hippocampus or amygdala. Secondly, in an elegant series of experiments by Arnsten and Goldman-Rakic (1985), aged monkeys, which are impaired on a test of spatial memory, showed significant improvement of performance with clonidine. The prefrontal cortex of monkeys (Goldman and Rosvold 1970) and specifically the catecholamine innervation of prefrontal cortex has been linked with spatial memory (Brozoski *et al.* 1979). Arnsten and Goldman-Rakic (1985) also report that clonidine improved performance on spatial memory in young monkeys with 60HDA induced depletion of catecholamines in prefrontal cortex. The effect of surgical ablation of the same area on spatial memory was not improved by the drug, suggesting that clonidine is acting on supersensitive post-synaptic $\alpha_2$receptors.

THEORETICAL CONSIDERATIONS AND CONCLUSIONS

Because much of the evidence presented above concerning neurotransmitter dysfunction in Korsakoff's syndrome is speculative, any conclusions at present must be considered tentative. The most substantial evidence suggests that there is decreased noradrenaline transmission in Korsakoff syndrome since:

(1) the neuropathological lesion of Korsakoff syndrome patients has been shown to frequently involve the site of origin of ascending noradrenergic pathways (Victor *et al.* 1971);

(2) noradrenaline metabolism has been shown to be decreased in Korsakoff's syndrome (McEntee and Mair 1978; McEntee *et al.* 1984);

(3) the degree of noradrenaline metabolite depletion is correlated with the severity of neuropsychological impairment in Korsakoff syndrome patients (Mair *et al.* 1985*a*); and

(4) the α_2-receptor agonist, clonidine, which possibly acts by enhancing noradrenergic transmission in the Korsakoff brain, improves performance on certain tests of memory in these patients (McEntee and Mair 1980; Mair and McEntee 1986).

However, two other transmitter systems should also be considered as candidates with links to Korsakoff's syndrome: serotonin and acetylcholine. Serotonin systems are implicated by way of their neuroanatomical position in the brain as described above, and also because in thiamine depleted animals there is altered serotonin metabolism (Plaitakis *et al.* 1982) and possible cell death within the raphe nuclei (Chan-Palay 1977). Furthermore, in Korsakoff's syndrome, serotonin metabolism has been shown to be decreased, although to a much lesser degree than noradrenaline (McEntee *et al.* 1984). Non-human primates, after thiamine depletion, demonstrate impairment on spatial-reversal learning and in recognition of highly familiar items (Witt and Goldman-Rakic 1983*a, b*), both of which may be analogous to similar deficits found in Korsakoff's syndrome (Oscar-Berman and Zola

Morgan 1980; see Witt and Goldman-Rakic 1983*b*). In this study the mamillary bodies, the mediodorsal nucleus of the thalamus and locus coeruleus were intact, but the site of the lesions corresponded closely to serotonergic cell bodies or axons.

The evidence supporting abnormalities in brain cholinergic systems is weak in that animal studies demonstrating decreased cholinergic turnover in thiamine deficiency have not shown whether this effect is permanent and the single study demonstrating pathology within the basal forebrain only studied the brains of three Korsakoff patients, the clinical manifestations of whom were not clear. There is still good reason to consider acetylcholine, however, because of the growing evidence implicating pathology within forebrain cholinergic systems in other human cognitive disorders, most notably Alzheimer's dementia (Sahakian 1987; Kopelman this volume). Butters (1985) suggests that in the alcoholic Korsakoff patient, the critical structures destroyed by a combination of thiamine deficiency and chronic alcohol consumption might be the cholinergic nuclei of the basal forebrain that innervate cortical and limbic structures subserving memory. He points out similarities between the cognitive deficits of Korsakoff's syndrome and early Alzheimer's dementia in which the mnemonic problems predominate and suggests that although there are also striking differences between the two populations, these may only reflect the degree of involvement of the nucleus basalis of Meynert, the latter group developing more severe and widespread destruction of cholinergic processes and hence more severe and widespread cognitive problems. Lishman (1986) has suggested that Wernicke-type pathology within the nucleus basalis of Meynert might account for cognitive deficits found in non-Korsakoff alcoholics. It has been well recognized that some alcoholics, often labelled as Korsakoff's syndrome, have in fact a global dementia (Horvath 1975; Cutting 1978*b*; Lishman 1981) and that the only underlying brain pathology in many of these is of the Wernicke type (Torvick *et al*. 1982). Although it has been argued that the memory disorder of alcoholics may reflect the diencephalic pathology of thiamine deficiency, plus the additional cognitive deficits due to direct neurotoxic action of alcohol on cortical cells, a more parsimonious argument is that damage within the nucleus basalis of Meynert can give rise to a spectrum of cognitive change depending on the site (as a reflection of the topographical organization) and/or the extent of involvement (see Lishman 1986).

From the above argument there are at least three neurotransmitter systems that might be implicated in the Wernicke–Korsakoff lesion and, as in Alzheimer's dementia, it is possible that a combination of noradrenergic, serotoninergic and cholinergic lesions may be involved (Adolfsson *et al*. 1979; Bondareff *et al*. 1981; Gottfries *et al*. 1969). These systems have in common projections to the entire neocortex and hippocampus suggesting non-specific roles in cognition. Although cognitive deficits in Korsakoff's syndrome may be more global than once thought, in the majority of cases amnesia is the most striking and significant impairment. Any neurochemical hypothesis of Korsakoff's syndrome must be able to explain this. One explanation is that only afferents to structures involved in learning and memory (e.g. hippocampus) may be damaged, as described above for the cholinergic system. Another explanation is that the pattern of *structural* damage is crucial to Korsakoff symptoms whereas neurochemical changes are merely epiphenomena. Finally, the explanation may lie in an understanding of the memory disorder itself in relation to cholinergic or monoaminergic function. For example, Mair and McEntee (1983, 1986) suggest that the amnesic symptoms of Korsakoff's syndrome can be explained by deficits in attention, and Robbins (1986) argues that, with respect to intact monoaminergic and cholinergic input to forebrain regions, 'converging lines of evidence point towards largely non-specific functions of the various systems, perhaps akin to arousal, optimum levels of which are necessary for the complex computational processes that are assumed to occur within their terminal regions'. He goes on to conclude that each neurochemical system may subserve different functions within the broad category of arousal. For example, noradrenaline may come into play during stressful conditions to preserve selective attention and possibly more associative and mnemonic processes, whereas the cholinergic system, he suggests, 'may function to maintain cortical arousal at an optimal level for cue discrimination'.

In conclusion, there are definite indications that the search for a neurochemical basis of Korsakoff's syndrome should proceed. Not only would this enlighten our understanding of memory and other cognitive processes, but it would also facilitate the study of possible therapeutic strategies for the treatment of the cognitive disorder in Korsakoff's syndrome, in which the dementing process is non-progressive.

ACKNOWLEDGEMENTS

I would like to thank Mrs Maja Fisher for preparing the manuscript and Dr Barbara Sahakian for comments. The author is a Wellcome Trust Lecturer in Mental Health.

REFERENCES

Adolfsson, R., Gottfries, C.G., Roos, B.E., and Winblad, B. (1979). Changes in brain catecholamines in patients with dementia of the Alzheimer type. *British Journal of Psychiatry* **135**, 216–23.

Anden, N.E., Corrodi, H., Fuxe, K., Hokfelt, B., Hokfelt, T., Rydin, C., and Svensson, T. (1970). Evidence for a central noradrenaline receptor stimulation by clonidine. *Life Sciences* **9**, 513–23.

Arendt, T., Bigl, V., Arendt, A., and Tennstedt, A. (1983). Loss of neurones in the Nucleus Basalis of Meynert in Alzheimer's disease, Paralysis Agitans, and Korsakoff's disease. *Acta Neuropathologica (Berlin)* **61**, 101–8.

Arnsten, A.F.T. and Goldman-Rakic, P. (1985). α_2-Adrenergic mechanisms in prefrontal cortex associated with cognitive decline in aged nonhuman primates. *Science* **230**, 1273–6.

Azmitia, E.C. (1978). The serotonin-producing neurons of the midbrain median and dorsal raphe nuclei. In *Handbook of Psychopharmacology*, Vol. 9 (ed. L.L. Iversen, S.D. Iversen, and S.H. Synder), pp. 233–314. Plenum Press, New York.

Barclay, L.L., Gibson, G.E., and Blass, J.P. (1981). Impairment of behaviour and acetylcholine metabolism in thiamine deficiency. *Journal of Pharmacology and Experimental Therapeutics* **217**, 537–43.

Boles, R.S. and Boles, R.S. (1951). Wernicke's disease: clinical and pathological study in nine cases. *Gastroenterology* **19**, 504–15.

Bondareff, W. Mountjoy, C.Q., and Roth, M. (1981). Selective loss of neurones of origin of adrenergic projections to cerebral cortex (nucleus locus coeruleus) in senile dementia. *Lancet* **i**, 783–4.

Botez, M.I., Young, S.N., Bachevalier, J., and Gauthier, S. (1982). Thiamine deficiency and cerebrospinal fluid 5-hydroxy indoleacetic acid: a preliminary study. *Journal of Neurology, Neurosurgery and Psychiatry* **45**, 731–3.

Brozoski, T.J., Brown, R.M., Rosvold, H.E., and Goldman, P.S. (1979). Cognitive deficit caused by regional depletion of dopamine in prefrontal cortex of rhesus monkey. *Science* **205**, 929–32.

Butters, N. (1984). Alcoholic Korsakoff's syndrome: an update. *Seminars in Neurology* **4**, 229–47.

—— (1985), Alcoholic Korsakoff's syndrome: some unresolved issues concerning aetiology, neuropathology and cognitive deficits. *Journal of Clinical and Experimental Neuropsychiatry* **7**, 181–210.

Chan-Palay, V. (1977). Indoleamine neurones and their processes in normal and in chronic diet-induced thiamine deficiency demonstrated by uptake of H-serotonin. *Journal of Comparative Neurology* **176**, 467–94

Cheney, D.L., Gubler, C.J., and Janussi, A.W. (1969). Production of acetylcholine in rat brain following thiamine deprivation and treatment with thiamine antagonists. *Journal of Neurochemistry* **16**, 1283–91.

Cruickshank, E.K. (1950). Wernicke's encephalopathy. *Quarterly Journal of Medicine* **19**, 327–38.

Cutting, J. (1978*a*). A cognitive approach to Korsakoff's syndrome. *Cortex* **14**, 485–95.

—— (1978*b*). The relationship between Korsakoff's syndrome and alcoholic dementia. *British Journal of Psychiatry* **132**, 240–51.

Dahlstrom A. and Fuxe, K. (1964). Evidence for the existence of monoamine-containing neurones in the central nervous system. 1. Demonstration of monoamines in the cell bodies of brainstem neurones. *Acta Physiologica Scandinavica* **62**, (Suppl. 232), 1–55.

De Wardener, H.E. and Lennox, B. (1947). Cerebral beri beri (Wernicke's encephalopathy). *Lancet* **ii**, 11–17.

Freund, G. (1973). Chronic central nervous system toxicity of alcohol. *Annual Review of Pharmacology* **13**, 217–27.

—— (1982). The interaction of chronic alcohol consumption and aging on brain structures and function. *Alcoholism: Clinical and Experimental Research* **6**, 13–21.

Frith, C.D., Dowdy, J., Ferrier, I.N., and Crow, T.J. (1985). Selective impairment of paired associate learning after administration of a centrally acting adrenergic agonist (clonidine). *Psychopharmacology* **87**, 490–3.

Gibson, G., Barclay, L., and Blass, J. (1982). The role of cholinergic systems in thiamine deficiency. In *Thiamine: Twenty Years of Progress.* (ed. H.Z. Sable and C.T. Gubler), Annals of the New York Academy of Sciences, Vol. 378, pp. 382–403.

Goldman, P.S. and Rosvold, H.E. (1970). Localisation of function within the dorsolateral prefrontal cortex of the rhesus monkey. *Experimental Neurology* **27**, 291–304.

Gothert, M., Huth, H., and Schlisker, E. (1981). Characterisation of the receptor subtypes involved in α-adrenoceptor-mediated modulation of serotonin release from rat brain cortex slices. *Naunyn Schmeiderberg's Archives of Pharmacology* **317**, 199–203.

Gottfries, C.G., Gottfries, I., and Roos, B.E. (1969). Homovanillic acid and 5-hydroxy indoleacetic acid in the cerebrospinal fluid of patients with senile dementia, presenile dementia and Parkinson's disease. *Journal of Neurochemistry* **16**, 1341–5.

Harper, C. (1979). Wernicke's encephalopathy: a more common disease than realised. *Journal of Neurology, Neurosurgery and Psychiatry* **42**, 226–31.

Iwata, H. (1982). Possible role of thiamine in the nervous system. *Trends in Pharmacological Sciences* **3**, 171–3.

Jollife, N., Wortis, H., and Fein, H.D. (1941). The Wernicke syndrome. *Archives of Neurology and Psychiatry* **46**, 569–97.

Leckman, J.F. and Maas, J.W. (1984). Plasma MHPG: relationship to brain noradrenergic systems and emerging clinical applications. In *Neurobiology of Mood Disorders* (ed. R.M. Post and J.C. Ballenger), Frontiers of Clinical Neuroscience, Vol. 1. Williams and Wilkins, Baltimore.

Lindvall, O. and Bjorklund, A. (1978). Organization of catecholamine neurones in the rat central nervous system. In *Handbook of Psychopharmacology*, Vol. 9 (ed. L.L. Iversen, S.D. Iversen, and S.H. Synder). Plenum Press, New York.

Lishman W.A. (1981). Cerebral disorders in alcoholism, syndromes of impairment. *Brain* **104**, 1–20.

—— (1986). Alcoholic dementia: A hypothesis. *Lancet* **i**, 1184–6.

Mair, R.G. and McEntee, W.J. (1983). Korsakoff's psychosis: noradrenergic systems and cognitive impairment. *Behavioural Brain Research* **9**, 1–32.

—— and —— (1986). Cognitive enhancement in Korsakoff's psychosis by clonidine: a comparison with L-Dopa and ephedrine. *Psychopharmacology* **88**, 374–80.

——, ——, and Zatorre, R.J. (1985*a*) Monoamine activity correlates with psychometric deficits in Korsakoff's disease. *Behavioural Brain Research* **15**, 247–54.

——, Anderson, C.D., Langlais, P.J., and McEntee, W.J. (1985*b*). Thiamine deficiency depletes cortical norepinephrine and impairs learning processes in the rat. *Brain Research* **360**, 273–84.

Mair, W.P.G., Warrington, E.K., and Weiskrantz, L. (1979). Memory disorder in Korsakoff's psychosis: a neuropathological and neuropsychological investigation of two cases. *Brain* **102**, 749–89.

Malamud, N. and Skillicorn, S.A. (1956). Relationship between the Wernicke and the Korsakoff syndrome. *Archives or Neurology and Psychiatry* **76**, 585–96.

Martin, P.R., *et al.* (1984) Central nervous system catecholamine metabolism in Korsakoff's psychosis. *Annual Review of Neurology* 15, 184–7.

McEntee, W.J. and Mair, R.G. (1978). Memory impairment in Korsakoff's psychosis: a correlation with brain noradrenergic activity. *Science* 202, 905–7.

—— and —— (1980). Memory enhancement in Korsakoff's psychosis by clonidine: further evidence for a noradrenergic deficit. *Annals of Neurology* 7, 466–70.

——, —— and Langlais, P.J. (1984). Neurochemical pathology in Korsakoff's psychosis: implication for other cognitive disorders. *Neurology* 34, 648–52.

Oscar-Berman, M. and Zola-Morgan, S.M. (1980). Comparative neuropsychology and Korsakoff's syndrome. 1. Spatial and visual reversal learning *Neuropsychologia* 18, 499–512.

Plaitakis, A., Chung Hwang, E., Van Woert, M.H., Szilagyi, P.I.A., and Berl, S. (1982). Effect of thiamine deficiency on rat neurotransmitter systems. In *Thiamine: Twenty Years of Progress* (ed. H.Z. Sable and C.J. Gubler). New York Academy of Sciences 378, 367–81.

Raiteri, M., Maura, G., and Versace, P. (1983). Functional evidence for two stereochemically different alpha-adrenoceptors regulating central norepinephrine and serotonin release. *Journal of Pharmacology and Experimental Therapeutics* 224, 679–84.

Robbins, T.W. (1986). Psychopharmacological and neurobiological aspects of the energetics of information processing. In *Energetics and Human Information Processing* (ed. R. Hockey, A. Gaillard, and M. Coles). Martini Nijhoff, Dordrecht, The Netherlands.

Sahakian, B.J. (1987). Cholinergic drugs and human cognition. In *Handbook of Psychopharmacology*, Vol. 20 (ed. L.L. Iversen, S.D. Iversen, and S.H. Snyder). Plenum Press, New York.

Svensson, T.H., Bunney, B.S., and Aghajanian, G.K. (1975). Inhibition of both noradrenergic and serotonergic neurones in brain by adrenergic agonist clonidine. *Brain Research* 92, 291–306.

Titeler, M., Tedesco, J.L., and Seeman, P. (1978). Selective labelling of presynaptic receptors by H-dopamine, H-apomorphine, H-clonidine labelling of postsynaptic sites by H-neuroleptics. *Life Sciences* 23 , 587–92.

Torvick, A., Lindboe, C.F., and Rogde, S. (1982). Brain lesions in alcoholics. A neuropathological study with clinical correlation. *Journal of Neurological Sciences* 56, 233–48.

Ungerstedt, U. (1971). Stereotaxic mapping of the monoamine pathways in the rat brain. *Acta Physiological Scandinavica* 367, 1–48.

U'Pritchard, D.C., Reisine, T.D., Mason, S.T., Fibiger, H.C., and Yamamura, H.I. (1980). Modulation of rat brain alpha- and beta-adrenergic receptor populations by lesions of the dorsal noradrenergic bundle. *Brain Research* 187, 143–54.

Victor, M., Adams, R.D., and Collins, G.H. (1971). *The Wernicke-Korsakoff Syndrome. Contemporary Neurology Series*, Vol. 7. F.A. Davis and Co., Philadelphia.

Vorhees, C.V., Schmidt, D.E., Barret, R.J., and Schenker, S. (1977). Effect of thiamine deficiency and acetylcholine levels and utilisation *in vivo* in rat brain. *Journal of Nutrition* 107, 1902–8.

——, ——, and —— (1978). Effect of pyrithiamine oxythiamine on acetylcholine levels and utilisation in rat brain. *Brain Research Bulletin* 3, 495–6.

Wernicke, C. (1981). *Lehrbuch der gehirnkrankheiten fur aerzte und studirende*, Vol. 2, Theodor Fischer, Kassel.

Wilkinson, D.A. (1982). Examination of alcoholics by computed tomographic (CT) scans: a critical review. *Alcoholism: Clinical and Experimental Research* 6, 31–48.

Winocur, G. and Weiskrantz, L. (1976). An investigation of paired associate learning in amnesia patients. *Neuropsychologia* 14, 97–110.

Witt, E.D. (1985). Neuroanatomical consequences of thiamine deficiency a comparative analysis. *Alcohol and Alcoholism* 20, 201–21.

—— and Goldman-Rakic, P.S. (1983a). Intermittent thiamine deficiency in the rhesus monkey. 1. Progression of neurological signs and neuroanatomical lesions. *Annals of Neurology* 13, 376–95.

—— and —— (1983b). Intermittent thiamine deficiency in the rhesus monkey. 2. Evidence for memory loss. *Annals of Neurology* 13, 396–401.

Young, W.S. and Kuhar, M.J. (1980). Noradrenergic alpha-1 and alpha-2 receptors: light microscopic autoradiographic localisation. *Proceedings of the National Academy of Sciences USA* 77, 1696–700.

18

Non-cholinergic pharmacology in human cognitive disorders

HARVEY J. ALTMAN, HOWARD J. NORMILE, AND
SAMUEL GERSHON

Recent estimates suggest that 12 per cent of the population of the United States or 24 million people are over the age of 65 (Cummings and Benson 1983). While the average life expectancy is not anticipated to increase substantially during the next 50 years, the percentage of the population aged 65 or older is expected to increase markedly. Estimates suggest that by the year 2030, 17–20 per cent of the population or 51 million people will be at least 65 years of age (Plum 1979; Wells 1981). The impact of this shift in the age distribution of the population is clearly beginning to make itself felt, particularly within the area of health care.

The elderly present with a unique profile of medically related needs and problems. One of these, dementia, accounts for more admissions and more hospital in-patient days than any other psychiatric illness of the elderly (Cummings and Benson 1983; Kachaturian 1985). In 1978 this translated into a cost of approximately $12 billion. By the year 2030 this figure is expected to more than double (Plum 1979).

Of the total population presenting with clinical symptons of dementia, approximately 22–39 per cent (average 29 per cent) of these can be attributed to the pathology associated with senile dementia of the Alzheimer's type (SDAT) (Marsden and Harrison 1972; Freemon 1976; Victoratos et al. 1977; Smith and Kiloh 1981; Hutton 1981; Heston 1981; Maletta et al. 1982; Benson et al. 1982). In addition, Ball et al. (1986) have reported a number of instances in which patients presented with significant signs of dementia without apparent SDAT-associated neuropathological and neurochemical deterioration, thus raising additional questions regarding the underlying bases of the dementias.

Behaviourally, SDAT is a devastating disease. The hallmark of the disease is an insidious and progressive loss of recent memory (Drachman et al. 1982; Reisberg 1983). In addition, patients frequently complain of disturbances in affect (Salzman and Shader 1979) and sleep (Feinberg et al. 1967; Lowenstein et al. 1928; Prinz et al. 1982; Reynolds et al. 1983; Vitiello et al. 1984). In later stages of the disease significant disturbances in activity (deLeon *et al.* 1984) and continence (Isaacs 1962; Hodkinson 1973; McLaren *et al.* 1981) also become apparent.

The average age of onset has been estimated to be approximately 75.6 years (Diesfeldt *et al.* 1986). The disease has an average duration of 7 years and inevitably ends in death (Heston *et al.* 1981; Barclay *et al.* 1985; Diesfeldt *et al.* 1986).

TREATMENT STRATEGIES

Initially, SDAT was thought to be primarily the result of cerebral ischaemia due to arteriosclerotic origins. As a result, early attempts to ameliorate the memory deficits associated with SDAT focused on increasing blood flow or the supply of oxygen to the brain.

Cerebral vasodilators

The drugs in this category essentially fall into one of two categories according to whether they act solely on the smooth musculature of the cerebrovascular system or whether they possess the additional capacity to stimulate cerebral metabolism. Drugs in the first class include: cyclandelate (Cyclospasmol), papaverine hydrochloride (Pavabid), isoxsuprine hydrochloride (Vaso-dilan), cinnerizine (Mitronal, Verta), buflomedil, nylidrin hydrochloride, betahistine hydrochloride, xanthinol niacinate, duxil (a mixture of almitrine and raubasine), indeloxazine (presently in clinical trials in Japan), and bency-clone. Drugs in the second class include: the vinca alkaloids (e.g. vincamine), dihydroergotoxin mesylate (DEM, Hydergine), nafronyl oxalate (Proxiline), pyritinol, and pentifylline.

In total, more than 105 controlled and uncontrolled studies have been conducted on the uses of cerebral vasodilators to ameliorate the memory deficits associated with senile dementia (for a detailed review see Yesavage *et al.* 1979). With only a few exceptions (Foster *et al.* 1955; Dhrymiotis and Whittier 1962; Baso 1973; Westreich *et al.* 1975) most of the studies report some degree of improvement on mental test performance. While some of the newer drugs in this category may prove useful for the treatment of dementia and should be explored further (e.g. nafronyl), Hydergine still remains the most widely used and prescribed drug of the cerebral vasodilators. However, at best, none of the drugs appear to produce clinically significant and consistent effects sufficient to warrant serious consideration as a research strategy for improving the treatment of SDAT.

Hyperbaric oxygen

An alternative method, proposed to increase the amount of oxygen getting to the brain, and therefore to increase brain metabolism, was to place subjects in an environment rich in oxygen (e.g. hyperbaric chambers). The first report of such an approach was encouraging (Jacobs *et al*. 1969). Thirteen elderly patients with chronic organic brain syndrome exhibited improved cognitive functioning following repeated exposure to hyperbaric oxygen. Age-matched controls, initially exposed to normal air, failed to improve. However, when shifted to hyperbaric oxygen, an improvement was observed in this latter group as well. Since this initial study, several additional controlled and uncontrolled studies of the effects of hyperbaric oxygen on cognitive functioning have been conducted. Five of these confirmed Jacobs' original observation (Jacobs *et al*. 1972; Boyle *et al*. 1974; Edwards and Hart 1974; Imai 1974; Ben-Yishay *et al*. 1974) and three did not (Goldfarb *et al*. 1972; Thompson 1974; Raskin *et al*. 1977). It is, therefore, questionable whether placing demented subjects in an oxygen rich environment will help consistently to ameliorate the memory impairments associated with SDAT.

Psychostimulants

A second early approach to the treatment of SDAT (psychostimulants) focused on the fact that people suffering from SDAT frequently exhibited increased fatiguability along with a significant decrease in motor activity, alertness, concentration and speed of CNS information processing, all of which were thought to contribute to a generalized slowing of cognitive processes (Botwinick 1977; Birren, 1974; Crook 1979). In addition, the elderly and especially those suffering from SDAT were often depressed (Smith and Kiloh 1981; Kral 1983; Straker 1984; Mahandra 1985). Psychostimulants seemed, therefore, to be a reasonable experimental therapeutic approach to improve cognitive processes in SDAT.

Drugs in the psychostimulant class include: methylphenidate (Ritalin), pentylenetetrazol (Metrazol), magnesium pemoline, procaine hydrochloride, and piradol. A substantial body of preclinical evidence suggests that psychostimulants can enhance learning and memory as well as attenuate some forms of experimentally induced amnesias (Quatermain and Altman 1982; Altman and Quartermain 1983; Quinton and Bloom 1977; Sara and Deweer 1982). However, none of the drugs in this category are currently approved by the U.S. Food and Drug Administration (FDA) for clinical use as cognitive enhancers. However, procaine is still marketed in Nevada. As a whole, the clinical results with psychostimulants are not encouraging; nearly all studies are negative. The positive data are principally derived from several uncontrolled studies. The clinical data do not, therefore, support the use of psychostimulants for the treatment of the memory impairments associated with SDAT. In addition, the side effects associated with the use of such compounds, including the potential for abuse, argue against prescribing them in any clinical population, especially the elderly.

Metabolic enhancers

Another class of compounds which has received considerable attention is the so-called 'metabolic enhancers' or 'nootropics'. Drugs in this class include: piracetam, etiracetam, aniracetam, pramiracetam, rolziracetam (CI-911), 3-phenoxypyridine (CI-844), dupracetam, fipexide (recently approved for use in Malaysia, Thailand, Hong Kong, and Singapore) and CI-933 (a propanoic acid of the methoxybenzoyl pyrrolidinones which is scheduled to go into phase-II clinical trials later this year).

There has been considerable speculation and debate regarding the mechanism of action of nootropics (Kopelevich *et al*. 1981; Dlabac *et al*. 1981; Wurtman *et al*. 1981; Poschel *et al*. 1983; Giurgea *et al*. 1983). Most studies have failed to demonstrate any direct interaction of nootropics with neurotransmitter systems. Indirect interactions, however, have not been entirely ruled out (Dlabac *et al*. 1981; Kopelevich *et al*. 1981).

Much of the evidence suggesting that this class of compounds may enhance cognitive processes is derived from preclinical research. Rather dramatic and consistent effects have been reported (Wolthuis 1971; Sara 1980; Wolthuis 1981). That is not to say that all of the animal studies report positive effects. There have been a few negative reports (Oglesby and Winter 1974; Buresora and Bures 1982). In addition, some evidence suggests that 'metabolic enhancers' may work best in cognitively impaired animals. The compounds appear least effective in normal young animals performing under optimal conditions (Gamzu 1985).

Unfortunately, much less dramatic and reliable effects have been observed clinically. In fact, there is a growing consensus that the metabolic enhancers will not prove to be effective in enhancing cognitive functioning in clinically diagnosed demented subjects. There does, however, appear to be one population of subjects that exhibits a modest, though reliable, improvement in performance: the 'normal' aged complaining of slight memory impairments. Nootropics, and piracetam in particular, may eventually be found to possess greater efficacy in human subjects when used in conjunction with other potentially active cognitive enhancers as shown for experimental animals (Bartus *et al*. 1983). Nootropics should still be seriously considered as potentially useful human cognitive enhancers, as a number of clinical trials with novel compounds are still in progress.

Neuropeptides

Using a radically different approach, a number of investigators began looking at compounds or classes of compounds interacting with neurotransmitter systems not implicated in the neuropathophysiology of Alzheimer's disease. In the animal literature there are an increasing number of reports of significant and reliable enhancement of learning and/or memory following administration of a variety of neuropeptides including ACTH (Rigter and Crabbe 1979; Bartus *et al.* 1983), vasopressin (de Wied 1965; Walter *et al.* 1975; de Wied and Gispen 1977; Bohus *et al.* 1978) and the opiate antagonists, naloxone and naltrexone (Carrassio *et al.* 1982; Gallagher 1982; Liang *et al.* 1983). It was reasoned that the cognitive impairments might be attenuated by addressing neurotransmitter systems suspected of playing a role in learning and memory. This approach would be particularly advantageous if the neurotransmitter systems in question did not appear to be significantly compromised as a result of the disease process.

ACTH and its analogues

Adrenocorticotropic hormone (ACTH) and its various analogues and fragments (particularly $ACTH_{4-10}$ and $ACTH_{4-9}$) have been studied extensively in animals. ACTH appears to stimulate learning and memory as well as to attenuate experimentally-induced amnesias (De Wied 1973; van Wimersma-Greidanus and De Wied 1976; De Wied and Bohus 1979). However, there is still some debate as to whether these effects are due to enhancement of associative or non-associative processes.

A similar problem in interpreting the effects of these compounds on cognition in man also exists. For example, Ferris (1983) evaluated the single dose and chronic effects of $ACTH_{4-9}$ (ORG-2766) using a double-blind, multiple cross-over design in 50 mildly to moderately impaired SDAT patients (aged 60–85). Single doses produced small improvements in visual recall, but slowed simple reaction time and increased fatigue. Chronic administration of $ACTH_{4-9}$ produced a small improvement in verbal memory and a consistent improvement in mood. However, others have failed to observe any indication that these compounds can significantly improve cognition (Branconnier and Cole 1980; Soininien *et al.* 1985). On the other hand, these compounds do appear to improve mood (Ferris *et al.* 1980), reduce fatigue (Ferris *et al.* 1980) and help maintain a high level of vigilance (Gaillard and Sanders 1975; O'Hanlon *et al.* 1978; Gaillard and Varney 1979), thus making them potentially useful for the treatment of depression and anxiety.

Vasopressin

A second neuropeptide which has received a good deal of attention in recent years is vasopressin. While it is unclear whether the levels of vasopressin are significantly reduced in either the CSF or brains of people suffering from SDAT (Rosser *et al.* 1980; Mazurek *et al.* 1986), it has frequently been shown that vasopressin facilitates learning and memory as well as antagonizes experimentally-induced amnesias in animals (Walter *et al.* 1975; Bohus *et al.* 1978). However, negative results have also been reported (Krejci *et al.* 1979) contributing to a growing controversy over the validity and mechanism(s) underlying the reported positive effects (van Ree *et al.* 1978).

Clinically, responses to vasopressin and its three analogues, lysine vaso-pressin (LVP), 1-desamino-8-D-arginine vasopressin (desmopressin, DDAVP) and desglycinamide arginine vasopressin (DGAVP) have been inconsistent. For example, some studies report enhancement of performance (Legros *et al.* 1978; Weingartner *et al.* 1981), while others report no effect (Jenkins *et al.* 1981, 1982; Tinklenberg and Thornton 1983; Tinklenberg *et al.* 1984; Peabody *et al.* 1986). Therefore, while clearly possessing CNS activity, the usefulness of these compounds for the treatment of the cognitive impairments associated with SDAT appears dubious at best.

The opiates

The opiates have received by far the most attention as possible treatments for the cognitive impairments in SDAT. This attention has arisen not because opiates are closely tied to learning and memory or because significant changes in the functional integrity of the opiate system have been reported (Kulmala 1986), but because positive clinical effects have been reported, initially as an open trial using 1.0 mg of naloxone administered intravenously to five subjects (Reisberg *et al.* 1983*a*) and then in a double-blind multiple-dose, placebo controlled study of naloxone in seven patients suffering from SDAT (Reisberg 1983). Unfortunately, a number of other laboratories have not been able to replicate these initial studies (Nasrallah *et al.* 1985; Pomara *et al.* 1985; Blass *et al.* 1983; Tariot *et al.* 1985; Steigler *et al.* 1985; Hyman *et al.* 1985). As a result, the efficacy of naloxone or naltrexone for the treatment of SDAT remains dubious at best.

The apparent lack of reproducible positive clinical effects in man, again, stands in marked contrast to the overall positive responses seen in animals. Numerous reports of facilitation of memory and/or reversal of experimentally-induced amnesias appear in the preclinical literature (Garzon *et al.* 1981; Carrasio *et al.* 1982; Gallagher 1982; Liang *et al.* 1983). In fact, research on the opiate system's participation in the neurochemical processes underlying learning and memory remains quite active today and will probably continue unabated for some time to come.

Somatostatin

As indicated earlier, a number of investigators have reported that the levels of somatostatin are significantly reduced in the brains of autopsied SDAT patients (Davies *et al.* 1980; Rossor *et al.* 1980). Somatostatin-like immuno-reactivity has also been shown to be significantly reduced in the cerebrospinal fluid of SDAT patients (Soininen *et al.* 1984). In addition, somatostatin-like immunoreactivity has been observed to be localized in senile neuritic plaques of SDAT brains (Morrison *et al.* 1985; Roberts *et al.* 1985). However, numerous other neuropeptides are also found in neurites associated with senile placques (Struble *et al.* 1985). Therefore, some caution should be exercised before concluding that somatostatin is idiosyncratically associated with senile plaques in SDAT. It should be noted, however, that an association between somatostatin and acetylcholine may exist. For example, based on studies in the rat, there is the suggestion that somatostatin and acetylcholine either coexist within neurons or that different neurons containing these two neuroactive substances lie in extremely close proximity to each other (Zhu *et al.* 1984; Delfs *et al.* 1984). Finally, the density, but not the affinity, of somatostatin receptors also appears to be significantly reduced in the brains of SDAT patients, particularly within the frontal and temporal cortex (Beal *et ql.* 1985).

Only one clinical study has been conducted to date which has attempted to augment somatostatin in patients suffering from SDAT (Cutler *et al.* 1985). In this study the cyclic somatostatin peptide analogue L-363, 586 (Merck, Sharp and Dohme), which is purported to be 50 times more potent than the naturally occurring peptide, was evaluated in a double-blind, placebo-controlled cross-over study in 10 patients with mild SDAT (age range 58–78). The compound was infused intravenously (120 μg/hour for 30 min) followed by a longer infusion (5.5 hours) at a slower rate (40 μg/hour). Neuropsychological assessment was conducted 3 and 5 hours after the start of the initial infusion period. No effect on either serial or paired associate learning was found. Clearly additional studies of the role of somatostatin in SDAT need to be conducted. The outcome of the Cutler *et al.* study may prove to be not all that surprising since the clinical significance of reductions in somatostatin in the brains of SDAT patients is still unknown.

Thyrotropin-releasing hormone

It is not yet clear whether the levels of TRH are actually affected as a result of SDAT. There is some evidence that TRH is reduced in CSF (Oram *et al.* 1981) and the amygdala (Biggins *et al.* 1983). However, others have failed to observe any change in this neuropeptide (Yates *et al.* 1983). While having no known direct involvement in learning and memory, there is limited evidence that this neuropeptide may modulate cholinergic neurotransmission (Yarbough 1979).

Based on the above, an initial double-blind trial of TRH (300 or 500 μg) *v.* placebo was conducted in four subjects. Drug and placebo were administered intravenously over a 1-minute period (Peabody *et al.* 1986). Neither dose produced an enhancement in performance.

Acetylcholine

Others in this volume (Kopelman, Agid, and Rossor) have reviewed the cholinergic neurochemistry of SDAT, so this topic will not be covered comprehensively here. Because of the intense interest in the cholinergic hypothesis of SDAT, we will comment on cholinergic mechanisms in the context of the potential importance of noncholinergic mechanisms in SDAT. Recently, the underlying cause of SDAT was hypothesized to be the result of a selective or preferential neuropathology associated with the cholinergic nervous system. Reductions in the activities of choline acetyltransferase (Bowen *et al.* 1976; Davies and Maloney 1976; Perry *et al.* 1977), acetylcholinesterase (Terry and Davies 1980; Bowen *et al.* 1976) as well as significant losses of cholinergic neurons within the ventral forebrain (Whitehouse *et al.* 1981, 1982; Price *et al.* 1982) have been observed in brains of Alzheimer's patients. Acetylcholine has long been suspected of playing an important role in learning and memory (Deutsch and Rogers 1979; Drachman and Leavitt 1974; Bartus *et al.* 1983). It was, therefore, no surprise that people suffering from SDAT presented with disturbances in cognition and intellect.

The course of action seemed obvious: pharmacologically stimulate the cholinergic nervous system with selective cholinergic agonists and this should restore intellectual functioning. A similar strategy has successfully been applied to the treatment of Parkinson's disease. Stimulated in part by the above and in part by the results of a number of key pharmacological studies conducted in animals (Bartus *et al.* 1980, 1983), a flurry of clinical trials were initiated in SDAT patients. However, the anticipated improvement in intellectual functioning has not been realized. While there are reports of significant improvement (Brinkman and Gershon 1983; Thal *et al.* 1983; Johns *et al.* 1985; Mohs *et al.* 1985; Harbourgh 1986), the number of subjects responding, compared to the total population sampled, is disappointingly small. In addition, improvement appears to be generally restricted to one or two measures of much larger test batteries. Furthermore, some investigators have failed to replicate their own findings. The panacea has, therefore, not yet been realized. Replenishment therapy (at least with respect to the cholinergic nervous system) may not be (in itself) the whole answer.

Most investigators and clinicians now realize that SDAT is not specifically a disease of the cholinergic nervous system, and attempts to reduce the behavioural symptoms of the disease to the effects of a single neurotransmitter (as has so often been tried in the past) is not likely to result in a solution to the problem. SDAT is now known to affect significantly the functional integrity of norepinephrine (Mann *et al.* 1982; Perry *et al.* 1981*a*), dopamine (Adolfsson *et al.* 1979; Gibson and Ball 1983), serotonin (Ishii 1966; Benton *et al.* 1982; Bowen *et al.* 1983; Reynolds *et al.* 1983; Yamamoto and Hirano 1985) and GABA (Davies 1979; Rossor *et al.* 1982; Zimmer *et al.* 1984; Mohr *et al.* 1986) as well as a number of neuropeptides (Oram *et al.* 1981; Biggins *et al.* 1983; Morley 1986), most notably somatostatin (Rossor *et al.* 1980;

Davies and Terry 1981; Perry *et al*. 1981*b*; Ferrier *et al*; 1983; Gomez *et al*. 1986; Mohr *et al*. 1986).

Dopamine

The dopaminergic nervous system has been suspected of playing a role in learning and memory for a number of years. For example, stimulation of dopaminergic neurotransmission has been shown to enhance memory as well as to attenuate experimentally-induced amnesias in animals (Quartermain and Altman 1982; Altman and Quartermain 1983; Quinton and Bloom 1977; Sara and Deweer 1982).

L-Dopa and dopaminergic agonists have also been tried clinically by several investigators. However, the results have been inconsistent. Half of the studies report a significant improvement in intellectual functioning (Lewis *et al*. 1978; Adolfsson *et al*. 1978; Schubert and Fleischhacker 1979), while the other half do not (Kristensen *et al*. 1977; Adolfsson *et al*. 1978; Fleischhacker *et al*. 1986). Clearly, additional studies need to be done in order to resolve this impasse.

GABA

An accumulating body of evidence suggests that CNS gamma-aminobutyric acid (GABA)-containing neurons may be adversely affected in SDAT. For example, CSF GABA levels of SDAT patients are significantly reduced when compared to age-matched controls (Enna *et al*. 1971; Manyam *et al*. 1980; Kuroda 1983; Zimmer *et al*. 1981; Zimmer *et al*. 1984). Furthermore, significant reductions in the levels of GABA, as well as in the activity of its synthetic enzyme glutamic acid hydroxylase have also been detected in SDAT brains (Perry *et al*. 1977; Davied 1979; Rossor *et al*. 1982).

Based in part on the above, as well as on the recent suggestion that cortical GABA interneurons may receive inputs from acetylcholine-containing neurons projecting from the basal forebrain (Coyle *et al*. 1983), a clinical trial was conducted in order to determine whether stimulation of GABAergic neurons would stimulate cognitive functioning in a population of mildly to moderately demented SDAT patients (Mohr *et al*. 1986). The GABAergic agonist used was 4,5,6,7,-tetrahydroisoxazolo (5,4,-c)pyridin-3-ol (THIP). A total of six male SDAT patients (mean age: 58 ± 2.0 yrs) and eight male normal age-matched controls (mean age: 61 ± 1.8 yrs) were included in this study. THIP was administered orally (QID), initially at a dose of 20 mg/day, which was then gradually increased until clinically significant side effects were observed (range: 40–160 mg) according to a double-blind, placebo-controlled, cross-over design. Although the investigators did observe slight, though non-significant improvements in some verbal test scores, 'clinically significant effects on cognitive function' were not observed.

Clearly, additional clinical trials need to be conducted (both with drugs that stimulate as well as block this system) before ruling out the possibility that modulating GABAergic neurons could potentially ameliorate cognitive deficits associated with SDAT.

Serotonin

Another neurotransmitter that appears to have received relatively little therapeutic attention with regard to SDAT is serotonin. However, there is a large body of preclinical evidence which suggests that the serotonergic nervous system may play an important role in information processing by the brain. Also, recent post-mortem studies suggest that the serotonergic nervous system is significantly affected in SDAT. The remaining discussion will focus on serotonin (5-HT), and we will review serotonergic changes in SDAT, as well as the results of clinical and preclinical trials using pharmacological interventions designed to modulate 5-HT.

Serotonin and SDAT

While the serotonergic nervous system has not been investigated systematically, several lines of evidence do suggest its involvement in SDAT. In fact, the involvement of 5-HT may be as global as that described for the cholinergic nervous system. Ishii (1966) was the first to report that cells in the dorsal and median raphe nuclei of SDAT patients contained dense neurofibrillary tangle formations. A more recent study has extended Ishii's initial observations, reporting cell loss and the presence of senile plaques in the dorsal raphe nucleus in SDAT (Yamamoto and Hirano 1985).

Several investigators have reported a reduction of serotonin within post-mortem SDAT brains. The reductions appear to be restricted to a number of nuclei or brain regions including the temporal lobe (Bowen *et al.* 1979, 1983), hippocampus, cingulate cortex, hypothalamus, and caudate nucleus (Winblad *et al.* 1982). Levels of 5-hydroxy-3-indoleacetic acid (5-HIAA) are reduced in the hippocampus and hypothalamus, as well as the cingulate, temporal, and frontal cortices (Adolfsson *et al.* 1979; Winblad *et al.* 1982; Cross *et al.* 1983). In a rather extensive study (Arai *et al.* 1984), concentrations of serotonin in nine regions (including the superior frontal gyrus, cingulate gyrus, amygdala, thalamus) and 5-HIAA in eight regions (including the amygdala, caudate, putamen, thalamus) were found to be significantly reduced in the brains of people suffering from SDAT.

The CSF from patients with a clinical diagnosis of SDAT has also been reported to show significant reductions in the levels of 5-HIAA (Gottfries *et al.* 1976; Gottfries and Roos 1976; Argentiero and Tavolato 1980). In addition, turnover of serotonin, as measured in CSF by probenecid loading, has been reported to be reduced in SDAT (Gottfries and Roos 1973; Gottfries *et al*. 1974).

Finally, markers of the serotonergic synapse also appear to be reduced in certain regions of SDAT brains. For example, the uptake of serotonin, as well as imipramine binding have been found to be reduced in biopsy samples taken from the temporal cortex of patients with SDAT (Bowen *et al*. 1983). In addition, a significant regional reduction in the number of serotonergic receptors (both S1 and S2) has also been observed (Bowen *et al*. 1979, 1983; Cross *et al*. 1984; Perry *et al*. 1984; Briley *et al*. 1986). These data, therefore, suggest that serotonergic neurotransmission may be impaired in SDAT. Accordingly, the most logical treatment strategy based on the above results would appear to be the use of pharmacological interventions thought to augment serotonergic activity.

Clinical trials

Only a small number of clinical trials have been conducted in an effort to determine whether manipulations of the serotonergic nervous system might be useful as a means of treating cognitive impairments in dementia. One study has attempted to assess the effects of blocking serotonergic neurotransmission by blocking serotonergic receptors with the selective serotonergic antagonist mianserin (Curran *et al*. 1986). In general, however, the approach has been to stimulate serotonergic neurotransmission. Overall, the results are rather discouraging (Cutler *et al*. 1985; Dehlin *et al*. 1985). Only one study (Passeri *et al*. 1985) reports observing significant improvements in cognitive performance; however, the results of this study are open to interpretation because the drug used in this study, minaprine, not only stimulates serotonergic neurotransmission, but dopaminergic and cholinergic neurotransmission as well (Biziere *et al*. 1982, 1983; Garattini *et al*. 1984). Drugs that have thus far proved unsuccessful in improving cognitive performance in SDAT include the two selective serotonergic reuptake blockers, zimeldine (Cutler *et al*. 1985) and alaproclate (Dehlin *et al*. 1985), and the 5-HT precursors tryptophan, and 5-hydroxytryptophan (5-HTP) (Meyer *et al*. 1977). However, another study reports that zimeldine attenuates the memory impairments associated with acute alcohol consumption (Weingartner *et al*. 1983).

On the other hand, numerous preclinical studies (summarized below) suggest not only that serotonin plays an important role in information processing, but also that some serotonergic drugs can significantly improve learning and memory as well as attenuate experimentally-induced amnesias in animals.

It may, therefore, be hasty to conclude that manipulations of the serotonergic nervous system will not prove useful in the treatment of the cognitive impairments associated with SDAT.

Pre-clinical data

Evidence for a serotonergic role in learning and memory derives both from studies of acute 5-HT stimulation and from studies of long-term disruption of serotonergic activity. For example, a variety of interventions thought to augment serotonergic activity including systemic administration of 5-hydroxytryptophan (Wooley and Van der Hoeven 1963; Joyce and Hurwitz 1964; Roffman and Lal 1972), intracerebral administration of serotonin (Essman 1978; Wetzel *et al.* 1980; Garzon *et al.*) electrical stimulation of the raphe nuclei (Fibiger *et al.* 1978), or *p*-chloroamphetamine (PCA)-stimulated release of serotonin (Ögren *et al.* 1977; Ögren 1982; Archer *et al.* 1981), all *impair* the performance of animals in various learning and memory tasks.

On the other hand, the effects observed following long-term disruption of serotonergic function (e.g. cytotoxic and electrolytic lesions) have been inconsistent. Such manipulations have been shown to impair (Stevens *et al.* 1969), facilitate (Stevens *et al.* 1967), or have no effect (Kohler and Lorens 1978) on learning and memory. There may be a number of reasons for these apparent inconsistencies, including differences in the experimental methods used to impair serotonergic activity and differences in behavioural tests resulting in varying degrees of serotonin depletion, and/or varying amounts of compensatory changes in non-serotonergic systems. The discrepancies may also reflect task specific, depletion-induced alterations in sensory processing, locomotor activity, and motivation (i.e. non-associative processes).

For the past few years, our laboratory has attempted to characterize the role of serotonin in learning and memory. In a series of studies employing acute pharmacological interventions, we have observed that either post-train or pre-test administration of serotonergic reuptake blockers (alaproclate and zimeldine) *facilitates* memory of a previously learned passive avoidance habit in mice (Altman *et al.* 1984; Altman 1985). Surprisingly, pretreatment with quipazine (a purported serotonergic agonist) completely *blocked* the facilitory effect induced by both reuptake blockers. Although quipazine has been reported to modify serotonin release via an autoreceptor mechanism (Martin and Sanders-Bush 1982), it has many other neuropharmacological properties which may account for its behavioural effect. We have observed that either post-train or pre-test administration of a variety of serotonergic receptor antagonists produced similar effects (Normile and Altman 1984, 1985; Altman and Normile 1986*a, b*). The combined data suggest that either stimulation or blockade of the serotonergic system may, in some instances, *facilitate* memory processes. However, how two classes of serotonergic compounds, thought to produce opposite effects on serotonergic function, can have a similar effect on performance remains to be explained. It should be noted that post-training administration of 5-HT has been shown to *impair* memory of an inhibitory avoidance task (Essman 1978). In addition, another laboratory has also reported that post-training administration of fluoxetine (a serotonergic reuptake blocker) *improved* memory using an active avoidance paradigm (Cherkin and Flood 1984).

Our laboratory has also examined the effects of PCA-induced 5-HT depletion on the performance of young and aged rats in a complex spatial discrimination task, the Stone 14-unit T-maze. Aged rats routinely make more errors and require more trials to reach criterion than young animals. Depletion of 5-HT by PCA facilitated performance in both young (Altman *et al.* 1985) and aged rats (Normile *et al.* 1986).

Finally, our laboratory is investigating possible functional relationships between the cholinergic and serotonergic neurotransmitter systems. We have observed that combined administration of alaproclate and oxotremorine (a muscarinic cholinergic agonist) *facilitates* the performance of mice in a one trial passive avoidance task at doses that were ineffective when either drug was administered separately (Altman *et al.* 1987). Noradrenaline mechanisms may also be involved since the facilitory effect of pre-test pirenperone (a serotonergic receptor blocker) on avoidance task performance has been reported to be completely blocked by pretreatment with the alpha-noradrenergic receptor antagonist, phenoxybenzamine (Normile and Altman 1984).

Miscellaneous

A number of other experimental approaches may prove useful in ameliorating the cognitive impairments associated with SDAT. These include: phosphatidylserine, a naturally occurring acidic phospholipid extracted from bovine brain and suspected of altering cerebral metabolism via an effect on membranes (Bruni and Toffano 1982; Toffano *et al.* 1983; Calderini *et al.* 1983), 'active lipid', a compound purported to restore membrane fluidity in aging neurons (Shinitski *et al.* 1983; Lyte and Shinitski 1984) and aluminium chelators (deBoni and McLachlan 1981).

Of the three, phosphatidylserine is currently receiving the greatest attention. Unfortunately, the results with phosphatidylserine are mixed and the improvements that are being reported are not convincing (Miceli *et al.* 1977; Castellani *et al.* 1978; Delwaide *et al.* 1986; Villardita *et al.*, submitted; Palmieri *et al.*, submitted). Phosphatidylserine may facilitate learning and memory in aged rats (Drago *et al.* 1983). A related compound, GM-1 (monosialotetrahexosylganglioside) will be tested in a group of mild-moderately demented SDAT subjects later this year (1986) in the USA.

DISCUSSION

Unfortunately no treatment strategy has proved strikingly effective in the treatment of SDAT. This may be due in part to the study of subjects too severely impaired to respond to the treatment(s) administered. Advanced SDAT patients may not have adequate functional integrity of critical neuronal systems to demonstrate a clinical response. Future trials should include subjects who are less severely affected, and who have an earlier stage of SDAT. Also, current drug treatment strategies focus on reversing or attenuating *present* cognitive impairments. Future strategies should also consider the possibility that certain drugs may be better at reversing or attenuating the *anticipated* cognitive decline than in treating *present* cognitive impairments. A major problem for future research strategies for SDAT therapy is the apparent lack of clinical vs pre-clinical correlation. Animal models of dementia predictive of clinical efficacy in SDAT are crucial and must be developed so that treatment strategies for SDAT can be formulated.

Cholinergically-based strategies for SDAT therapy have already proven to be somewhat promising. Better methods of delivering cholinergic drugs to the brain (e.g. infusions of drugs via permanently indwelling catheters) as well as more selective and longer acting cholinergic drugs (e.g. preferential M1 agonists) may prove extremely rewarding. However, the magnitude of the overall clinical response as well as the total proportion of responders versus non-responders suggests that pharmacologic manipulations restricted to the cholinergic nervous system may fall short of projected expectations.

SDAT is more than a cholinergic deficiency syndrome and must be regarded as a multifactoral disease involving many neurotransmitters and neuropeptidergic systems.

For example, serotonin is significantly affected in SDAT and may play an important role in neuronal processes underlying learning and memory. Serotonin and acetylcholine neurons may interact with a number of other neurotransmitter systems which are essential to information processing. Thus, future research should attempt to clarify how various neurotransmitter systems interact in the processing of information and how dysfunctions in these interactions might contribute to the symptomatology associated with SDAT. Such an understanding may lead to the development of rational 'combination' drug therapy strategies which prove to be efficacious for SDAT.

REFERENCES

Adolfsson, R., Gofftries, C.G., Oreland, L., Roos, B.E., and Winblad, B. (1978). Reduced levels of catecholamines in the brain and increased activity of monoamine oxidase in platelets in Alzheimer's disease: therapeutic implications. In *Alzheimer's Disease: Senile Dementia and Related Disorders*, Aging Vol. 7 (ed. R. Katzman, R.D. Terry, and K.L. Bick), pp. 441–51. Raven Press, New York.

——, ——, Roos, B.E., and Winblad, B. (1979). Changes in the brain catecholamines in patients with dementia of Alzheimer type. *British Journal of Psychiatry* **135**, 216–23.

Altman, H.J. (1985). Mediation of storage and retrieval with two drugs that selectively modulate serotonergic neurotransmission. In *Memory Dysfunctions: An Integration of Animal and Human Research from Preclinical and Clinical Perspectives* (ed. D.S. Olton, E. Gamzu, and S. Corkin), pp. 497–8. The New York Academy of Sciences, New York.

——, Nordy, D.A., and Ogren, S.O. (1984). Role of serotonin in memory: Facilitation by alaproclate and zimeldine. *Psychopharmacology* **84**, 496–502.

—— and Normile, H.J. (1986*a*). Serotonin, learning and memory: Implications for the treatment of dementia. In *Treatment Development Strategies for Alzheimer's Disease* (ed. T. Crook, R.T. Bartus, S. Ferris, and S. Gershon), pp. 361–86. Mark Powley Associates, Inc., New York.

—— and —— (1986*b*). Enhancement of the memory of a previously learned aversive habit following pre-test administration of a variety of serotonergic antagonists in mice. *Psychopharmacology* **90**, 24–7.

—— and Quartermain, D. (1983). Facilitation of memory retrieval by centrally administered catecholamine stimulating agents. *Behavioural Brain Research* **7**, 51–63.

——, Normile, H.J., and Ogren, S.O., (1985). Facilitation of discrimination learning in the rat following cytotoxic lesions of the serotonergic nervous system. *Society of Neuroscience (Abstracts)* **11**, 874.

——, Stone, W.S., and Ogren, S.O. (1987) Evidence for a functional interaction between serotonergic and cholinergic mechanisms in memory retrieval. *Behavioural Neurology and Biology* (in press).

Arai, H., Kosaka, K., and Iizuka, R. (1984). Changes of biogenic amines and their metabolites in postmortem brains from patients with Alzheimer-type dementia. *Journal of Neurochemistry* **43**, 388–93.

Archer, T., Ogren, S.O., and Johansson, C. (1981). The acute effect of p-chloroamphetamine on the retention of fear conditioning the rat: evidence for a role of serotonin in memory consolidation. *Neuroscience Letters* **25**, 75–81.

Argentiero, V. and Tavolato, B. (1980). Dopamine (DA) and serotonin metabolic levels in the cerebrospinal fluid (CSF) in Alzheimer's presenile dementia under basic conditions and after stimulation with cerebral cortex phospholipids. *Journal of Neurology* **224**, 53–8.

Ball, M.J., *et al.* (1986). A new definition of Alzheimer's disease: A hippocampal dementia. *Lancet* **i**, 14–16.

Barclay, L.L., Zemcov, A., Blass, J.P., and Sansone, J. (1985). Survival in Alzheimer's disease and vascular dementias. *Neurology* **35**, 834–40.

Bartus, R.T., Dean, R.L.,Goas, J.A., and Lippa, A.S. (1980). Age-related changes in passive avoidance retention: Modulation with dietary choline. *Science* **209**, 301–3.

——, ——, and Beer, B. (1983). An evaluation of drugs for improving memory in aged monkeys: Implications for clinical trials in humans. *Psychopharmacology Bulletin* **19**, 168–84.

Baso, A.J. (1973). Ergot preparation (Hydergine) vs papaverine in treating common complaints of the aged: Double-blind study. *Journal of the American Geriatrics Society* **10**, 63–71.

Beal, M.F., Mazurek, M.F., Tran, V.T., Chattha, G., Bird, E.D., and Martin, J.B. (1985). Reduced numbers of somatostatin receptors in the cerebral cortex in Alzheimer's disease. *Science* **229**, 289–91.

Benson, D.F., Cummings, J.L., and Tsai, S.Y. (1982). Angular gyrus syndrome simulating Alzheimer disease. *Archives of Neurology* **39**, 616–20.

Benton, J.S., *et al.* (1982). Alzheimer's disease as a disorder of isodendritic core. *Lancet* **i**, 456.

Ben-Yishay, Y., *et al.* (1974). The alleviation of cognitive and functional impairments in senility by hyperbaric oxygenation combined with systemic cuing. In *Fifth International Hyperbaric Conference, Vol. 2* (ed. W.G. Trapp, E.W. Bannister, A.J. Davison, and P.A. Trapp) pp. 424–31. Simon Fraser University Press, Burnaby, Canada.

Biggins, J.A., *et al.* (1983). Post-mortem levels of thyrotropin-releasing hormone and neurotensin in the amygdala in Alzheimer's disease, schizophrenia and depression. *Journal of Neurological Science* **58**, 117–22.

Birren, J.E. (1974). Translation in gerontology—from lab to life: Psychophysiology and speed of response. *American Psychology* **29**, 808–15.

Biziere, K., Kan, J.P., Muyard, J.P., and Roncucci, R. (1982). Pharmacological evaluation of miniprine dihydrochloride, a new psychotropic drug. *Arzneimittel Forschung (Drug Res.)* **32**, 824–31,

——, ——, Goniot, C., Garattini, S., and Wermuth, C.G., (1983). Neurochemical profile of minaprine, a novel psychotropic drug. *Journal of Neurochemistry* **41**, 78.

Blass, J.P., Drachman, D., Katzman, R., and Spar, J.E. Letter to the editor. *New England Journal of Medicine* **211**, 556.

Bohus, N., Kovacs, G.L., and de Wied, D. (1978), Oxytocin, vasopressin and memory: Opposite effects on consolidation and retrieval processes. *Brain Research* **157**, 414–17.

Botwinick, J.P. (1977). Intellectual abilities. In *Handbook of the Psychology of Aging* (ed. J.E. Birren and K.W. Schaie), pp. 580–605. Van Nostrand, New York.

Bowen, D.M., Smith, C.B., White, P., and Davison, A.N. (1976). Neurotransmitter-related enzymes and indices in senile dementia and other abiotrophies. *Brain* **99**, 459–96.

——, *et al.* (1979). Accelerated ageing or selective neuronal loss as an important cause of dementia? *Lancet* **i**, 11–14.

——, *et al.* (1983). Biochemical assessment of serotonergic and cholinergic dysfunction and cerebral atrophy in Alzheimer's disease. *Journal of Neurochemistry* **41**, 266–72.

Boyle, E., *et al.* (1974). Hyperbaric oxygen and acetazolamide in the treatment of senile cognitive functions. In *Fifth International Hyperbaric Conference, Vol. 2* (ed. W.G. Trapp, E.W. Bannister, A.J. Davison, and P.A. Trapp), pp. 432–8. Simon Fraser University Press, Burnaby, Canada.

Branconnier, R.J. and Cole, O.J. (1980). The effects of Org 2766 in mild senile organic brain syndrome. Report to Organon.

Briley, M., Chopin, P., and Moret, C. (1986). New concepts in Alzheimer's disease. *Neurobiology of Aging* **7**, 57–62.

Brinkman, S.D. and Gershon, S. (1983). Measurement of cholinergic drug effects on memory in Alzheimer's disease. *Neurobiology of Aging* **4**, 139–45.

Bruni, A. and Toffano, G. (1982). Lysophosphotidylserine, a short-lived intermediate with plasma membrane regulatory properties. *Pharmacological Research Communication* **14**, 469–84.

Buresova, O. and Bures, J. (1982). Radial maze as a tool for assessing the effect of drugs on the working memory of rats. *Psychopharmacology* **77**, 268–71.

Calderini, G., Bonnetti, A.C., Battistella, A., Crews, F.T., and Toffano, G. (1983).

Biochemical changes of rat brain membranes with aging. *Neurochemistry Research* **8**, 483–91.

Carrasio, M.A., Dias R.D., and Izquierdo I. (1982). Naloxone reverses retrograde amnesia induced by electroconvulsive shock. *Behavioural Neurology and Biology* **34**, 352–7.

Castellani, A., *et al*. Clinical experiments with brain cortex phospholipids in psychogeriatrics. *Acta Neurologica* **33**, 217–29.

Cherkin, A. and Flood, J.F. (1984). Fluoxetine, a serotonin uptake blocker, enhances memory processing in mice. *Society of Neuroscience* (Abstr.) **10**, 255.

Coyle, J.D., Price, D.L., and Delong, M.R. (1983). Alzheimer's disease: A disorder of cortical cholinergic innervation. *Science* **219**, 1184–9.

Crook, T. (1979). Central-nervous-system stimulants: Appraisal of use in geropsychiatric patients. *Journal of the American Geriatrics Society* **27**, 476–7.

Cross, A.J., *et al*. (1983). Monoamine metabolism in senile dementia of Alzheimer type. *Journal of Neurological Science* **60**, 383–93.

——, *et al*. (1984). Studies on neurotransmitter receptor systems in cortex and hippocampus in senile dementia of the Alzheimer-type *Journal of Neurological Science* **64**, 109–17.

Cummings, J.L. and Benson, D.F. (1983). *Dementia: A Clinical Approach*. Butterworths, Boston, Mass.

Curran, H.V., Shine, P., and Lader, M. (1986). Effects of repeated doses of fluvoxamine, mianserin and placebo on memory and measures of sedation. *Psychopharmacology* **89**, 360–3.

Cutler, N.R., *et al*. (1985). Evaluation of zimeldine in Alzheimer's disease: Cognitive and biochemical measures. *Archives of Neurology* **42**, 744–8.

Davies, P. (1979). Neurotransmitter-related enzymes in senile dementia of the Alzheimer's type. *Brain Research* **171**, 319–27.

—— and Maloney, A.J.F. (1976). Selective loss of central cholinergic neurons in Alzheimer's disease. *Lancet* **ii**, 1403.

——, Katzman, R., and Terry, R.D. (1980). Reduced somatostatin-like immunoreactivity in cerebral cortex from cases of Alzheimer's disease and Alzheimer type dementia. *Nature* **228**, 279–80.

—— and Terry, R.D. (1981). Cortical somatostatin-like immunoreactivity in cases of Alzheimer's disease and senile dementia of the Alzheimer type. *Neurobiology of Aging* **2**, 9–14.

DeBoni, V. and McLachlan, D.R.C. (1981). Biochemical aspects of SDAT and aluminium as a neurotoxic agent. In *Strategies for the Development of an Effective Treatment for Senile Dementia* (ed. T. Crook and S. Gershon), pp. 215–30. Mark Powley Associates, New Canaan, CT.

Dehlin, O., Hedenrud, B., Jansson P., and Norgard, J. (1985). A double-blind comparison of alaproclate and placebo in the treatment of patients with senile dementia. *Acta Psychiatrica Scandinavica* **71**, 190–6.

deLeon, M.J., Potegal, M., and Gurland, B. (1984). Wandering and parietal signs in senile dementia of the Alzheimer's type. *Neuropsychobiology* **11**, 155–7.

Delfs, J.R., Zhu, C-H., and Dichter, M.A., (1984). Coexistence of acetylcholinesterase and somatostatin-immunoreactivity in neurons cultured from rat cerebrum. *Science* **223**, 61–3.

Delwaide, P.J., Gyselynck-Mambourg, A.M., Hurlet, A., and Ylieff, M. (1986). Double-blind randomized controlled study of phosphatidylserine in senile demented patients. *Acta Neurologica Scandinavica* **73**, 136–40.

Deutsch, J.A. and Rogers, J.B. (1979). Cholinergic excitability and memory: Animal studies and their clinical implications. In *Brain Acetylcholine and Neuropsychiatric Disease* (ed. K.L. Davis and P.A. Berger), pp. 175–204. Plenum Press, New York.

De Wied, D. (1965). The influence of posterior and intermediate lobe of the pituitary and pituitary peptides on the maintenance of a conditioned avoidance response in rats. *International Journal of Neuropharmacology* **4**, 157–67.

—— (1973). The role of the posterior pituitary and its peptides on the maintenance of conditioned avoidance behavior. In *Hormones and Brain Function* (ed. K. Lissas), pp. 391–7. Plenum Press, New York.

—— and Bohus, B. (1979). Modulation of memory processes by neuropeptides of hypothalamic-neurohypophyseal origin. In *Brain Mechanisms in Memory and Learning: From the Single Neuron to Man* (ed. M.A.B. Brazier), pp. 139–49. Raven Press, New York.

—— and Gispen, W.H. (1977). Behavioural effects of peptides. In *Peptides in Neurobiology* (ed. H. Gainer), pp. 418–48. Plenum Press, New York.

Dhrymiotis, A.D. and Whittier, J.R. (1962). Effects of a vasodilator, isoxsuprine, on cerebral ischemic episodes. *Current Therapy Research* **4**, 124–9.

Diesfeldt, H.F.A., van Houte, L.R., and Moerkens, R.M. (1986). Duration of survival in senile dementia. *Acta Psychiatrica Scandinavica* **73**, 366–71.

Dlabac, A., Krelci, I., and Kupkova, B. (1981). The interaction of nootropic drugs with anticonvulsants. *Activitas Nervosa Superior*. (Praha), **23**, 218–19.

Drachman, D.A. and Leavitt, J. (1974). Human memory and the cholinergic system. *Neurology* **30**, 113–21.

—— *et al.* (1982). Memory decline in the aged: Treatment with lecithin and physostigmine. Paper presented at the American Academy of Neurology.

Drago, F., Toffano, G., Catalano Rossi Danielli, L., Continella, G., and Scapagnini, U. (1983). Phosphotidylserine facilitates learning and memory processes in aged rats. In *Aging of the Brain* (ed. D. Samuel, S. Algeri, S. Gershon, V.E. Grimm and G. Toffano), pp. 309–16. Raven Press, New York.

Edwards, A.E. and Hart, G.M. (1974). Hyperbaric oxygenation and the cognitive functioning of the aged. *Journal of the American Geriatrics Society* **22**, 376–9.

Enna, S.J., Stern, L.Z., Wastek, G.J., and Tamamura, H.I. (1971). Cerebrospinal fluid gamma-aminobutyric acid variations in neurologic disorders. *Archives of Neurology* **34**, 683–5.

Essman, W.B. (1978). Serotonin in learning and memory. In *Serotonin in Health and Disease*: Vol. 3. *The Central Nervous System* (ed. W. Essman), pp. 69–143. Spectrum, New York.

Feinberg, I., Koresko, R.L., and Heller, N. (1967). EEG sleep patterns as a function of normal and pathological aging in man. *Journal of Psychiatric Research* **5**, 107–44.

Ferrier, I.N., *et al.* (1983). Neuropeptides in Alzheimer's type dementia. *Journal of Neurological Science* **62**, 159–70.

Ferris, S. (1983). Neuropeptides in the treatment of Alzheimer's disease. In *Alzheimer's Disease: The Standard Reference* (ed. B. Reisberg), pp. 369–73. The Free Press, New York.

——, ——, Reisberg, B., and Gershon, S. (1980). Neuropeptide effects on cognition in the elderly. In *Aging in the 1980's: Selected Contemporary Issues in the Psychology of Aging* (ed. Poon). pp. 239–55. American Psychological Association, Washington D.C.

Fibiger, H.C., Lepiane, F.G., and Phillips, A.G. (1978). Disruption of memory produced by stimulation of the dorsal raphe nucleus: mediation by serotonin. *Brain Research* **155**, 380–6.

Fleischhacker, W.W., Buchgeher, A., and Schubert, H. (1986). Memantine in the treatment of senile dementia of the Alzheimer type. *Progress in Neuropsychopharmacology Biology and Psychiatry* **10**, 87–93.

Forster, W., Schultz, S., and Henderson, A.L. (1955). Combined hydrogenated alkaloids of ergot in senile dementia and arteriosclerotic psychoses. *Geriatrics* **10**, 26–30.

Freemon, F.R. (1976). Evaluation of patients with progressive intellectual deterioration. *Archives of Neurology* **33**, 658–9.

Gaillard, A.W.K. and Sanders, A.F. (1975). Some effects of ACTH 4–10 on performance during a serial reaction task. *Psychopharmacology (Berlin)*, **42**, 201–8.

—— and Varey, C.A. (1979). Some effects of an ACTH 4–9 analogue (ORG 2766) on human performance. *Psychology and Behaviour* **23**, 79–84.

Gallagher, M. (1982). Naloxone enhancement of memory processes: Effects of other opiate antagonists. *Behaviour, Neurology, and Biology* **35**, 375–82.

Gamzu, E. (1985). Animal behavioural models in the discovery of compounds to treat memory dysfunction. In *Memory Dysfunctions: An Integration of animal and Human Research From Preclinical and Clinical Perspectives* (ed. D.S. Olton, E. Gamzu, and S. Corkin), pp. 370–93. The New York Academy of Sciences, New York.

Garattini, S., Forloni, G.L., Tirelli, A.S., Landinski, H., and Consolo, S. (1984). Neurochemical effect of minaprine a novel psychotropic drug on the central cholinergic system of the rat. *Psychopharmacology* **82**, 210–14.

Garzon, J., Rubio, J., and del Rio, J. (1981). Naloxone blocks the effects of 5-hydroxy-tryptophan on passive avoidance learning in rats: Implication of endogenous opioid peptides. *Life Sciences* **29**, 17–25.

Gibson, C.J. and Ball, M.J. (1983). Hippocampal monoamine deficits in Alzheimer's disease. *Journal of Neurochemistry* **41**, S-20.

Giurgea, C.E., Greindl, M.G., and Preat, S. (1983). Nootropic drugs and aging. *Acta Psychiatrica Belgia* **83**, 349–58.

Goldfarb, A.I., Hochstadt, N.J., Jacobson, J.H., and Weinstein, W.A. (1972). Hyperbaric oxygen treatment of organic mental syndrome in aged persons. *Journal of Gerontology* **27**, 212–17.

Gomez, S., Puyirat, J., Valade, P., Davous, P., Rondot P., and Cohen, P. (1986). Patients with Alzheimer's disease show an increased content of 15K dalton somatostatin precursor and a lowered level of tetradecapeptide in their cerebrospinal fluid. *Life Sciences* **39**, 623–7.

Gottfries, C.G. and Roos, B.E. (1973). Acid monoamine metabolites in cerebrospinal fluid from patients with presenile dementia (Alzheimer's disease). *Acta Psychiatrica Scandinavica* **49**, 257–63.

——, Kjallquist, A., Ponten, U., Roos, B.E., and Sundbarg, G. (1974). Cerebrospinal fluid pH and monoamine and glucolytic metabolites in Alzheimer's disease. *British Journal of Psychiatry* **124**, 280–7.

—— and Roos, B.E. (1976). Monoamine metabolites in cerebrospinal fluid (CSF) in patients with organic presenile and senile dementia. *Aktuel Gerontologie* **6**, 37–42.

——, ——, and Winblad, B. (1976). Monoamine and monoamine metabolites in the human brain post mortem in senile dementia. *Aktuel Gerontologie* 6, 429–35.

Harbourgh, R.E. (1986). Drug delivery to the brain by central infusion: Clinical Application of a chemical paradigm of brain function. In *Treatment Development Strategies for Alzheimer's Disease* (ed. T. Crook, R. Bartus, S. Ferris, and S. Gershon), pp. 553–66. Mark Powley Associates, Inc., New York.

Heston, L.L. (1981). Genetic studies of dementia: With emphasis on Parkinson's disease and Alzheimer's neuropathology. In *The Epidemiology of Dementia* (ed. J.A. Mortimer and L.M. Schuman), pp. 101–14. Oxford University Press, New York.

——, Mastri, A.R., Anderson, V.E., and White, J. (1981). Dementia of the Alzheimer type: Clinical genetics, natural history and associated conditions. *Archives of General Psychiatry* 38, 1085–1090.

Hodkinson, H.M. (1973). Mental impairment in the elderly. *Journal of the Royal College of Physicians in London* 4, 305–17.

Hutton, J.T. (1981). Results of clinical assessment for the dementia syndrome: Implications for epidemiologic studies. In *The Epidemiology of Dementia* (ed J.A. Mortimer and L.M. Schuman) pp. 62–9. Oxford University Press, New York.

Hyman, B.T., Eslinger, P., and Damasio, A. (1985). Effect of naltrexone on senile dementia of the Alzheimer type. *Journal of Neurological and Neurosurgical Psychiatry* 48, 1169–71.

Imai, Y. (1974). Hyperbaric oxygen (OHP) therapy on memory disturbances. In *Fifth International Hyperbaric Conference*, Vol. 2 (ed. W.G. Trapp, E.W. Bannister, A.J. Davison, P.A. Trapp), pp. 402–8. Simon Fraser University Press, Burnaby, Canada.

Isaacs, B. (1962). A preliminary evaluation of a paired-associate verbal learning test in geriatric practice. *Gerontology Clinics* 4, 43–55.

Ishii, T. (1966). Distribution of Alzheimer's neurofibrillary changes in the brain stem and hypothalamus of senile dementia. *Acta Neuropathologica* 6, 181–7.

Jacobs, E.A., Alvis, H.J., and Small, S.M. (1972). Hyperoxygenation: A central nervous system activator? *Journals of Geriatric Psychiatry* 5, 107–21.

——, Winter, P.M., Alvis, H.J., and Small, S.M. (1969). Hyperoxygenation effects on cognitive functioning in the aged. *New England Journal of Medicine* 281, 753–75.

Jenkins, J.S., Mather, H.M., Coughlan, A.K., and Jenkins, D.G. (1981). Desmopressin and desglyanamide vasopressin in post-traumatic amnesia. *Lancet* i, 39.

——, ——, and —— (1982). Effect of desmopressin on normal and impaired memory. *Journal of Neurological and Neurosurgical Psychiatry* 45, 830–1.

Johns, C.A., Haroutunian, V., Davis, B.M., Horvath, T.B., Mohs, R.C., and Davis, K.L. (1983). Acetylcholine inhibitors in Alzheimer's disease and animal models. In *Alzheimer's Disease: Advances in Basic Research and Therapies* (ed. R.J. Wurtman, S.H. Corkin, and J.H. Growdon) pp. 349–73. Center for Brain Sciences and Metabolism Charitable Trust, Cambridge, Mass.

Joyce, D. and Hurwitz. H.M.B. (1964). Avoidance behavior in the rat after 5-hydroxytryptophan (5-HTP) administration. *Psychopharmacologia* 5, 424–30.

Khachaturian, Z. (1985). Progress of research on Alzheimer's disease. In *Senile Dementia of the Alzheimer Type* (ed. J.T. Hutton and A.D. Kenny), pp. 379–84. Alan R. Liss, Inc., New York.

Kohler C. and Lorens, S.A. (1978). Open field activity and avoidance behavior following serotonin depletion: a comparison of the effects of parachlorophenyl-alanine and electrolytic midbrain raphe lesions. *Pharmacology, Biochemistry, and Behaviour* **8**, 223–33.

Kopelevich, V.M., Sytinsky, I.A., and Gunar, V.I. (1981). Current approach to the design of nootropic agents on the basis of gamma-aminobutyric acid. *Khim. Farm.* **15**, 27–38.

Kral, V.A. (1983). The relationship between senile dementia (Alzheimer type) and depression. *Canadian Journal of Psychiatry* **28**, 304–10.

Krejci, I., Kupova, B., Metys, J., Barth T., and Jost, K. (1979). Vasopressin ana-logues: Sedative properties and passive avoidance behavior in rats. *European Journal of Pharmacology* **56**, 347–53.

Kristensen, V., Olsen, M., and Theilgaard, A., (1977). Levodopa treatment of pre-senile dementia. *Acta Psychiatrica Scandinavica* **55**, 41–51.

Kulmala, H.K. (1986). Some enkephalin- or VIP-immunoreactive hippocampal pyramidal cells contain neurofibrillary tangles in the brains of aged humans and persons with Alzheimer's disease. *Neurochemistry and Pathology* **3**, 41–51.

Kuroda, H. (1983). Gamma-aminobutyric acid (GABA) in cerebrospinal fluid. *Acta Medica Okayama* **37**, 167–77.

Legros, J.J., *et al.* (1978). Influence of vasopressin in learning and memory. *Lancet* i, 41–2.

Lewis, C., Ballinger, B.R., and Presly, A.S. (1978). Trial of levodopa in senile dementia. *British Medical Journal* **1**, 550.

Liang, K.C., Messing R.B., and McGaugh, J.L. (1983). Naloxone attenuates amnesia caused by amygdaloid stimulation: The involvement of a central opioid system. *Brain Research* **271**, 41–9.

Lowensein, R.J., Weingartner, H., Gillin, J.C., Kaye, W., Ebert, M., and Mendelson, W.B. (1982). Disturbances of sleep and cognitive functioning in patients with dementia. *Neurobiology of Aging* **3**, 371–7.

Lyte, M. and Shinitzky, M. (1984). Possible reversal of tissue aging by a lipid diet. In *Alzheimer's Disease: Advances in Basic Research and Therapies* (ed. R.J. Wurtman, S.H. Corkin, and J.H. Growdon), pp. 126–42. Center for Brain Sciences and Metabolism Trust, Cambridge, MA.

Mahendra, N. (1985). Depression and dementia: The multi-faceted relationships. *Psychology and Medicine* **15**, 227–34.

Maletta, G.J., Pirozzolo, F.T., Thompson, G., and Mortimer, J.A. (1982). Organic mental disorders in a geriatric outpatient population. *American Journal of Psychiatry* **139**, 521–3.

Mann, D.M.A., Yates, P.O., and Hawkes, J. (1982). The noradrenergic system in Alzheimer and multi-infarct dementias. *Journal of Neurological and Neuro-surgical Psychiatry* **45**, 113–19.

Manyam, N.V.B., Katz, L., Hare, T.A., Gerber, J.C., and Grossman, M.H. (1980). Levels of gamma-aminobutyric acid in cerebrospinal fluid in various neurologic disorders. *Archives of Neurology* **37**, 352–5.

Marsden, C.D. and Harrison, M.J.G. (1972). Outcome of investigation of patients with presenile dementia. *British Medical Journal* **2**, 249–52.

Martin, L.L. and Sanders-Bush E. (1982). The serotonin autoreceptor: Antagonism by quipazine. *Neuropharmacology* **21**, 445–50.

Mazurek, M.F., Growdon, J.H., Beal, M.F., and Martin, J.B. (1986). CSF vaso-pressin concentration is reduced in Alzheimer's disease. *Neurology* **36**, 1133–6.

McLaren, S.M., McPherson, F.M., Sinclair, R., and Ballinger, G.R. (1981). Prevalence and severity of incontinence among hospitalized female psychogeriatric patients. *Health Bulletin* **39**, 157-61.

Meyer, J.S., *et al.* (1977). Neurotransmitter precursor amino acids in the treatment of mutli-infarct dementia and Alzheimer's disease. *Journal of American Geriatrics Society* **25**, 289-98.

Miceli, G., Caltagirone, C., and Gainotti, G. (1977). Gangliosides in the treatment of mental deterioration. *Acta Psychlogica Scandinavica* **55**, 123-8.

Mohs, R.C., *et al.* (1985). Clinical studies of the cholinergic deficit in Alzheimer's disease: psychopharmacological studies (II). *Journal of the American Geriatrics Society* **33**, 749-57.

Mohr, E., *et al.* (1986). GABA-agonist therapy for Alzheimer's disease. *Clinical Neuropharmacology* **9**, 257-63.

Morley, J.E. (1986). Neuropeptides, behavior, and aging. *Journal of the American. Geriatrics Society* **34**, 52-62.

Morrison, J.H., Rogers, J., Scherr, M., Benoit R., and Bloom, F.E. (1985). Somatostatin immunoreactivity of neuritic placques of Alzheimer's patients. *Nature* **314**, 90-2.

Nasrallah, H.A., Varney, N., Coffman, J.A., Bayliss, J., and Chapman, S. (1985). Effects of naloxone on cognitive deficits following electroconvulsive therapy. *Psychopharmacology Bulletin* **21**, 89-90.

Normile, H.J. and Altman, H.J. (1984). Facilitation of retrieval following pre-test administration of pirenperone in mice. *Society of Neuroscience (Abstr.)* **10**, 255.

—— and —— (1985). The effects of serotonergic receptor blockade on learning and memory in mice. *Society of Neuroscience* (Abstr.) **11**, 875.

——, ——, and Galloway, M. (1986). Facilitation of discrimination learning in aged rats following depletion of brain serotonin. *American Geriatrics Society* (Abstr.) **43**, 549.

Oglesby, M.M. and Winter, J.C. (1974). Strychnine sulfate and piracetam: Lack of effect on learning in the rat. *Psychopharmacology* **77**, 268-71.

Ögren, S.O. (1982). Forebrain serotonin and avoidance learning. Behavioral and biochemical studies on the acute effect of p-chloroamphetamine on one-way active avoidance learning in the male rat. *Pharmacology, Biochemistry, and Behaviour* **16**, 881-95.

——, Ross, S.B., Holm, A.C., and Baumann, L. (1977). 5-Hydroxytryptamine and avoidance performance in the rat. Antagonism of the acute effect of p-chloroamphetamine by zimeldine, an inhibitor of 5-hydroxytryptamine uptake. *Neuroscience Letters* **7**, 331-6.

O'Hanlon, J.F., Fussler, C., Sancia, E., and Grandjean, E.P. (1978). Efficacy of an ACTH 4-9 analogue, relative to that of a standard drug (amphetamine) for blocking the 'vigilance decrement' in man. Report to Organon, 1978.

Oram, J., Edwardson, J., and Millard, P. (1981). Investigation of cerebrospinal fluid neuropeptides in idiopathic senile dementia. *Gerontology* **27**, 216-23.

Palmieri, G., *et al.* (1987). Double-blind controlled trial of phosphotidylserine in patients with senile mental deterioration. *Clinical Trials Journal* **24**(1), 73-83.

Passeri, M., Cucinotta, D., deMello, M., Storchi, G., and Biziere, K. (1985). Minaprine for senile dementia. *Lancet* **i**, 824.

Peabody, C.A., Deblois, T.E., and Tinklenberg, J.R. (1986). Thyrotropin-releasing hormone (TRH) and Alzheimer's disease. *American Journal of Psychiatry* **143**, 262-3.

Perry, E.K., Gibson, P.H., Blessed, G., Perry, R.H., and Tomlinson, B.E. (1970). Neurotransmitter enzyme abnormalities in senile dementia: Choline acetyltransferase and glutamic acid decarboxylase activities in necropsy brain tissue. *Journal of Neurological Science* **34**, 247-65.

——, Tomlinson, B.E., Blessed, G., Perry R.H., and Cross, A.J. (1981*a*). Neuropathological and biochemical observations on the noradrenergic system in Alzheimer's disease. *Journal of Neurological Science* **51**, 279-87.

——, Dockray, G.J., Dimaline, R., Perry, E.K., Blessed, G., and Tomlinson, B.E. (1981*b*). Neuropeptides in Alzheimer's disease, depression and schizophrenia: a post mortem analysis of vasoactive intestinal peptide and cholecystokinin in cerebral cortex. *Journal of Neurological Science* **51**, 465-72.

Perry, E.K., *et al.* (1984). Cortical S2 receptor binding abnormalities in patients with Alzheimer's disease: comparison with Parkinson's disease. *Neuroscience Letters* **51**, 353-7.

Plum, F. (1979). Dementia: An approaching epidemic. *Nature* **279**, 372-3.

Pomara, N., Roberts, R., Rhiew, H.B., Stanley, M., and Gershon, S. (1985). Multiple, single-dose naltrexone administrations fail to affect overall cognitive functioning and plasma cortisol in individuals with probable Alzheimer's disease. *Neurobiology of Aging* **6**, 233-6.

Poschel, B.P.H., Marriott, J.G., and Gluckman, I. (1983). Pharmacology of the cognition activator pramiracetam (CI-879). *Drugs: Experimental and Clinical Research* **IX**, 853-71.

Price, D.L., *et al.* (1982). Basal forebrain cholinergic system in Alzheimer's disease and related dementias. *Neuroscience Communications* **I**, 84-92.

Prien, R.F. (1983). Psychostimulants in the treatment of senile dementia. In *Alzheimer's Disease: the Standard Reference* (ed. B. Reisberg), pp. 381-6. The Free Press, New York.

Prinz, P.N., *et al.* (1982). Sleep, EEG, and mental function changes in senile dementia of the Alzheimer's type. *Neurobiology of Aging* **3**, 361-70.

Quartermain, D. and Altman, H.J. (1982). Facilitation of retrieval by *d*-amphetamine following anisomycin-induced amnesia. *Physiology and Psychology* **18**, 283-92.

Quinton, E.E. and Bloom, A.S. (1977). Effects of *d*-amphetamine and strychnine on cycloheximide and diethyldithiocarbamate-induced amnesia in mice. *Journal of Comparative Physiology and Psychology* **91**, 1390-8.

Raskin, A., *et al.* (1977). The effects of hyper- and normobaric oxygen on cognitive impairment in the elderly. *Psychopharmacology Bulletin* **13**, 45-6.

Reisberg, B. (1983). Clinical presentation, diagnosis and symptomatology of age-associated cognitive decline and Alzheimer's disease. In *Alzheimer's Disease: the Standard Reference* (ed. B. Reisberg) pp. 173-97. Free Press, New York.

——, Ferris, S.H., Anand, R., Mir, P., DeLeon, M.J., and Roberts, E. (1983*a*). Naloxone effects on primary degenerative dementia (PDD). *Psychopharmacology. Bulletin* **19**, 45-7.

——, *et al.* (1983*b*). Effects of naloxone in senile dementia: A double-blind trial. *England Journal of Medicine* **308**, 721-2.

Reynolds, C.F., Spiker, D.G., Hanin, I., and Kupfer, D. (1983). Electroencephalographic sleep, aging and psychopathology: New data and state of the art. *Biological Psychiatry* **18**, 139-55.

Rigter, H. and Crabbe, J.C. (1979). Modulation of memory by pituitary hormones and related peptides. *Vitamins and Hormones* **337**, 153–241.

Roberts, G.W., Crow, T.J., and Polak, J.M. (1985). Location of neuronal tangles on somatostatin neurons in Alzheimer's disease. *Nature,* **314**, 92–4.

Roffman, M. and Lal, H. (1972). Role of brain amines in learning associated with 'amphetamine-state'. *Psychopharmacologia* **25**, 195–204.

Rosser, M.N., Fahrenkrug, J., Emson, P.C., Mountjoy, C.Q., Iversen, L.L., and Roth, M. (1980). Reduced cortical choline acetyltransferase activity in senile dementia of Alzheimer type is not accompanied by changes in vasoactive intestinal polypeptide. *Brain Research* **201**, 249–53.

——, Garrett, N.J., Johnson, A.L., Mountjoy, C.Q., Roth, M., and Iverson, L.L. (1982). A post-mortem study of the cholinergic and GABA systems in senile dementia. *Brain* **105**, 313–30.

Salzman, C. and Shader, R.I. (1979). Clinical evaluation of depression in the elderly. In *Psychiatric Symptoms of Depression in the Elderly* (ed. A. Raskin and L.J. Jarvik), pp. 39–63. Hemisphere, Washington. DC.

Sara, S.J. (1980). Memory retrieval deficits: Alleviation by etiracetam, a nootropic drug. *Psychopharmacology* **68**, 235–41.

—— and David-Remacle, M. (1974). Recovery from electroconvulsive shock-induced amnesia by exposure to the training environment: Pharmacological enhancement by piracetam. *Psychopharmacology* **36**, 59–66.

—— and Deweer, B. (1982). Memory retrieval enhanced by amphetamine after a long retention interval. *Behavioural Neurology and Biology* **36**, 146–60.

Schubert, H. and Fleischhacker, W. (1979). Therapeutische ansatze bei dementiellen syndromen. Ergebnisse mit amantadin-sulfat unter stationaren bedingungen. *Arztliche* Praxis **46**, 2157–60.

Shinitzky, M., Heron, D.S., and Samuel, D. (1983). Restoration of membrane fluidity and serotonin receptors in the aged mouse brain. In *Aging of the Brain* (ed. D. Samuel, S. Algeri, S. Gershon, V.E. Grimm, and G. Toffano), pp. 329–36. Raven Press, New York.

Smith, J.S. and Kiloh, L-G. (1981). The investigation of dementia: Results in 200 consecutive admissions. *Lancet* i, 824–6.

Soininen, H.S., Jolkkonen, J.T., Reinikainen, K.J., Halonen, T.O., and Reikkinen, P.J. (1984). Reduced cholinesterase activity and somatostatin-like immunoreactivity in the cerebrospinal fluid of patients with dementia of the Alzheimer type. *Journal of Neurological Science* **63**, 167–72.

——, Koskinen, T., Helkala, E-L., Pigache, R., and Riekkinen, P.J. (1985). The treatment of Alzheimer's disease with a synthetic ACTH 4–9 analogue. *Neurology* **35**, 1348–51.

Steiger, W.A., Mendelson, M., Jenkins, T., Smith, M., and Gay, R. (1985). Effects of naloxone in treatment of senile dementia. *Journal of the American Geriatrics Society* **33**, 155.

Stevens, D.A., Resnick, O., and Krus, D.M. (1967). The effects of *p*-chlorophenylalanine, a depletor of brain serotonin on behavior: I. Facilitation of discrimination learning. *Learning Sciences* **6**, 2215–20.

——, Fletcher, L.D., and Resnick, O. (1969). The effects of *p*-chlorophenylalanine, a depletor of brain serotonin on behavior: II. Retardation of passive avoidance learning. *Life Sciences* **8**, 379–85.

Straker, M. (1984). Depressive pseudodementia. *Psychiatric Annals* **14**, 93–102.

Struble, R.G., Powerss, R.E., Casanova, M.F., Kitt, C.A., O'Connor, D.T., and

Price, D.L. (1985). Multiple transmitter-specific markers in senile placques in Alzheimer's disease. *Journal of Neuropathology and Experimental Neurology* **44**, 325.

Tariot, P.N., Sunderland, T., Weingartner, H., Murphy, D.L., Cohen, M.R., and Cohen, R.M. (1985). Low- and high-dose naloxone in dementia of the Alzheimer's type. *Psychopharmacology Bulletin* **21**, 680-2.

Terry, R.D. and Davies, P. (1980). Dementia of the Alzheimer type. *Annual Review of Neuroscience* **3**, 77-95.

Thal, L.J., Masur, D.M., Fuld, P.A., Sharpless, N.S., and Davies, P. (1983). Memory improvement with oral physostigmine and lecithin in Alzheimer's disease. In *Banbury Report: Biological Aspects of Alzheimer's Disease* (ed. R. Katzman) pp. 461-70. Cold Spring Harbor Laboratory, New York.

Thompson, L.W. (1974). Effects of hyperbaric oxygen on behavioural functioning in elderly persons with intellectual impairment. Paper presented at the meeting of the American College of Neuropsychopharmacology, San Juan, Puerto Rico, 12/74.

Tinklenberg, J.R. and Thornton, J.E. (1983). Neuropeptides in geriatric psychopharmacology. *Psychopharmacological Bulletin* **91**, 198-211.

——, Peabody, C.A., and Berger, P.A. (1984). Vasopressin effects on cognition and affect in the elderly. In *Neuropeptide and Hormone Modulation of Brain Function and Homeostasis* (ed. J.M. Ordy, J.R. Sladek, and B. Reisberg), pp. 301-19. Raven Press, New York.

Toffano, G., *et al.* (1983). Experimental evidence supporting the use of phosphotidylserine liposomes in aging brain. In *Neuroradiological and Neurophysiological Correlations* (ed. A. Cecchini, G. Nappi, and A. Arrigo), pp. 312-21. International Multidisciplinary Seminar.

van Ree, J.M., Bohus, B., Versteeg, D.H.G., and de Wied, D. (1978). Commentaries: Neurohypophyseal principles and memory processes. *Biochemistry and Pharmacology* **27**, 1793-800.

van Wimersma-Greidanus, Tj.B. and de Wied, D. (1976). Dorsal Hippocampus: A site of action of neuropeptides on avoidance behavior? *Pharmacology, Biochemistry, and Behaviour* **5** (Supp.) 29-33.

Victoratos, G.C., Lenman, J.A.R., and Herzberg, L. (1977). Neurological investigation of dementia. *British Journal of Psychiatry* **130**, 131-3. 1977.

Villardita, C., *et al.* (1987). Multicenter clinical trial of brain phosphotidylserine in elderly subjects with intellectual deterioration. *Clinical Trials Journal* **27**(1), 84-93.

Vitiello, M.V., Bokan, J.A., Kukull W.A., Muniz, R.L., Smallwood, R.G., and Prinz, P.N. (1984). Rapid eye movement sleep measures of Alzheimer's type dementia patients and optimally healthy aged individuals. *Biological Psychiatry* **19**, 721-34.

Walter, R., Hoffman, P.L., Flexner, J.A., and Flexner, L.B. (1975). Neurohypophyseal hormones, analogues, and fragments: their effect on puromycininduced amnesia. *Proceedings of the National Academy of Sciences* **72**, 4180-4.

Weingartner, H., *et al.* (1981). Vasopressin enhances memory consolidation in man. *Science* **211**, 601-3.

——, Rudorfer, M.V., Buchsbaum, M.S., and Linnoila, M. (1983). Effects of serotonin on memory impairments produced by ethanol. *Science* **221**, 472-4.

Wells, C.E. (1981). A deluge of dementia. *Psychosomatics* **2**, 837-8.

Westreich, G., Alter, M., and Lundgren, S. (1975). Effects of cyclandelate on dementia. *Stroke* **6**, 535-8.

Wetzel, W., Getsova, V.M., York, R., and Matthies, H. (1980). Effect of serotonin on Y-maze retention and hippocampal protein synthesis in rats. *Pharmacology, Biochemistry, and Behaviour* 12, 319–22.

Whitehouse, P.J., Price, D.L., Clark, A.W., Coyle, J.T., and DeLong, J.T. (1981). Alzheimer disease: Evidence for selective loss of cholinergic neurons in the nucleus basalis. *Annals of Neurology* 10, 122–6.

——, ——, Struble, R.G., Clark, A.W., Coyle, J.T., and DeLong M.R. (1982). Alzheimer's disease and senile dementia: Loss of neurons in the basal forebrain. *Science* 215, 1237–9.

Winblad, B., Adolfsson, R., Carlsson, A., and Gottfries, C.G. (1982). Biogenic amines in brains with Alzheimer's disease. In *Alzheimer's Disease: A Report of Progress, Vol. 19. Aging* (ed. S. Corkin, K.L. Davis, J.H. Growdon, E. Usdin, and R.J. Wurtman), pp. 25–33. Raven Press, New York.

Wolthuis, O.L. (1971). Experiments with UCB 6215, a drug which enhances acquisition in rats: its effects compared with those of methamphetamine. *European Journal of Pharmacology* 16, 283–97.

—— (1981). Behavioural effects of etiracetam in rats. *Pharmacology, Biochemistry, and Behaviour* 15, 247–55.

Woolley, D.W. and Van der Hoeven, T. (1963). Alteration in learning ability caused by changes in cerebral serotonin and catecholamines. *Science* 139, 610–11.

Wurtman, R.J., Magil, S.G., and Reinstein, D.K. (1981). Piracetam diminishes hippocampal acetylcholine levels in rats. *Life Science* 28, 1091–3.

Yamamoto, T. and Hirano, A. (1985). Nucleus raphe dorsalis in Alzheimer's disease: Neurofibrillary tangles and loss of large neurons. *Annals of Neurology* 17, 573–7.

Yarbrough, G. (1979). On the neuropharmacology of thyrotropin-releasing hormone (TRH). *Progressive Neurobiology* 12, 291–312.

Yates, C.M., *et al.* (1983). Thyrotropin-releasing hormone, leutinizing hormone-releasing hormone and substance-P immunoreactivity in post-mortem brain from cases of Alzheimer-type dementia and Down's syndrome. *Brain Research* 258, 45–52.

Yesavage, J.A., *et al.* (1979). Vasodilators in senile dementia: A review of the literature. *Archives of General Psychiatry* 36, 220–3.

Zhu, C-H., Delfs, J., Mufson, E. Dichter, M., and Mesulam, M-M (1984). Acetylcholinesterase and somatostatin-like-immunoreactivity coexist in rat cerebral cortex and hippocampus, but not in CH4 cholinergic neurons of the basal forebrain. *Society of Neuroscience Abstracts* 10, 696.

Zimmer, R., Teelken, A.W., Trieling, W.B., Weber, W., Weihmayer, T., and Lauter, H. (1980), Gamma-aminobutyric acid in cerebrospinal fluid in various neurologic disorders. *Archives of Neurology* 37, 352–5.

——, ——, —— and ——. (1984). Gamma-aminobutyric acid and homovanillic acid concentration in CSF of patients with senile dementia of Alzheimer's type. *Archives of Neurology* 41, 602–4.

19

Dementia praecox: cognitive function and its cerebral correlates in chronic schizophrenia

T.J. CROW

INTRODUCTION

The form of cognitive disturbance in schizophrenia, and its relationship to the pathology of the disease, has long been a focus of interest. The symptoms are polymorphic and the question arises whether they can all be related to a single underlying dysfunction. The issue has relevance to the definition of the disease and its demarcation from other conditions. Thus, while Kraepelin (1919) distinguished dementia praecox from manic-depressive insanity mainly on the basis of outcome, others have sought to identify psychopathological characteristics which are specific to the syndrome. The most successful of such attempts is Schneider's (1957) description of the nuclear or 'first rank' symptoms. These include types of auditory hallucination (e.g. hearing one's thoughts spoken aloud and hearing voices discussing one in the third person) and delusions concerning thoughts (e.g. that thoughts are removed from or inserted into one's head, or broadcast to other people) which it is suggested occur only in schizophrenia and certain organic psychoses. These symptoms are of a somewhat striking nature; their occurrence gives support to the view that in schizophrenia there is a loss of ability to discriminate mental contents which arise from stimuli in the external environment from those which arise internally from ongoing mental activity. However, while they have achieved utility in recent diagnostic systems, e.g. the DSMIII and the Catego system, their basis in cerebral activity and relation to normal functions remains obscure. While related in some way to the schizophrenic process they are by no means always present. Recent studies suggest they do not necessarily predict a poor outcome.

Of greater interest with respect to prognosis and with implications for the nature of the disease process is the occurrence of intellectual impairment. Although Kraepelin's use of the term dementia praecox suggests to the contemporary reader a degree of loss of intellectual function it is far from clear that this was his intention. The term dementia seems to have had a less specific connotation than it subsequently acquired. Both Kraepelin and

E. Bleuler, who introduced the term schizophrenia, appear to have believed that true intellectual impairment was not characteristic of the condition.

Bleuler stressed the point, perhaps to emphasize the demarcation between the functional and the organic psychoses. Thus, he wrote 'In contrast to the organic psychoses, we find in schizophrenia . . . that sensation, memory, consciousness . . . are not directly disturbed' (Bleuler 1950, p. 55) and 'The integration of perceptions concerning spatial and temporal orientation is quite good; even delirious schizophrenics are for the most part well orientated as to place and even to time' (p. 58), 'memory as such does not suffer in this disease' (p. 59) and '. . . consciousness . . . is not altered in the chronic conditions of schizophrenia' (p. 62). In a similar vein Kraepelin (1919) states that in dementia praecox 'perception . . . is not usually lessened' (p. 5), 'orientation is not usually disordered', 'consciousness . . . if we leave out of account the terminal condition of dementia, is in many cases clear throughout' (p. 17) and 'memory is comparatively little disordered. The patients are able, when they like, to give a correct detailed account of their past life, and often know accurately to a day how long they have been in the institution' (p. 18). The comment about terminal states introduces a degree of ambiguity. Later in his book Kraepelin includes in this category *simple weak mindedness* in which patients are 'clear about time, place and person, also about their position, and give reasonable and connected information' (p. 189), but they 'lose a great part of their knowledge; they become impoverished in thought, monotonous in their mental activities' (p. 190) and *drivelling dementia*, in which there is 'a general decay of mental efficiency' and patients 'are often not clear about their position and surroundings' but 'they often answer simple questions quite correctly' (p. 198). Thus, it appears that while Bleuler viewed the psychopathological changes of schizophrenia as quite distinct from those of the dementias Kraepelin acknowledged features in the defect states compatible with true intellectual impairment but strove to explain them in other ways.

AGE DISORIENTATION

The views of both Kraepelin and Bleuler are challenged by the phenomenon of 'age disorientation'. Aspects of this phenomenon have been noted separately by a number of workers who have examined chronic schizophrenic patients. Lanzkron and Wolfson (1958) described a 'perceptual distortion of temporal orientation' in 50 chronic patients, and Dahl (1958) replicated the finding in a population of 500 institutionalized patients and claimed that this 'singular distortion of temporal orientation' occurred only in schizophrenia. Ehrenteil and Jenney (1960) found that a number of hospitalized schizophrenic patients gave ages younger than their true age and asked 'Does time stand still for some psychotics?'. Michelson (1968) also apparently unaware of previous findings, described systematic errors made by 62 patients in estimating their ages over a 6-year period.

In the course of a survey of in-patients with schizophrenic illnesses of long-standing Crow and Mitchell (1975) were impressed by the frequency with which such subjects believed themselves to be an age widely different from their true age. In 237 patients with chronic schizophrenia in the wards of four mental hospitals in Scotland they found that approximately 25 per cent believed themselves to be 5 or more years younger than they really were (Fig. 19.1). Twelve per cent believed themselves to be within 5 years of their age on admission although on average they were 28 years older than this. This figure was significantly ($P < 0.001$) greater than would be expected if the subjects were merely guessing their age. Five per cent of the sample thought their age was within 1 year of their age on admission. The findings were interpreted as indicating that these patients had a failure of new learning capacity.

Stevens *et al.* (1978) surveyed the population of patients with a diagnosis of schizophrenia in Shenley Hospital in North West London and found that age disorientation (defined as a 5-year discrepancy between subjective and true age) was present in 25 per cent of this sample, a figure closely similar to that reported in the less representative Scottish sample of Crow and Mitchell

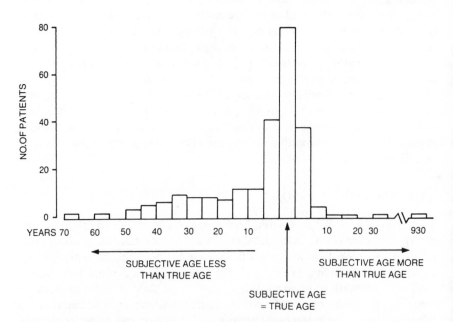

FIG. 19.1.   Distribution of subjective age (in years) in relation to true age in a population of 237 chronic schizophrenic patients in Scottish mental hospitals. Sixty (25.3 per cent) patients believed themselves to be 5 or more years younger than they were (from Crow and Mitchell 1975).

(1975) and that noted by Smith and Oswald (1976) in patients in the Harlem Valley Psychiatric Center in New York. Furthermore, Stevens *et al.* (1978) found that schizophrenic patients *without* age disorientation were younger at first admission and had a longer duration of stay than age-matched patients with age disorientation. They suggested that age disorientation was a feature of a type of schizophrenic illness with early onset and poor prognosis.

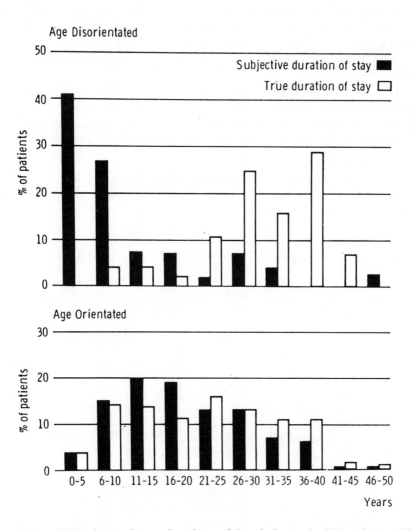

FIG. 19.2. Subjective and true durations of hospital stay in 299 patients with chronic schizophrenia in Shenley Hospital subdivided according to whether (above) or not (below) they showed age disorientation (from Crow and Stevens 1978).

THE NATURE OF THE COGNITIVE DEFICIT

The question arises as to whether age disorientation is an isolated psycho-pathological deficit, part of a wider impairment of temporal orientation, or part of a more general loss of intellectual function. Crow and Stevens (1978) found that patients with age disorientation (n = 77) were much less likely than those without age disorientation (n = 227) to be able to give correct answers to simple questions about dates and the passage of time (e.g. their date of birth, the present year, and the duration of their hospital stay). The age disorientated patients systematically underestimated the present year and their duration of hospital stay; the errors made by individual patients in estimating these figures were consistent with their concept of their own age (Fig. 19.2 and 19.3).

Thus, age disorientation may be part of a constellation of defects of temporal orientation. For these patients 'time stands still'. However, in some patients an incorrect appreciation of their own age coexists with correct awareness of the present year, and between these patients and those for

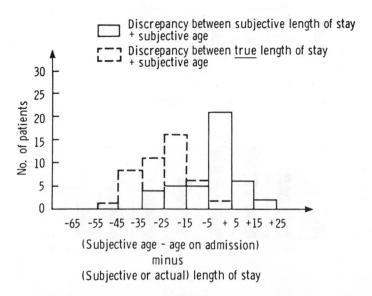

FIG. 19.3.   The discrepancy between subjective length of stay and subjective age is much less than between true length of stay and subjective age in 44 schizophrenic patients with age disorientation (from Crow and Stevens 1978).

whom subjective time stands still there appears to be a continuum of increasing temporal disorientation. This spectrum of temporal disorientation may correlate with severity of intellectual impairment.

In order to determine the effect of past physical treatments on the prevalence of age disorientation, we compared age disorientated and age orientated patients for frequency of having received electroconvulsive therapy, insulin coma therapies (or a combination of the two), or leucotomy (Table 19.1). Since the age disorientated patients were slightly *less* likely to have received one of these typical treatments, age disorientation cannot be attributed to these physical treatments.

TABLE 19.1. *Past physical treatments in relation to age disorientation*

	Age-orientated $n = 222$	Age disorientated $n = 88$
ECT	96 (43%)	35 (40%)
Insulin coma	80 (40%)	27 (31%)
Insulin coma and ECT	48 (22%)	16 (18%)
Leucotomy	22 (10%)	6 (7%)

Liddle and Crow (1984) addressed the question of whether the temporal disorientation is associated with general deficits of intellectual function. A series of 21 patients with age disorientation (each of whom believed himself to be within 5 years of his age on admission) were compared on a battery of cognitive function tests with a group of 21 patients without age disorientation who were matched for age and duration of hospital stay. On all tests the age disorientated patients were more impaired (Fig. 19.4). This was true of tests of orientation and general knowledge and estimation of numerical quantities as well as tests of new learning (face-name learning, the digit symbol test of the WAIS), problem solving (Raven's matrices) and tests of previous learning (the Peabody picture vocabulary and the famous personalities test), and dysphasia (the Boston naming test). On those tests in which a random guessing rate could be estimated, age disorientated patients performed little above this level, and on the mental test score (which assesses several aspects of intellectual function) the two groups were separated with an overlap of only two patients ($P < 0.001$).

Thus, it seems that age disorientation is but one aspect of a global impairment of intellectual function. This pattern of intellectual deficits in schizophrenic patients is difficult to distinguish from that of patients with organic dementias such as Alzheimers Disease. Indeed, the performance of these schizophrenic patients on the Peabody vocabulary test (which is intended to tap premorbid intellectual function) did not establish that these patients had

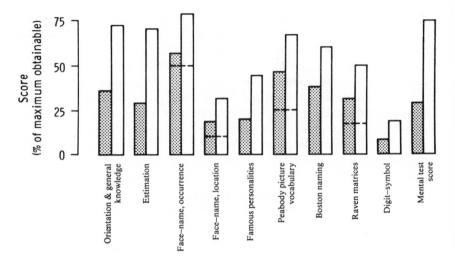

Fig. 19.4.   Mean scores of 21 age-disorientated schizophrenic patients (dotted bars) and 21 non-age-disorientated schizophrenic controls on nine tests of cognitive function. Dotted lines indicated expected score for random guessing (from Liddle and Crow 1984).

previously functioned at an adequate intellectual level.

This possibility of inadequate premorbid intellectual function is of interest in relation to the concept of 'childhood asociality' as formulated by Quitkin *et al.* (1976). According to these authors there is a group of patients who fail to develop interpersonal relationships, and have poor educational attainments, but later progress to develop typical episodes of schizophrenia. Other studies have identified groups of patients with intellectual impairments preceding the onset of psychosis. Pollack *et al.* (1966) found school performance was worse among those who subsequently developed schizophrenia by comparison with their non-schizophrenic siblings and that such deficits were greater in those with an earlier onset of illness. Lane and Albee (1964) found that intelligence assessed during the second grade at school (at about the age of 7 or 8) of boys who subsequently developed schizophrenia was lower than that of their siblings. Mason (1956) found lower premorbid intelligence scores at army induction procedures among those who subsequently developed hebephrenia and simple schizophrenia when compared to non-schizophrenic army inductees.

To test whether the educational level of patients who subsequently developed age disorientation was particularly impaired, Buhrich *et al.* (in press) attempted a retrospective comparison of school performance in age disoriented patients with that of age oriented patients. Age disoriented patients did *not* perform more poorly, and as far as could be ascertained, most patients had achieved an average performance at school.

Thus, it appears that intellectual function declines with progression of illness. This has been well documented by Garside (1969). In a group of 101 schizophrenic patients, 99 hospitalized non-schizophrenic psychiatric patients, and 50 normal controls, he compared performance on the Wechsler–Bellevue scales with the school grading examination these subjects had obtained years earlier. Schizophrenic patients showed significant intellectual deterioration by comparison with the other groups, this deterioration being greatest in those diagnosed as hebephrenic and least in those suffering from paranoid schizophrenia. Somewhat similar findings were reported by Rappaport and Webb (1950). From these studies and from the existence of a group of patients with apparently severe intellectual deficits, it appears safe to conclude that schizophrenia is sometimes associated with a progressive decline in intellectual function.

WHAT IS THE CEREBRAL BASIS?

Following the introduction of computerized tomography, we (Johnstone *et al.* 1976, 1978) were able to demonstrate that a group of institutionalized patients with chronic schizophrenia had greater lateral cerebral ventricular area than a group of age and premorbid occupation-matched controls. The change was unrelated to past physical treatment. Within the patient group, ventricular enlargement was associated with intellectual impairment. Subsequent work (e.g. Owens *et al.* 1985) has confirmed the finding of ventricular enlargement, although the relationship with intellectual impairment remains unclear.

To account for the discrepancy between apparent responsiveness of some symptoms (e.g. delusions and hallucinations) to neuroleptic medication, and the relative resistance and irreversibility of others (e.g. affective flattening and poverty of speech) a two syndrome concept (Crow 1980) was advanced according to which the former (the positive symptoms) represent one dimension of pathology in schizophrenia and the latter (the negative symptoms) another. The positive symptoms (the type I syndrome) are generally responsive to medication while the type II syndrome of negative symptoms are less so. The type II syndrome is closely related to intellectual impairments and might be expected to be associated with structural changes in the brain.

A recent post-mortem study (Brown *et al.* 1986) supports the view that there are structural brain changes and suggests that these are in the temporal lobe. In comparison to the brains of patients with affective illness, the brains of patients with schizophrenia were 5–6 per cent lighter, had significantly increased temporal horn area, and thinner parahippocampal gyri, especially on the left side. Although detailed information on the pre-mortem clinical state of this group of patients was not available, one can speculate that these changes in the temporal lobe might be relevant to the intellectual deficits.

Some neurochemical findings are also relevant. In a study of six neuropeptides in post-mortem brain (Roberts *et al.* 1983), differences between schizophrenic patients and controls were confined to temporal lobe structures, and were seen particularly in patients with negative symptoms. Thus, the cholecystokinin and somatostatin content of hippocampus and cholecystokinin content of amygdala were reduced in patients with negative symptoms. Since these peptides are located in interneurones in these structures, the losses may represent destruction or aplasia of interneurones which may be related in turn to the structural changes. Cholecystokinin binding was found reduced in hippocampus and also in frontal cortex (Ferrier *et al.* 1985). A parallel change in temporal structures and frontal cortex was found for monoamine oxidase activity. A reduction in type B but not type A MAO activity was seen in amygdala, hippocampus and frontal cortex, but again was confined to patients with negative symptoms (Owen *et al.* in press). Together, these findings suggest that a pathological process occurring predominantly in the temporal lobe but also in areas of frontal cortex is present particularly in patients with the defect state or type II syndrome. These changes may be reflected in the intellectual impairments which form an integral part of the syndrome.

Whatever the nature of these changes, it is clear that they do not resemble those seen in Alzheimer's disease. Neither dopamine β-hydroxylase nor choline acetyl transferase, markers of adrenergic and cholinergic neurones respectively, are lost in patients with schizophrenia in general, nor in particular in patients with negative symptoms (Crow *et al.* 1981). One possibility is that the structural changes in schizophrenia represent some type of developmental arrest, i.e. that they occur at an early stage of the disease process. A recent study (Colter N. *et al.*, unpublished data) suggests that structural changes may occur particularly in schizophrenic patients with early onset of disease and that these changes affect the development of asymmetries in the brain. This could account for the apparent predilection of the disease process for the left hemisphere (Crow 1986).

CONCLUSIONS

Temporal disorientation is seen in some patients with chronic schizophrenia and reflects the presence of a global impairment of intellectual function (a true 'dementia praecox'). When present, it is unrelated to past physical treatments, but is associated with other features of intellectual loss and behavioural deterioration. It is a component of the defect state or type II syndrome. Although this temporal disorientation is associated with early onset of illness, it is not a direct consequence of poor premorbid intellectual function; there is also evidence of progression of intellectual loss with activity of the disease process. Structural brain changes in schizophrenia are now well established in CT scan and post-mortem studies and have a degree of selectivity for the temporal lobe. While it is tempting to relate such changes to the intellectual impairments which are sometimes present, this relationship is not yet clearly established. The disease process underlying such changes differs from that which occurs in Alzheimer's disease. In schizophrenia the process may arrest brain development, but there may also be an element of neuronal loss. The process has a degree of selectivity for the left hemisphere.

REFERENCES

Bleuler, E. (1950). *Dementia Praecox or the Group of Schizophrenias*. International Universities Press, New York (Translated by J. Zinkin).

Brown, R., *et al.* (1986). Post-mortem evidence of structural brain changes in schizophrenia. *Archives of General Psychiatry* **43**, 36–42.

Buhrich, N., Crow, T.J., Johnstone, E.C., and Owens, D.G.C. Age disorientation in chronic schizophrenia is not associated with premorbid intellectual impairment or past physical treatment. *British Journal of Psychiatry* (in press).

Crow, T.J. (1980). Molecular pathology of schizophrenia: More than one dimension of pathology? *British Medical Journal* **280**, 66–8.

—— (1986). Left brain, retrotransposons and schizophrenia. *British Medical Journal* **293**, 3–4.

—— and Mitchell, W.S. (1975) Subjective age in chronic schizophrenia: evidence for a sub-group of patients with defective learning capacity? *British Journal of Psychiatry* **126**, 360–3.

—— and Stevens, M. (1978). Age disorientation in chronic schizophrenia: the nature of the cognitive deficit. *British Journal of Psychiatry* **133**, 137–42.

——, *et al.* (1982). The search for changes underlying the type II syndrome of schizophrenia. In *Biological Psychiatry* (1981) (ed. C. Perris, G. Struwe, and B. Jansson), pp. 727–31. Elsevier/North Holland, Amsterdam.

Dahl, M. (1958). A singular distortion of temporal orientation. *American Journal of Psychiatry* **115**, 146–9.

Ehrenteil, O.F. and Jenney, P.B. (1960). Does time stand still for some psychotics? *Archives of General Psychiatry* **3**, 1–3.

Ferrier, I.N., *et al.* (1985). Reduced cholecystokinin levels in the limbic lobe in schizophrenia: a marker for the defect state? *Annals of the New York Academy of Sciences* **448**, 495–506.

Garside, R.F. (1969). The relationship between schizophrenia and intelligence. PhD thesis, University of Newcastle.

Johnstone, E.C., Crow, T.J., Frith, C.D., Husband, J., and Kreel, L. (1976). Cerebral ventricular size and cognitive impairment in chronic schizophrenia. *Lancet* **ii**, 924–7.

——, ——, ——, Stevens, M., Kreel, L., and Husband, J. (1978). The dementia of dementia praecox. *Acta Psychiatricia Scandinavica* **57**, 305–24.

Kraepelin, E. (1919). *Dementia Praecox and Paraphrenia*. R.E Krieger, New York. (Translated by R.M. Barclay and G.M. Robertson.)

Lane, E. and Albee, G. (1964). Early childhood intellectual differences between schizophrenic adults and their siblings. *Journal of Abnormal and Social Psychology* **68**, 193–5.

Lanzkron, J. and Wolfson, W. (1958). Prognostic value of perceptual distortion of temporal orientation in chronic schizophrenics. *American Journal of Psychiatry* **114**, 744–6.

Liddle, P. and Crow, T.J. (1984). Age disorientation in chronic schizophrenia is associated with global intellectual impairment. *British Journal of Psychiatry* **144**, 193–9.

Mason, C.F. (1956). Pre-illness intelligence of mental hospital patients. *Journal of Consulting Psychology* **20**, 297–300.

Michelson, M. (1968). A note on age confusion in psychosis. *Psychiatric Quarterly* **42**, 331–8.

Owen, F., *et al.* Selective decrease in MAO-B activity in post-mortem brains from schizophrenic patients with the type II syndrome. *British Journal of Psychiatry* (in press).

Owens, D.G.C., Johnstone, E.C., Crow, T.J., Frith, C.D., Jagoe, J.R., and Kreel, L. (1985). Cerebral ventricular enlargement in schizophrenia. Relationship to the disease process and its clinical correlates. *Psychological Medicine* **15**, 27–41.

Pollack, M., Woerner, M.G., Goodman, W., and Greenberg, I.M. (1966). Childhood developmental patterns of hospitalized adult schizophrenic and non-schizophrenic patients and their siblings. *American Journal of Orthopsychiatry and Psychiatry* **36**, 510–17.

Quitkin, F., Rifkin, A., and Klein, D.F. (1976). Neurologic soft signs in schizophrenia and character disorders. *Archives of General Psychiatry* **33**, 845–53.

Rappaport, S.R. and Webb, W.B. (1950). An attempt to study intellectual deterioration by premorbid testing. *Journal of Consulting Psychology* **14**, 95–8.

Roberts, G.W., *et al.* (1983). Peptides, the limbic lobe and schizophrenia. *Brain Research* **228**, 199–211.

Schneider, K. (1957). Primare und sekundare symptome bei der Schizophrenie. *Fortschritte fur Neurologie und Psychiatrie* **25**, 487–90.

Smith, J.W. and Oswald, W.T. (1976). Subjective age in chronic schizophrenia. *British Journal of Psychiatry* **128**, 100.

Stevens, M., Crow, T.J., Bowman, M.J., and Coles, E.C. (1978). Age disorientation in schizophrenia: a constant prevalence of 25 per cent in a chronic mental hospital population? *British Journal of Psychiatry* **133**, 130–6.

20

Neurochemical approaches to cognitive disorders—a discussion

M. TRAUB AND S. FREEDMAN

The discovery that there is a profound loss of cholinergic activity in the cerebral cortex and hippocampus in Alzheimer's disease, gave rise to the hope that, just as for Parkinson's disease, palliative replacement therapy might prove effective. Such treatments have included substrate loading with lecithin and administration of anticholinesterases (e.g. physostigmine) which enhance cholinergic activity, or the use of directly acting muscarinic agonists. However, as discussed by Collerton and by Altman (this volume), such attempts at boosting central cholinergic transmission in patients with Alzheimer's disease have generally met with only modest clinical results at best (see Summer's *et al.* 1986 for a contrary view). As discussed by Rossor (this volume) it is now known that there are multiple neurotransmitter deficits in Alzheimer's disease, although it is generally accepted that pronounced cholinergic depletion is a consistent feature of this disorder. Authors in this section of *Cognitive neurochemistry* have re-evaluated the role of this cholinergic deficit in the light of current knowledge.

In his review, Collerton (this volume) provides stern criticism not only of the 'cholinergic hypothesis', but also of various experimental approaches that have been employed in the study of Alzheimer's disease. Certain obstacles to research remain until a specific *in vivo* diagnostic test for the condition is developed. In the mildest cases where treatment might have most to offer, there is usually a considerable element of diagnostic uncertainty. Indeed, our knowledge of the neurochemistry of the disorder is almost entirely based on the study of post-mortem tissue from patients with end-stage disease when, perhaps, the loss of crucial neurotransmitter systems is obscured by more generalized secondary changes, and correlations with behavioural tests performed during life must inevitably be crude.

Given the methodological difficulties of studying Alzheimer's disease itself, considerable attention has been devoted to the study of pharmacologically induced models of dementia both in animals and man. Altman (this volume) has highlighted an important limitation of animal studies by stressing that a number of drugs that appear to improve memory in rodents are totally ineffective clinically. In man the most thoroughly investigated 'pharmacological dementia' has been that induced by anticholinergic drugs. The most obvious appeal of such models is that in some respects amnesia caused by pharmacological intervention mimics memory loss, a cardinal feature of Alzheimer's disease. However, as Collerton points out, most investigations have been deficient in a number of respects. For example, the majority have ignored the non-specific effects of the drugs on arousal, which may contribute to amnesia, and have failed to carry out proper dose-response studies. In his review, Kopelman (this volume) has examined the effects of scopolamine over a whole range of mnemonic processes in order to make a more meaningful comparison between a pharmacological model of central cholinergic blockade and Alzheimer's disease. He argues that, in contrast to primary dementia, acute scopolamine administration spares the more passive aspects of primary memory (e.g. forward digit span) as well as the remote component of secondary memory; functions which he attributes to other neurotransmitter systems. Kopelman's detailed examination on the effects of scopolamine on human memory serves to underline the naivity of certain earlier studies, which tended to exaggerate the similarities between Alzheimer's disease and the pharmacological model on the basis of relatively crude observations. Nevertheless, such evidence as that presented by Kopelman cannot amount to total rejection of the 'cholinergic hypothesis' given the limitations of an acute study.

In view of the heterogeneity of Alzheimer's disease as well as the ethical limitations of collecting cerebral biopsy material during life, clinico-pathological correlations are fraught with difficulty. An alternative approach is to compare the clinical features, neuropathology, and neurochemical changes in a variety of dementing conditions. Both Rossor and Agid (this volume) have examined the distinction between subcortical and cortical dementias with the aim of delineating the separate role of cortical and subcortical structures in cognitive processing. Such an approach has served to emphasize the importance of diencephalic-cortico projections in high intellectual functions. For example, the combination of pathological and PET scanning data suggests that frontal lobe symptomatology in progressive supranuclear palsy is not due to cortical damage, but to frontal deafferentation. If, as Agid suggests, cortical 'programmes' are indeed preserved in progressive supranuclear palsy, effective therapy may be achieved by amplification of signals in the damaged pathways projecting to cortex or by the replacement of an absent neuromodulator. By contrast, in the case of Alzheimer's disease and the dementia of Parkinson's disease, it is unclear to what extent cognitive failure is attributable to damage to subcortico-cortical projections and how much is caused by intrinsic cortical pathology. Clearly, as Agid points out, palliative replacement therapy is most unlikely to be effective in Alzheimer's disease, if there is significant disorganization of cortical 'hard-wiring'.

Before one takes this pessimistic view, there are a number of unanswered questions which need to be addressed as to why cholinomimetic therapy has been so disappointing. The narrow therapeutic window observed with anticholinesterases, such as physostigmine (see Bartus 1979), means that unless precise dosing information is obtained, treatment may actually result in a cognitive impairment. It is likely that individual titration of dosage in every patient may be the only way to overcome this. It is perhaps surprising that even under these circumstances physostigmine therapy has been so consistently therapeutically efficacious. The most recent report of anticholinesterase therapy with tetrahydroaminoacridine (THA) (Summers *et al.* 1986) has raised a great deal of interest, particularly with reference to the long-term treatment of patients and the lack of serious cholinergic side effects. It remains, however, to determine whether these beneficial affects are due to cholinesterase inhibition or to the acetylcholine releasing properties associated with aminopyridines. Wessling and Agostan (1984) have previously reported that 4-aminopyridine is effective at improving memory in clinical trials.

The use of directly acting muscarinic agonists has the attraction of acting postsynaptically at receptors which have been reported not to change in Alzheimer's disease. Yet the clinical results with arecoline, RS86, and BM-5, and similar cholinergic agonists have been very disappointing. It should be noted, however, that many of these compounds have extremely short half lives, penetrate into the central nervous system very poorly, and have the pharmacological profile of being weak partial agonists in functional studies. It may be that the muscarinic agonist of choice has yet to be tested, and even if such a compound exists there is still the problem of side effects. Many of the clinical trials of the past have reported severe side effects including salivation, flushing, nausea, and vomiting, all of which have limited the doses at which cholinergic compounds can be administered. Perhaps a novel way of overcoming these problems would be the use of alaproclate, a selective 5HT uptake blocker, to potentiate the responses of muscarinic agonists (Ogren *et al.* 1985). It remains to be seen whether the peripheral and central effects of muscarinic agents can be separated by this mechanism.

The new generation of selective muscarinic agents e.g. pirenzepine, AFDX-116, hexahydrosilaprocyclidine (Hammer *et al.* 1980; Giachetti *et al.* 1986; Eglen and Whiting 1986) have been important in our understanding of multiple muscarinic receptors and the next few years may reveal a new generation of selective muscarinic agents which penetrate into the central nervous system.

Recent advances in molecular biology have made this goal appear more realistic. The sequence of the M-1 muscarinic receptor was recently published in *Nature* (Kubo *et al*. 1986) and it is believed that this will be closely followed by the sequence of the cardiac M-2 receptor. It is hoped that these developments will lead directly to more selective muscarinic agonists and therefore improve the therapeutic potential of cholinergic agents by reducing their adverse side-effects.

REFERENCES

Bartus, R.T. (1979). Physostigmine and recent memory effects in young and aged nonhuman primates. *Science* **206**, 1087–9.

Eglen, R.M. and Whiting, R.L. (1986). Differential affinities of muscarinic antagonists at ileal and atrial receptors. *British Journal of Pharmacology* **87**, Supp. 33P.

Giachetti, A., Micheletti, R., and Montayne, E. (1986). Cardioselective profile of AF-DX 116, a muscarinic M-2 receptor antagonist. *Life Sciences* **38**, 1663–77.

Hammer, R., Berrie, C.P., Birdsall, N.J.M., Burgen, A.S.V., and Hulme, E.C. (1980) Pirenzepine distinguishes between different subclasses of muscarinic receptors. *Nature* **283**, 90–2.

Kubo, T., *et al*. (1986). Cloning, sequencing and expression of complementary DNA encoding the muscarinic acetylcholine receptor. *Nature* **323**, 411–16.

Ogren, S.O., Nordstrom, O., Danielsson, E., Peterson, L.L., and Bartfail, T. (1985). *In vivo* and in vitro studies on the potentiation of muscarinic receptor stimulation by Alaproclate, a selective 5HT uptake blocker. *Journal of Neurology and Transmission* **61**, 1–20.

Summers, W.K., Majorski, L.V., Marsh, G.M., Tachiki, K., and Kling, A. (1986). Oral Tetrahydroaminoacridine in long term treatment of Senile dementia, Alzheimer type. *New England Journal of Medicine* **315**, 1241–5.

Wessling, H. and Agoston, S. (1984). Effects of 4-amino-pyridine in elderly patients with Alzheimer's disease. *New England Journal of Medicine* **310**, 988–9.

Index